Neurobiology of Spinal Cord Injury

Contemporary Neuroscience

Neurobiology of Spinal Cord Injury

Edited by

Robert G. Kalb, MD

Stephen M. Strittmatter, MD, PHD

School of Medicine, Yale University
New Haven, CT

Humana Press Totowa, New Jersey

© 2000 Humana Press Inc.
999 Riverview Drive, Suite 208
Totowa, New Jersey 07512

For additional copies, pricing for bulk purchases, and/or information about other Humana titles, contact Humana at the above address or at any of the following numbers: Tel.: 973-256-1699; Fax: 973-256-8341; E-mail: humana@humanapr.com or visit our Website: http://humanapress.com

This publication is printed on acid-free paper. (∞)

ANSI Z39.48-1984 (American Standards Institute) Permanence of Paper for Printed Library Materials.

Cover design by Patricia F. Cleary.

Cover illustration: A motor neuron and its synaptic inputs. Immunohistology for synaptophysin (green puncta), a marker for presynaptic terminals, is combined with DiI labeling of a motor neuron axon, proximal dendrites, and cell body. Optimizing motor function after spinal cord injury will depend on improving the ability of excitatory synapses to drive motor neuron activation. Photograph by Drs. Laising Yen and Robert Kalb.

Photocopy Authorization Policy:

Printed in the United States of America. 10 9 8 7 6 5 4 3 2 1

Library of Congress Cataloging-in-Publication Data

Neurobiology of spinal cord injury / edited by Robert G. Kalb, Stephen M. Strittmatter.
    p.   cm. -- (Contemporary neuroscience)
    Includes bibliographical references and index.
    ISBN 0-89603-672-3 (alk. paper)
    1. Spinal cord--Wounds and injuries--Pathophysiology. I. Kalb, Robert G. II. Strittmatter, Stephen M. III. Series.
    [DNLM: 1. Spinal Cord Injuries--physiopathology. 2. Neurobiology.
    3. Spinal Cord Injuries--therapy.   WL 400 N4938 2000]
    RD594.3.N4686 2000
    617.4'82044--dc21
    DNLM/DLC
    for Library of Congress
                   99-23563
                   CIP

Preface

Neurobiological Research in SCI
Suggests a Multimodality Approach to Therapy

Neuroscientists representing a wide variety of disciplines are drawn to the problem of spinal cord injury (SCI). One of the many reasons for this is humanistic: the severity of neurologic deficits can have a devastating impact on most aspects of an individual's functioning. As compassionate individuals we cannot help but believe that a small lesion, tragically localized, can be amenable to a therapeutic intervention. Interest in spinal cord injury research also stems from the view that the overall problem can be broken down into a number of understandable and experimentally tractable smaller problems (*see* Table 1). The clearness and discreteness of the issues to be addressed help focus the research effort. The hope is that if one could amalgamate the progress made in these smaller arenas, a significant overall benefit would be available for patients. This book highlights the major areas of basic science research in which progress is being made today in our battle against the problem of spinal cord injury.

Important advances in developing effective intervention to promote functional recovery after spinal cord injury depends on animal models. The utility of complete or partial spinal cord transection models is complemented by studies in which the spinal cord is injured by dropping a weight of known mass from a fixed distance onto the cord. This has led to reproducible lesions and functional deficits. There is a remarkable concordance between this experimentally induced lesion and that seen in postmortem specimens from injured human spinal cords. In their chapter, Drs. Beattie and Bresnahan describe a number of important insights gleaned from this model system. First, there is a significant amount of delayed apoptotic death of neurons and glia both at the lesion site and remotely. It stands to reason that prevention of this cell death will have important consequences for recovery of function. Second, anatomical studies of the cystic lesion induced by the trauma reveal a complex mix of astrocytes, Schwann cells, inflammatory cells, as well as axons of dorsal root ganglion cells and spared descending axons in various states of demyelination. Some of the cellular components of the contusion lesion matrix appear to arise from ependymal progenitor cells. If new neurons or glial cells were incorporated to the contusion site, they might integrate into functional circuitry and/or help form tissue bridges for regrowing axons.

Though the glial component of the injury is important (as we will see), one of the major reasons for functional impairment after spinal cord injury is the death of neurons. Effective neuroprotective strategies after spinal cord insult will therefore depend on understanding the mechanisms of neuronal death. A major advance in this field has come from the focus on intracellular Ca^{2+}. Intracellular Ca^{2+} is a fundamental regulator of cellular physiology and, as such, its concentration and subcellular distribution is highly regulated. Derangement of this regulation plays a central role in neuronal death. In their chapter, Chu et al. discuss the controversies surrounding the mechanism by which rises in intracellular Ca^{2+} lead to cell death. In particular they focus on the extent to which pathogenic deregulation of Ca^{2+} homeostasis is a function of the ion's intracellular concentration, its spatial distribution, or the route of entry into cells.

After injury to the cervical or thoracic spinal cord, the neural elements in the lumbar enlargement are deprived of inputs from higher command centers (brainstem and cortex), but are in a certain sense otherwise intact. Can the isolated lumbar enlargement generate the patterned muscle activation required for locomotion or is this capacity dependent on information encoded by the descending inputs? The functional capacity of the deafferented lumbar cord has been investigated in lamprey, rodents, and cats and under certain experimental conditions exhibits a remarkable ability to generate the neural activity subserving locomotion. Thus, there is great interest in understanding the functional capacity of the isolated lumbar cord. Rossignol et al. describe a set of experiments using cats to define the intrinsic capacity of the isolated lumbar cord to generate locomotion. Through a combination of lesion studies and pharmacological manipulations, this work is beginning to outline the cellular and biochemical mechanisms involved. A particularly exciting set of experiments provides compelling evidence for spinal learning—the ability of the isolated cord to make adaptations to new environmental demands (such as stepping over an obstruction or walking on a tilted surface). The circuitry inherent within the isolated lumbar spinal cord and its ability to adapt to changing demands may form an important substrate for therapeutic intervention.

A more detailed anatomical and electrophysiological analysis of the neural substrate underlying the locomotor generating ability of the isolated lumbar spinal cord requires a more experimentally accessible system. Dr. Cazalets has pioneered the use of an ex vivo neonatal rat spinal cord preparation and details work on the location and properties of the major central pattern generator for fictive locomotion. Though many regions within the spinal cord can be experimentally manipulated to generate oscillatory firing of neurons, the neural circuitry of the lower thoracic and upper lumbar spinal cord are likely to be the dominant central pattern generator for hindlimb locomotor activity. The region around the central canal, particularly ventrally, appears to be key. If this region is undamaged after a spinal cord insult, then stimulation by regrowing axons or pharmacological means might form the basis for functional restitution.

Assuming sufficient numbers of neurons and glia can be coaxed to survive the primary and secondary wave of cell death and that the pattern generators for locomotion are intact, a central concern focuses on restoring the continuity between the brain and the distal spinal cord. Steeves and Tetzlaff provide an overview of molecules known to promote axonal regeneration in SCI and those which inhibit regeneration. Growth-promoting molecules include proteins expressed within regenerating neurons, such as GAP-43 and Tα1-tubulin. In addition, a number of neurotrophins can induce axonal sprouting, and the effects of these molecules in SCI are reviewed. The ability of certain axonal guidance molecules, such as CAMs, semaphorins, netrins, and ephrins, to promote axonal regeneration is also discussed. The major inhibitory factors are those derived from CNS myelin and from astrocytes. In SCI, the growth inhibitors exert a predominant effect over the growth-promoting factors. To increase functional recovery after SCI, intervention to both decrease growth-inhibiting effects and increase growth-promoting effects must be considered.

The extension of axons is primarily regulated from their distal end, a specialization termed the growth cone. A number of endogenous macro-molecules, including some present in injured spinal cords, can repel axons and prevent axonal regeneration. We review the molecular events that transduce the presence of extracellular signals into a cessation of axonal extension. Particular emphasis is placed on the mechanism of action of the semaphorins, since they are well studied, and on components of CNS myelin, because they are likely to underlie the failure of axonal regeneration after SCI. As the normal functioning of the growth cone becomes understood, pharmacological methods to overcome the failure of axonal growth cone advance in SCI might become obvious.

If the milieu that a growing axon encounters is permissive or even promotes growth, ultimately the axon elaboration machinery must be engaged. Peter Baas reviews the role of microtubules in forming axon structure and determining axonal elongation rates. The biochemistry and cell biology of microtubules are reviewed with special focus on their contribution to axonal extension. Evidence suggests that microtubules are nucleated from a site near the centrosome and then transported into axons. The protein dynein is the motor responsible for transporting microtubules into and down the axon. A number of microtubule associated proteins (MAPs) are likely to regulate microtubule formation and transport. Although the functional organization of these proteins in developing axons is becoming clear, their role in facilitating or preventing axonal regeneration remains less so. Further investigation of such mechanisms in SCI might provide novel opportunities to promote axon regeneration and recovery of function.

From the perspective of problems incurred by spinal cord injury that may be amenable to therapeutic intervention in the near future, three have received the most attention: (1) How do we keep the maximal number of neurons and glial cells alive after the injury? (2) How do we promote the extension of axons past the lesion site? (3) How can we promote myelination of surviving or newly formed axons so that they can efficiently transmit action potentials distally? Remarkably, research from a number of different labs has indicated that cell transplants and neurotrophic factors may have utility for each of these problems. Work from Barbara Bregman's lab indicates that transplanting fetal spinal cord tissue and similtaneously providing neurotrophic factors can have a beneficial effect on both neuronal survival and axon growth. Though it has long been clear that trophic factors are important for the survival and differentiation of immature neurons, it has only been recently that evidence has been accrued that they also play a role in mature neuronal survival. One important advance is the recognition that different neurons (i.e., brainstem, cortical, etc.) have distinct trophic factor dependence for survival and axon elaboration. When combined with a suitable substrate for growth (such as fetal spinal cord transplants), trophic factors remarkably enhance neuronal survival and promote axonal growth and sprouting. These interventions lead to functionally significant improvement in animal behavior, particularly when applied to immature animals. The challenge will be to adapt this strategy to maximize the benefits for mature animals.

Since cell transplantation strategies hold great hope for functional recovery after SCI, what other choices exist beyond the use of fetal spinal cord tissue? Bartolomei and Greer consider the various cell transplantation strategies now being developed. Transplanted tissue has included fetal nervous system, peripheral nervous system Schwann cells, and more recently olfactory ensheathing cells. The olfactory ensheathing cell appears to hold

Table 1
Therapeutic Approaches in SCI

Mechanism	Chapter	Site of action	Timing	Current clinical therapy	Animal studies	Potential future approaches
Limit cell death	1, 2, 8, 11	Injury	Acute	Prednisone	Growth factors, cytokines	Specific anti-inflammatory and anti-apoptotic
Distal intrinsic circuits	3, 4	Distal	Chronic	Physical therapy	—	Training directed by electrophysiology
Environment for regeneration	8, 9, 10, 11	Injury	Chronic	—	Olfactory ensheathing cell transplants	Human cell transplants
Enhance axon growth	5, 6, 7, 8, 9	Injury	Chronic	—	Olfactory ensheathing cell transplants	Specific axon growth regulation
Enhance remyelination	9, 10	Injury	Chronic	—	Olfactory ensheathing cell transplants	Specific regulators of myelination

much promise for a variety of reasons: it is easily obtained in large numbers; it can bridge the border between a glial scar at the site of SCI and the surrounding spinal cord; it is capable of remyelinating any surviving but demyelinated axons; and it may secrete trophic factors. The lineage and properties of olfactory ensheathing cells are discussed in detail. Recent experiments demonstrating the efficacy in ensheathing cell transplantation in SCI are reviewed.

The loss or dysfunction of oligodendrocytes has increasingly become an area of interest and Drs. Waxman and Kocsis focus on the existence of demyelinated axons in SCI. They discuss data indicating that, after SCI, a number of axons remain anatomically intact but lose their myelin sheaths and exhibit conduction block. The distribution of ion channels in demyelinated axons indicates that conduction block might be relieved by potassium channel blockers. The authors discuss experiments showing that the function of demyelinated axons is improved in the presence of 4-AP, a potassium channel blocker. Remyelination will probably turn out to be the optimal method for improving axonal impulse conduction. The authors' data indicate that transplanted olfactory ensheathing cells can provide functional remyelination in spinal cord demyelination models.

Wise Young considers the more practical aspects of various therapeutic interventions after spinal cord injury. The clinical NACSIS trials have validated the ability of high dose steroid therapy immediately after SCI to improve outcome measures. Potential mechanisms of methylpredinisolone action are discussed. A number of other compounds, including cyclooxygenase inhibitors, cyclosporine A, and opiate antagonists, possess anti-inflammatory effects and improve outcomes in animal models of SCI. Paradoxically, some proinflammatory agents, such as interleukins, gangliosides, and bacterial toxins, may also enhance recovery after SCI. Reducing inflammation immediately after SCI may limit tissue damage, while at later times proinflammatory effects may facilitate tissue repair. Neurotransmitter receptor blockers also appear to reduce tissue damage after SCI. There is some evidence that naloxone therapy may improve outcomes in human SCI, and NMDA receptor antagonists reduce neuronal degeneration in animal models. Efforts to increase regeneration of axons are less well developed.

Our hope is that this book will delineate the major areas in which therapeutic interventions may alleviate the neurological deficit of the spinally injured patient. It is clear from the extensive research described in these pages that the basic neurobiology underlying spinal cord dysfunction is rapidly advancing. Already, this has led to novel cellular and molecular approaches to improving function after spinal cord injury in laboratory animals. The progress in animal research bodes well for the prospect of clinically successful interventions to achieve restoration of neurological performance in humans. We note that research is advancing along several independent, but equally promising, avenues. As each modality to improve function after SCI is perfected, combinations of therapies are likely to produce strongly synergistic benefit. Even small degrees of anatomical recovery after human SCI are likely to reap enormous rewards. The difference between an SCI victim being respirator-dependent and taking his or her own breath, or being wheelchair bound or standing without assistance, may hinge on the recovery of function in just a small fraction of injured axons. There is every reason to believe that neurobiological research will provide the opportunity for SCI victims to walk again.

Robert Kalb, MD
Stephen M. Strittmatter, MD, PHD

Contents

Contributors

PETER W. BAAS, PHD, *Department of Anatomy, The University of Wisconsin Medical School, Madison, WI*

HUGHES BARBEAU, PHD, *École de Physiothérapie et d'Ergothérapie, Université McGill, Montréal, Québec, Canada*

JUAN C. BARTOLOMEI, MD, *Department of Neurosurgery, Yale University School of Medicine, New Haven, CT*

MICHAEL S. BEATTIE, PHD, *Department of Neuroscience, College of Medicine and Public Health, The Ohio State University Medical Center, Columbus, OH*

MARC BÉLANGER, PHD, *Département de Kinanthropologie, Université du Québec, Montréal, Québec, Canada*

LAURENT BOUYER, PHD, *Département de Physiologie, Centre de Recherche en Sciences Neurologiques, Faculté de Médecine, Université de Montréal, Montréal, Québec, Canada*

BARBARA S. BREGMAN, *Division of Neurobiology, Department of Cell Biology, Georgetown University School of Medicine, Washington, DC*

JACQUELINE C. BRESNAHAN, PHD, *Department of Neuroscience, College of Medicine and Public Health, The Ohio State University Medical Center, Columbus, OH*

EDNA BRUSTEIN, PHD, *Centre de Recherche en Sciences Neurologiques, Département de Physiologie, Faculté de Médecine, Université de Montréal, Montréal, Québec, Canada*

JEAN-RENÉ CAZALETS, *CNRS Laboratoire de Neurobiologie et Mouvements, Marseille, France*

CONNIE CHAU, PHD, *Département de Physiologie, Centre de Recherche en Sciences Neurologiques, Faculté de Médecine, Université de Montréal, Montréal, Québec, Canada*

GORDON K. T. CHU, MD, *Division of Neurosurgery, University of Toronto, Toronto, Canada*

TREVOR DREW, PHD, *Département de Physiologie, Centre de Recherche en Sciences Neurologiques, Faculté de Médecine, Université de Montréal, Québec, Canada*

ALYSON FOURNIER, PHD, *Department of Neurology, Yale University School of Medicine, New Haven, CT*

NATHALIE GIROUX, MSC, *Département de Physiologie, Centre de Recherche en Sciences Neurologiques, Faculté de Médecine, Université de Montréal, Québec, Canada*

CHARLES A. GREER, PHD, *Department of Neurosurgery and Section of Neurobiology, Yale University School of Medicine, New Haven, CT*

CLAUDE-ANDRÉ GRENIER, DVM, *Département de Physiologie, Centre de Recherche en Sciences Neurologiques, Faculté de Médecine, Université de Montréal, Montréal, Québec, Canada*

ROBERT G. KALB, MD, *Departments of Neurology and Pharmacology, Yale University School of Medicine, New Haven, CT*

J. D. KOCSIS, PHD, *Department of Neurology, Yale University School of Medicine, New Haven, and PVA/EPVA Neuroscience Research Center, VA Hospital, West Haven, CT*

FUMIO NAKAMURA, MD, PHD, *Department of Neurology, Yale University School of Medicine, New Haven, CT and PVA/EPVA Neuroscience Research Center, VA Hospital, West Haven, CT*

TOMAS A. READER, MD, PHD, *Département de Physiologie, Université de Montréal, Centre de Recherche en Sciences Neurologiques, Faculté de Médecine, Montréal, Québec, Canada*

SERGE ROSSIGNOL, MD, PHD, *Département de Physiologie, Centre de Recherche en Sciences Neurologiques, Faculté de Médecine, Université de Montréal, Montréal, Québec, Canada*

JOHN D. STEEVES, *CORD (Collaboration on Repair Discoveries), The University of British Columbia (UBC), Vancouver, British Columbia, Canada*

STEPHEN M. STRITTMATTER, MD, PHD, *Department of Neurology, Yale University School of Medicine, New Haven, CT*

TAKUYA TAKAHASHI, MD, *Department of Neurology, Yale University School of Medicine, New Haven, CT*

CHARLES H. TATOR, MD, PHD, FRCS(C), *Division of Neurosurgery, University of Toronto, Toronto, Ontario, Canada*

WOLFRAM TETZLAFF, *CORD (Collaboration on Repair Discoveries), Departments of Zoology, Anatomy, and Surgery, The University of British Columbia (UBC), Vancouver, British Columbia, Canada*

MICHAEL TYMIANSKI, MD, PHD, FRCSC, *Division of Neurosurgery and Playfair Neuroscience Unit, Toronto Western Hospital, Toronto, Ontario, Canada*

LI-HSIEN WANG, PHD, *Department of Neurology, Yale University School of Medicine, New Haven, CT*

STEPHEN G. WAXMAN, MD, PHD, *Department of Neurology, Yale University School of Medicine, New Haven, and PVA/EPVA Neuroscience Research Center, VA Hospital, West Haven, CT*

WISE YOUNG, MD, PHD, *Neuroscience Center, Rutgers, The State University, Nelson Biological Laboratories, Piscataway, NJ*

Cell Death, Repair, and Recovery of Function after Spinal Cord Contusion Injuries in Rats

Michael S. Beattie and Jacqueline C. Bresnahan

1. INTRODUCTION: MODELS OF SCI

Many attempts to model human spinal cord injury in animals have been made through the years. Recently, some consensus has been reached in terms of the goals of such models. They need to be reliable, consistent, and reproducible from laboratory to laboratory *(1)*. They need to replicate some of the important pathologic features of human spinal cord injury (SCI). The models should allow assessment of some of the mechanistic features of damage and recovery after injury. When large treatment trials are needed, the models need to be efficient, and outcome measures need to be relatively simple. These considerations have led to a number of new models in rats, cats, and other species over the past several years. Our laboratory has worked to develop models in the rat *(2–8)* and most recently, we have joined a consortium of other laboratories to help standardize a weight-drop model with chronic neurologic and histologic outcome measures *(3,9,10)* The Multicenter Animal Spinal Cord Injury Study (MASCIS) was formed to combine the expertise of multiple laboratories, to provide for interlaboratory reproducibility in assessing preclinical treatments for SCI (MASCIS was initiated by Dr. Wise Young). As a result of this effort, and the work of many other laboratories, it is now possible to predict outcomes and test therapies efficiently in contusion injury models. The lesions produced by the MASCIS contusion device and by other similar models, for example, the Ohio State University electromechanical model *(7,11,12)*, are characterized by the development of central hemorrhagic necrosis that spreads radially, as well as rostrocaudally over time. The end result is an ellipsoidal, loculated cystic cavity or cell-filled injury site *(7,8,13–15)*. Contusion lesions in rats respond to early pharmacologic interventions similar to human cord injuries *(6,16–18)*, suggesting that this model is useful for assessing clinical treatments. Some of the features of these lesions are shown in Figure 1.

Detailed data on the pathology of human SCI, are few, but the reports available suggest that the rat contusion lesion is similar to a large number of human

From: *Neurobiology of Spinal Cord Injury*
Edited by: R. G. Kalb and S. M. Strittmatter © Humana Press Inc., Totowa, NJ

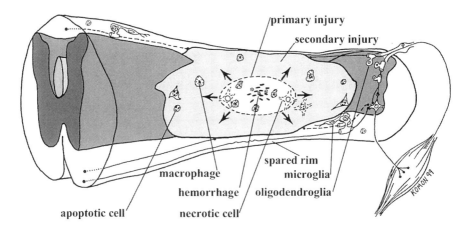

Fig. 1. Schematic showing the features of spinal cord contusion lesions. Contusion lesions are characterized by a central region of hemorrhagic necrosis that expands over time due to activation of secondary injury processes including programmed cell death (apoptosis). Characteristically, at the center of the injury, a spared rim of tissue remains at the periphery. Besides blood elements that escape from the broken vessels, recruitment of immune mediators occurs early after injury. Large numbers of macrophages are present in the lesion site within a few days, and activated microglia extend along the edges of the lesion rostrocaudally within the fiber tracts undergoing wallerian degeneration. The microglia may participate in the apoptotic death of oligodendrocytes, which also occurs along the degenerating tracts, extending into regions far remote from the lesion. Intact segmental circuitry caudal to the injury and circuitry rostral to the injury are denervated from loss of descending and ascending axons, respectively, initiating reorganization in those systems.

SCIs *(19–21)*. A critical aspect of these lesions is that even after severe injuries, a small peripheral rim of spared tissue and axons remains *(3,7,10,15)*. Furthermore, the amount of spared tissue, in the thoracic cord, correlates highly with locomotor function *(3,4,6,7,15)*. This spared rim of axons has also been observed in human SCI, in which even neurologically complete patients often exhibit some peripheral tissue sparing. Such spared tissue can be targeted for both acute and chronic treatments. Acute treatments could reduce the spread of secondary injury, leaving more fibers and myelin intact and thereby increasing function; chronic treatments could use rehabilitative strategies to enhance residual function or enhance conduction, thereby increasing function (e.g., ref. *22*). Alternatively, regenerative strategies might also be employed to add fibers to those surviving the initial and secondary insults *(23–28)*.

Many studies of recovery and regeneration after SCI have used transections or partial transections to produce localized, reproducible injuries that unequivocally sever specific spinal tracts. For example, studies of corticospinal tract

(CST) regeneration using antibodies to inhibitory molecules *(26,27)*, or applications of trophic factors *(28)*, have used a dorsal hemisection that completely severs the dorsal CST but spares ventral tracts. These injuries produce only mild locomotor deficits but are useful because they allow for quantitative assessment of the regeneration of completely severed tracts like the dorsal CST. Complete transections of the cord are used in bridging studies, in which complete destruction of all tracts is desired as a baseline for demonstration of the effects of small numbers of regenerating fibers *(23,24)*. Transection lesions are different from the contusion lesions in that they have a limited rostrocaudal spread of secondary injury *(29)*, and because contusion lesions have regions of partially damaged white matter that is subject to demyelination of intact axons (discussed in detail in Chapter 10). In addition, transections or hemisections are made with the meninges opened, which may lead to differences in the invasion of peripheral cells. There are important differences between lesion models that may relate to the biology underlying the mechanisms of repair and recovery, and studies that use both are particularly useful for comparisons between the occurrence and time-course of pathologic events. In this chapter, we review some of the recent work from our laboratory and those of others on cell death, repair, and recovery after contusion injuries, noting some comparisons with other models.

2. DEGENERATION AFTER SPINAL CORD INJURIES

2.1. Primary and Secondary Injury and Cell Death

While clearly direct damage is induced by contusion of the cord (e.g., membrane disruption, vascular damage, and hemorrhage), the final pathologic picture is far greater than that identifiable in the first few hours after injury. The spread of damage is thought to be due to activation of biochemical events leading to frank cellular breakdown (necrosis) and excitotoxic damage. The idea that secondary biochemical events contribute to SCI, especially contusion SCI, has been the driving force behind the development of therapies for acute injuries in humans *(16,17,30)*. Recent data suggest that in addition to secondary necrosis, the active processes of programmed cell death may be important in the production of damage after ischemia and central nervous system (CNS) trauma *(31)*. Studies in rats have shown that both neurons and glia can die by apoptosis after SCI and that the time-course of these events can extend for weeks *(32–35)*. Much of the apoptosis that occurs long after injury is related to wallerian degeneration; oligodendrocytes undergo apoptosis in the white matter regions containing degenerating axons. In monkeys with SCI, the same pattern is observed, with apoptotic cells located all along the degenerating ascending and descending tracts *(33)*. These observations have been extended to human SCI as well *(36)*. Wallerian degeneration is accompanied by the activation of resident microglia, which are in intimate contact with apoptotic oligodendrocytes *(37)*. It is suggested that these microglia (which also undergo apoptosis) may be responding to the death of the oligodendrocytes, or, alternatively, may be inducing apoptosis in oligodendrocytes via the release of cytokines or other factors. The end result may contribute to chronic demyelination and dysfunction after partial SCI. If so, then therapies aimed at apoptosis may provide novel means to affect outcome by reducing both neuronal death and demyelination.

2.2. Immune Response

Although the CNS has been presumed to be an immunologically privileged organ, it is now clear that there is an important immune component to CNS ischemia and injury *(38–42)*. This is easy to understand in traumatic SCI, in which the blood-brain barrier is damaged and peripheral cells can invade. In addition, however, an important response of resident cells, microglia, may also contribute to secondary damage through both apoptosis and necrosis and to repair. However, cytokines may also be involved in stimulating regeneration *(43)*. Growth factors such as fibroblast growth factor (FGF) can be released in response to injury by astrocytes, suggesting that there is an endogenous neuro-protective response to injury *(44,45)*.

Shuman et al. *(37)* showed that the pattern of microglial activation follows the pattern of wallerian degeneration and that microglia were closely apposed to apoptotic oligodendrocytes, suggesting that they may be involved in the activation of cell death programs. In models of multiple sclerosis, it has been shown that tumor necrosis factor-α can induce apoptosis in oligodendrocytes *(46)*. Activated microglia may secrete cytokines that can induce cell death in oligodendrocytes after spinal cord injury as well. A number of other cytokines and growth or neurotrophic factors are also candidates for the induction of cell death in oligodendrocytes. For example, nerve growth factor (NGF) has been considered to be neuroprotective, but studies in oligodendendrocytes have shown that NGF can induce apoptosis if the oligodendrocytes express the p75 NGF receptor *(47)*. Furthermore, Yoon et al. *(48)* have shown that this p75-mediated apoptosis may be opposed by cell-protective effects of NGF mediated through the trk A tyrosine kinase pathway. This provides an interesting example of balance between injury and repair mediated by a single ligand and suggests that the endogenous response to injury may depend on a complicated interplay among cytokines, growth factors, and neurotrophins induced by injury and also that the response could be very different in different cell types. Secretion of NGF by microglia apparently regulates cell death in the retina during development *(49)*. If microglia express NGF after SCI, then their close apposition to apoptotic oligodendrocytes could provide a pathway for cell death mediated by NGF. On the other hand, neurotrophins may induce repair, including the addition of new oligodendrocytes from the precursor population *(50)*.

2.3. Remote Effects on Cell Death and Degeneration

Axotomized neurons in remote nuclei may also undergo atrophy or cell death after injury, and this too may be regulated by endogenous responses to injury such as the production of neurotrophic factors *(51,52)*. Although cell death has not been studied systematically after contusion injuries, it is clear that spared descending systems play an important role in recovery *(3)*. Therefore, treatments aimed at sparing tissue in the lesion area might also affect the sparing of cells whose axons form long descending or ascending tracts. Bregman and colleagues have shown that transplants of fetal tissue, as well as the application of neurotrophins, can spare axotomized red nucleus neurons in neonatal and adult rats *(53)*. Neurotrophins can both spare neurons and promote regeneration after SCI *(51,54,55)*. In the adult cat, Goldberger, Murray, Tessler, and collaborators have shown that axotomy of Clarke's nucleus neurons results in death of about 30% of these cells. Interestingly, this cell death, which appears to be apoptotic, is prevented by sectioning the dorsal roots of the segments that innervate Clarke's nucleus cells *(56)*. This suggests that primary afferent activity may contribute to cell death in axotomized neurons. The death is also preventable by blocking glutamate receptors, and by applying neurotrophic factors *(57,58)*. Thus, a complicated series of interactions can determine whether cells live or die after axotomy. This situation is surely true also after contusion lesions, but may be even more complicated by the addition of the secondary events associated with lesion spread, for example, free radical stress.

3. REPARATIVE EVENTS AFTER INJURY

3.1. Compensation by Sprouting

Removal of circuits and cells by anterograde degeneration and retrograde cell death results in the availability of new sites for synaptic replacement or growth. It is likely that the final functional outcome is determined in part by reorganization of residual systems, for example, by synaptic sprouting *(58)*. Indeed, one of the earliest descriptions of plasticity in the nervous system was by Liu and Chambers *(59)*, who described reorganization in the primary afferent system after lesions in the spinal cord. These observations were followed by numerous studies of this phenomenon including those by Goldberger and Murray and their colleagues *(60,61)* in the spinal cord.

We have studied changes in inputs to spinal motoneurons and autonomic preganglionic neurons after cord lesions in our laboratory. The motoneurons that innervate the pudendal musculature including the external anal sphincter and the bulbospongiosus muscles are innervated by descending fibers from the brainstem *(62–64)*, which are lost following spinal cord transection. Evidence for compensatory sprouting of segmental systems was obtained in electron microscopic studies in cats *(65)*. Some of the synaptic sprouting appears to be γ-amino butyric acid (GABA) ergic *(66)*. Apparent increases were also seen in the number of calcitonin gene-related peptide-containing terminals in contact with sacral parasympathetic nucleus neurons, suggesting that dorsal root afferents sprout after transection *(67)*. Similar results have been reported in the rat thoracic intermediolateral cell column after hemisection *(68)*. Plasticity of synapses onto pudendal motoneurons in the rat has also been demonstrated with alterations in testosterone

(69,70). Together, these and other studies show convincingly that changes in segmental circuitry can occur in response to denervation and injury. Such changes might be important in recovery by restoring excitability to denervated neurons. For example, rapid changes in synaptic inputs to phrenic motoneurons are associated with return of respiratory function after spinal hemisection *(71,72).* However, as suggested originally by Liu and Chambers *(59),* recent evidence also suggests that sprouting might also play a role in dysfunction by increasing inappropriate inputs, thus causing spasticity or aberrant pain *(73,74).* Sprouting, of course, is not the only way in which synaptic organization can change after injury, and it is likely that many cellular synaptic changes underlie both recovery and the production of aberrant neural activity following injury.

There is also evidence that descending systems may sprout in cases of partial injury. Goldberger and Murray showed that serotonin (5-HT) fibers decreased and then reappeared in the dorsal horn after spinal hemisection in the cat *(75).* Sprouting and reorganization of the serotonergic system is thought to contribute to recovery of locomotor function after spinal hemisection *(76).*

The evidence for sprouting and synaptogensis after CNS injury in the adult mammal is now overwhelming, although its precise role in recovery remains to be seen. It may be that regulation of circuit reorganization will be an important means of therapy. For example, Schwab and collaborators have recently shown that antibodies raised against inhibitory proteins in myelin can be used to enhance sprouting of rostral systems after injury to descending spinal axons. This sprouting is correlated with return of function *(77).* Since synaptic plasticity and reorganization may also be regulated by endogenous growth and trophic factors *(78),* it appears again that the endogenous balance between growth-enhancing and growth-inhibiting molecules may determine the functional outcome.

Studies of reorganization after injuries to the spinal cord have used cord sections almost exclusively. Studies of remote reorganization after contusion lesions are needed to determine whether such plasticity is involved in recovery of function. In our contusion model in the rat, changes in spasticity of pudendal motoneuron reflexes are noted after contusion lesions, and this hyperreflexia is reduced over time, coincident with partial recovery of motor function *(79).* It is possible that this is due to sprouting of spared descending systems or of segmental systems.

3.2. Tissue Repair after Contusion Lesions

It has been recognized for years that glial cells, i.e., astrocytes, can proliferate or hypertrophy in response to injury, and that this reparative event, or the formation of scar, might actually retard regeneration. In addition, invasion of the CNS by peripheral cells, especially Schwann cells, is well documented, and this might provide the substrate for axonal growth *(80,81).* More recent evidence shows that there is a proliferative response in the adult CNS after injury, and that this response may actually involve the production of new neurons and glia that might be useful in repair.

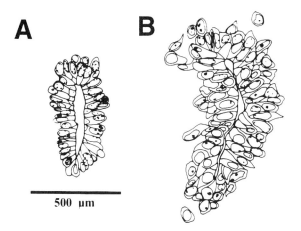

Fig. 2. Drawing from a 1-μm-thick plastic section showing the ependymal zone from a normal rat **(A)** and 48 hr after a contusion spinal cord injury **(B)** at the caudal end of the developing lesion.

3.2.1. Proliferation of Ependymal/Subependymal Cells After Injury

Work on amphibians in our laboratory and those of others *(82–84)* has shown that during metamorphosis, the site of a spinal transection is reconstituted, apparently by ependymal cell proliferation. Robust regeneration of descending fibers through this reconstituted cord has been documented. In mammals, very early injury to the cord may allow for cord reconstitution and substantial regeneration, as has recently been reported in the rat *(85)* and in the opossum *(86)*. This is likely to be related to the presence of a population of progenitor or stem cells that can be activated to reconstitute the cord matrix, and perhaps neurons as well. Our studies in tadpoles, as well as recent studies showing the presence of subependymal progenitor or stem cells in the adult rat brain and spinal cord *(87–89),* led us to examine the region of the central canal in normal rats and in rats with contusion lesions. We found that at several time points after injury, there was swelling and proliferation of cells in the region of the central canal at the rostral and caudal edges of the expanding secondary injury *(10;* Fig. 2). Bromodeoxyuridine labeling could be seen in ependymal cells in normal rats, and in many periependymal cells after injury (unpublished observations). These cells may provide a route of entry for CNS axons into the lesion cavity and may also contribute to the cellular matrix that includes invading Schwann cells and peripheral axons. A recent study by Frisen et al. *(87)* reports that both nestin and vimentin appear to be induced in cells near the central canal after a knife-cut injury of the dorsal columns in rats, as well as in astrocytes within long tracts undergoing degeneration. Our collaborators at UCLA have also documented nestin-positive cells apparently proliferating after contusion lesions in our rats;

some of these cells express the Id transcription factor genes related to astrocyte differentiation *(90)*. More recently, progenitor cells isolated from the postnatal rat cord have been shown to produce neuronal and glial progeny in vitro *(91)*. These findings support our hypothesis that the contusion lesion matrix is formed in part through the development of cellular trabeculae emerging from the region of the central canal at the ends of the expanding lesion cavity.

Additional new evidence shows that ependymal cells are themselves neural stem cells in the rat and mouse and that they respond to injury *(92)*. Although pluripotent cells may exist and may be stimulated to proliferate after injury, it is clear that they are not able to reconstitute the cord in the adult mammal the way they appear to do in amphibians and fetal mammals. It may be that the proliferation and differentiation of these cells is regulated by factors expressed in the injured adult nervous system. Information on that regulation, whether it be by cytokine-induced apoptosis *(93)*, or by growth factors, could serve to provide strategies for enhancing the reparative response using manipulation of endogenous precursors. Epidermal growth factor and basic FGF have been shown to enhance the production of progeny from adult progenitor cells *(89)*. Furthermore, a very recent study shows that FGF-2 is sufficient to induce proliferation of progenitor cells in adult rat spinal cord *(94)*. Thus, therapeutic effects of this growth factor might derive in part from effects on cellular proliferation.

After lesions made with the NYU/MASCIS weight drop contusion injury device, cells in the ependymal zone proliferate as the lesion expands *(10,32;* Fig. 3). The proliferating cells appear to provide cellular trabeculae that form a matrix on which axons can grow. Many of the axons that invade the lesion can

Fig. 3. *(Right)* Schematic (center) of a longitudinal section through a chronic contusion lesion with micrographs illustrating different features of the injury site. The lesion (center schematic, indicated in gray) extends rostrally and caudally, but spares a thin rim of tissue at the periphery (**A**; 20-μm-thick paraffin section, luxol fast blue stain). This rim surrounds a cavitated region that frequently contains a cellular matrix consisting of aggregates of macrophages (**D**, short arrow), blood vessels, and large numbers of axon fascicles myelinated by Schwann cells (**D**, long arrow). The lesion site fills with macrophages early (**B**; 1-μm-thick plastic section stained with toluidine blue, 5 days after injury), and a loose matrix of cells develops, providing paths for ingrowing axons and Schwann cells (**E**; the large arrow indicates an axon myelinated by a Schwann cell, and the small arrow indicates a fascicle of growing unmyelinated axons; 1-μm-thick plastic section, toluidine blue stain, 3 weeks after injury). The ependymal cells in the central canal regions at the edges of the lesion site appear to proliferate and can be observed to contribute to the cellular matrix in the lesion site (**C** and **F**; the region of the expanded canal in **C** is shown at higher power in **F**; 1-μm-thick plastic section, toluidine blue stain, 3 weeks after injury). Many of the axons in the lesion site appear to be from the dorsal roots (**A** and schematic), but some may be of CNS origin (schematic, arrows on corticospinal tract regenerates; see text).

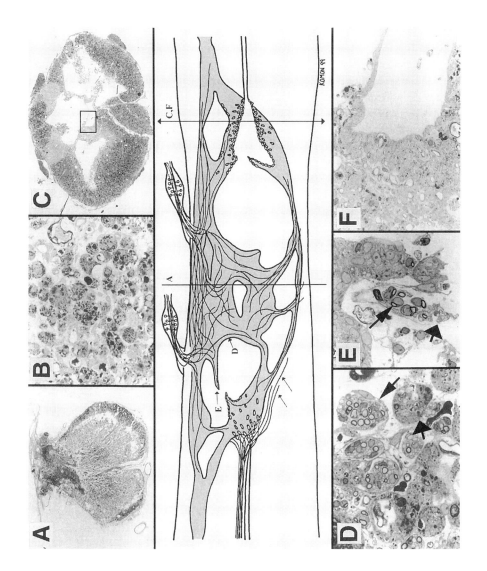

be traced to the dorsal roots; however, there is evidence that some may come from the CST. The possible presence of fibers from this and other CNS regions is under investigation in our laboratory. The chronic lesion at 6 weeks post injury is filled with fibers that are myelinated by Schwann cells, providing a favorable environment for axonal growth. Figure 3 shows many of the features of the repair process in the chronic contusion lesion.

Evidence for regeneration after contusion SCI comes from a large series of studies done by the MASCIS consortium and analyzed in the Beattie-Bresnahan labs *(10)*. The large lesion cavities produced by contusion lesions can become filled with cells and axons, as has been observed in human postmortem cases of SCI *(20)*. Aggregates of macrophages float in the cavities, which are lined by astrocytes (Fig. 3C–E). Cells appear to be born at the ependymal zone of the central canal at the rostral and caudal margins of the lesion (Fig. 3C, F) and to contribute to trabeculae, which divide the cavity rooms and provide a matrix for invading Schwann cells and axons (Fig. 3E). At 3 weeks after SCI, a peripheral rim of spared axons can be seen in the ventral and lateral funiculi. The center of the cavity is, however, filled with a loose matrix. Cellular elements that resemble immature astrocytes, Schwann cells, and large numbers of macrophages are present. Many growing axons are being wrapped by Schwann cells, and many fascicles of small myelinated and unmyelinated axons are interspersed with macrophages engorged with membranous whorls (Fig. 3D). At earlier times, this area is occupied by granular debris, myelin fragments, macrophages, and invading blood vessels (Fig. 3B).

The contusion lesions, filled with cellular elements and axons (Fig. 3A), are quite different from the typical picture seen after transection or hemisection, in which few if any axons cross the lesion site. It should be pointed out that the lesion sites in the contusions are frequently cavities that fill with peripheral nerve elements, so they are likely to be very different from the CNS compartment of the distal segment of the spinal cord.

There is evidence that some of the axons within the lesion are from the CNS. Injections of FluoroRuby (Molecular Probes, Eugene, OR) have been made into the motor region of the cerebral cortex or into the brainstem in rats with chronic spinal cord contusions *(95)*. The CST within the dorsal columns is completely disrupted by these lesions. Labeled fibers can be seen approaching the rostral end of the lesion and branching into the adjacent gray matter. Many axons exhibit retraction bulbs or "dystrophic" endings *(96,97)*. However, some appear to go around the rostral edge of the lesion and track along its lateral edge. Some of these reach quite far (several millimeters) beyond the disrupted tract. In addition, some of these axons enter the lesion cavity proper via the tissue bridges described above. The same pattern is seen for axons labeled by injections into the dorsal portion of the nucleus paragigantocellularis lateralis at the pontomedullary junction. It is therefore likely that at least some axons can regenerate and either grow alongside or enter the contusion lesion cavity. It seems unlikely, however, that these fibers contribute very much if at all to functional recovery. Their presence rather suggests that the capacity for regrowth is retained and that appropriate therapies with growth or trophic factors, or other treatments, might be used to enhance the regeneration to the extent that functional effects might be evident. The contusion lesion model is a unique way in which to investigate regeneration and recovery; it is more difficult to prove that fibers crossing the thoracic cord are actually regenerated than if the cord had been completely transected. On the other hand, since this lesion resembles many human cord injuries, it seems useful to determine what therapies are useful when some residual tissue remains intact. The lesion environment is different in contusions than in transections, and it may be that useful regeneration will be more easily enhanced in this situation. Indeed, enhancement of the number of neurofilament-positive axons within a contusion lesion cavity have been reported with application of fibroblasts engineered to produce neurotrophins *(50)*. Our laboratory will continue to examine therapies that may increase regeneration in the contusion model.

4. BEHAVIORAL AND NEUROLOGIC RECOVERY FOLLOWING SPINAL CORD LESIONS

4.1. Measures of Recovery of Function

Treatments can be designed based on our understanding of the biology of SCI, but for evaluation, relevant and reliable outcome measures are needed. The features and time-course of behavioral and neurologic recovery after SCI in rats have been documented by many laboratories. Here we will describe some measures of motor function that appear to be useful for evaluation of treatment and that also might give insights into which neural systems are involved in mediating recovery. We use individual tests in our laboratory; some investigators have developed combined scores that provide an overall assessment of sensorimotor function by adding individual test scores in a weighted fashion *(98)*. These generally correlate quite well with scores on the locomotor scale described below.

4.1.1. Locomotion

Walking has been a traditional outcome measure for SCI studies. Since most lesion models disrupt fibers descending in the thoracic cord, the segmental circuits and motoneurons necessary for locomotion are spared. Complete transection of the cord at thoracic levels yields complete paraplegia without recovery of overground locomotion *(3,4)*. Complete transection thus provides a good baseline to test for functional regeneration. Indeed, successful regeneration accompanied by some return of locomotion has been recently claimed *(24)*. It is also known that the central pattern generators for locomotion are present in the lumbosacral cord and can be trained and used in animals with complete spinal cord transections *(99; see chaps. 3 and 4)*. However, we have never seen any effective spontaneous (volitional) overground locomotion in rats with complete cord lesions (but see ref. *(100)* for the results of lesions early in development).

Partial sections of the descending tracts using either surgical sections or contusion spare descending fibers and also spare some return of locomotor function. We have extended the locomotor scales traditionally used in SCI studies to provide a scoring system based on the natural history of recovery of locomotor elements in rats with various severities of contusive injury. Rats are tested for spontaneous locomotion in an open field (actually, a child's swimming pool). This "Basso, Beattie, and Bresnahan" locomotor scale has been shown to be useful for describing the level of locomotor abilities in rats with different lesions and also for charting the course of locomotor recovery over time *(2,3,9)*. The scale does not provide the kind of detail obtained by kinematic or multiple muscle electromyographic (EMG) studies, but it does yield a specific profile of locomotor function that has proved to be "scalable". The scores are based on locomotor elements that correspond to both severity of injury and stage of recovery. Thus the lower third of the scale scores hindlimb movements, from none (0) to slight movements of single joints, to movements of all three joints in the limb. The next portion of the scale scores the presence of weight support, and then weight-supported stepping, and finally, coordination between the forelimbs and hindlimbs. The uppermost part of the scale scores limb position, ground clearance, and stability while walking, which are the final elements that recover in rats with only mild injuries.

In studies on many hundreds of animals, the outcomes have proved to be quite reliable and consistent even between laboratories *(9)*. It is useful in part because it takes only 4 min to obtain a score for each animal, allowing for rapid testing conducive to large scale preclinical trials. The scale has now been tested by a large number of laboratories, as well as in the MASCIS consortium laboratories, and is proving useful for communicating functional status between laboratories. It has also been shown to be sensitive to several treatments that appear to alter outcome. It is also now being used for locomotor analyses in the mouse.

Rats with complete transections of the cord can still exhibit hindlimb movements in the open field. Rats with slight injuries score a 21, the highest point on the scale. These extremes of the scale are best supplemented with other locomotor analyses. It is difficult to separate effects on segmental excitability from the effects of descending activation at the low end of the scale; evidence that descending influences are mediating recovery may require retransection of the cord *(3)*. Rats with good recovery (to 21 on the scale) may still show deficits if challenged in more difficult tasks such as walking on grids *(7)* or beam walking *(101)*. Because the scoring sheets provide more information than that used to generate the scores, it may be useful to devise subscaling strategies or other analysis techniques to increase the scale's sensitivity and reliability further.

4.1.3. Tests of Eliminative and Sexual Function

Bladder and bowel function are disrupted after SCI in rats as they are in humans. Much of the difficulty in maintaining healthy experimental animals involves bladder infections (and, more rarely, fecal impactions) that require extensive nursing care. A relationship between severity of injury and bladder dysfunction has been observed by many investigators and has been used by some as an index of recovery or response to treatment after partial lesions *(102,103)*. The dysfunction involves a bladder contraction with urethral sphincter dyssynergia in which the sphincter retards voiding. Rats with moderate injuries to the thoracic cord usually develop functional urination after several weeks. Data from the literature on sexual reflexes shows that in rats, *ex copula* penile reflexes are released from descending inhibition after thoracic cord lesions *(104,105)*. Much less attention has been given to bowel function in experimental SCI, although bowel dysfunction is a considerable problem in patients with SCI *(106,107)*. The external anal and urethral sphincters (EAS and EUS), as well as the somatic muscles that affect penile erection, are innervated by the pudendal motoneurons in the sacral cord of cats and the lumbosacral cord of rats. Our laboratory has been studying these neurons and alterations to their inputs after spinal lesions for some time *(65–67,69)*. More recently, we have studied the recovery of these reflexes after complete transection and partial, contusion lesions.

Immediately after complete spinal cord transection at thoracic levels, the external anal sphincter is completely flaccid and unresponsive to stimuli that elicit a rectoanal reflex. By 48 hr after the injury, however, stimuli that simulate passage of a fecal bolus, or tactile stimulation of the perianal region, elicit bursts of muscle contractions. By 4 days, the EAS is hyperreflexic, and minimal stimulation induces a sustained burst of EMG activity and visible contractions closing the anal orifice *(108)*. This hyperreflexia is maintained chronically and is unchanged at 6 weeks post transection. Penile reflexes are also chronically hyperreflexic after transection, as measured by erection magnitude and latency in *ex copula* tests. We also see hyperreflexia of pudendal reflexes in our contusion injury model. However, the hyperreflexia of both anal sphincter and penile reflexes is reduced over time, coincident with the restoration of partial locomotor function *(79)*. It is possible that similar reparative processes, such as sprouting of residual descending systems, may be responsible for these concurrent changes in different motor systems. We believe that this is another example of the potential usefulness of the contusion model in studying mechanisms of recovery, and are in the process of examining the putative role of descending 5-HT fibers in the restoration of function in our model.

5. USING ANIMAL MODELS TO PLAN STRATEGIES FOR CLINICAL INTERVENTION

Contusion lesion models in rodents and cats have proved to be useful in preclinical studies of neuroprotective agents. Methylprednisolone (MP), the current acute therapy for SCI, was tested in a long series of animal studies *(30)*, notably a compression model of partial SCI in cats *(109)*. Our contusion model *(6)* and the MASCIS trials (unpublished data) show a small, but significant protective effect of MP. Thus, models are now in place for evaluating acute therapies for SCI. We believe that contusion injuries can also contribute to the development of chronic therapies, first by providing insights into the biology of SCI, and second by providing a method for evaluating long-term recovery of function. Transplantation of human fetal tissue into cavities in the injured spinal cord has begun *(110,111)*, based on prior work in rat and cat models *(112)*. It is likely that strategies for tissue replacement and repair will be tested in a number of models of SCI, including contusion lesions. With the advent of a multitude of new strategies aimed at both acute and chronic repair and protection, it will be important to use a variety of models to tease out the multiple mechanisms of neurologic recovery. The exciting prospects for rehabilitation training in human SCI, again based on animal studies *(113,114)*, suggest that both long tracts and segmental circuits can be engaged in the recovery process. Thus, therapies that can affect wide-ranging aspects of recovery may be most useful *(115)*. Animal models will continue to improve and to provide important biologic and preclinical data that will contribute to success in treating SCI in humans.

REFERENCES

1. Beattie, M. S., Stokes, B. T., and Bresnahan, J. C. (1988) Experimental spinal cord injury: strategies for acute and chronic intervention based on anatomic, physiologic, and behavioral studies, in *Pharmacological Approaches to the Treatment of Brain and Spinal Cord Injury,* Stein, D. and B. Sabel, eds., Plenum, New York, pp. 43–74.
2. Basso, D. M., Beattie, M. S., and Bresnahan, J. C. (1995) A new sensitive locomotor rating scale for locomotor recovery after spinal cord contusion injuries in rats. *J. Neurotrauma* **12,** 1–21.
3. Basso, D. M., Beattie, M. S., and Bresnahan, J. C. (1996) Histological and locomotor studies of graded spinal cord contusion using the NYU weight-drop device versus transection. *Exp. Neurol.* **139,** 224–256.
4. Behrmann, D. L., Bresnahan, J. C., and Beattie, M. S. (1992) Graded controlled displacement lesions in the rat spinal cord: behavioral and histological analyses. *J. Neurotrauma* **9,** 197–217.
5. Behrmann, D. L., Beattie, M. S., and Bresnahan, J. C. (1993) A comparison of nalmefene, U-50488, and YM-14673 after controlled displacement spinal cord injury in the rat. *Exp. Neurol.* **119,** 258–267.
6. Behrmann, D. L., Beattie, M. S., and Bresnahan, J. C. (1994) Modeling spinal cord injury in the rat: neuroprotection and enhanced recovery with methylprednisolone and YM-14673. *Exp. Neurol.* **126,** 61–75.
7. Bresnahan, J. C., Beattie, M. S., Todd, F. D. III, and Noyes, D. H. (1987) A behavioral and anatomical analysis of spinal cord injury produced by a feedback-controlled impaction device. *Exp. Neurol.* **95,** 548–570.
8. Bresnahan, J. C., Beattie, M. S., Stokes, B. T., and Conway, K. M. (1991) Three-dimensional computer-assisted analysis studies of graded contusion lesions in spinal cord of the rat. *J. Neurotrauma* **8,** 91–101.
9. Basso, D. M., Beattie, M. S., Bresnahan, J. C., et al. (1996) Multicenter analysis of open field test locomotory scores: a test of training effects and reliability. *J. Neurotrauma* **13,** 343–359.

10. Beattie, M. S., Bresnahan, J. C., Koman, J., et al. (1997) Endogenous repair after spinal cord contusion injuries in the rat. *Exp. Neurol.* **148,** 453–463.
11. Noyes, D. L. (1987) An electromechanical impactor for producing experimental spinal cord injury in animals. *Med. Biol. Eng. Comp.* **25,** 335–340.
12. Stokes, B. T., Behrmann, D. L., and Noyes, D. H. (1992) An electromechanical spinal injury device with dynamic sensitivity. *J. Neurotrauma* **9,** 187–195.
13. Bresnahan, J. C., King, J. S., Martin, G. F., and Yashon, D. (1976) A neuroanatomical analysis of spinal cord injury in the Rhesus monkey *(Macaca mulatta). J. Neurol. Sci.* **28,** 521–542.
14. Guizar-Sahagun, G., Grijalva, O., Madrazo, E., Oliva, E., and Zepeda, A. (1994) Development of post-traumatic cysts in the spinal cord of rats subjected to severe spinal cord contusion. *Surg. Neurol.* **41,** 241–249.
15. Noble, L. and Wrathall, J. (1985) Spinal cord contusion in the rat: morphometric analyses of alterations in the spinal cord. *Exp. Neurol.* **88,** 135–149.
16. Bracken, M. B., Shepard, M. J., Collins, W. F., et al. (1990) A randomized controlled trial of methylprednisolone or naloxone in the treatment of acute spinal-cord injury. *N. Engl. J. Med.* **322,** 1405–1461.
17. Bracken, M. B., Shepard, M. J., Holford, T. R., et al. (1997) Administration of methylprednisolone for 24 or 48 hours or tirilazad mesylate for 48 hours in the treatment of acute spinal cord injury—results of the Third National Acute Spinal Cord Injury Randomized Controlled Trial. *JAMA* **277,** 1597–1604.
18. Constantini, S. and Young, W. (1994) The effects of methylprednisolone and the ganglioside GM1 on acute spinal cord injury in rats. *J. Neurosurg.* **80,** 97–111.
19. Bunge, R. P. (1994) Clinical implications of recent advances in neurotrauma research, in *The Neurobiology of CNS Trauma* (Salzman, S. A. and Faden, A. I., eds.), Oxford University Press, New York, pp. 329–339.
20. Bunge, R. P., Puckett, W. R., Becerra, J. L., Marcillo, A., and Quencer, R. M. (1993) Observations of the pathology of human spinal cord injury. A review and classification of 22 new cases with details from a case of cord compression with extensive focal demyelination. *Adv. Neurol.* **59,** 75–89.
21. Kakulas, B. A. (1984) Pathology of spinal injuries. *CNS Trauma* **1,** 117–129.
22. Hayes, K. C., Potter, P. J., Wolfe, D. L., Hsieh, J. T. C., Delaney, G. A., and Blight, A. R. (1994) 4-aminopyridine-sensitive neurologic deficits in patients with spinal cord injury. *J. Neurotrauma* **11,** 433–446.
23. Xu, X. M., Guenard, V., Kleitman, N., Aebischer, P., and Bunge, M. B. (1995) Axonal regeneration into Schwann cell seeded guidance channels grafted into transected adult rat spinal cord. *J. Comp. Neurol.* **351,** 145–160.
24. Cheng, H., Cao, Y., and Olson, L. (1996) Spinal cord repair in adult paraplegic rats: partial restoration of hindlimb function. *Science* **273,** 510–513.
25. Bernstein-Goral, H. and Bregman, B. S. (1993) Spinal cord transplants support the regeneration of axotomized neurons after spinal cord lesions at birth: a quantitative double-labeling study. *Exp. Neurol.* **123,** 118–132.
26. Bregman, B., Kunkel-Bagden, E., Schnell, L., Dai, N. Gao, D., and Schwab, M. E. (1995) Recovery from spinal cord injury mediated by antibodies to neurite growth inhibitors. *Nature* **378,** 498–501.

27. Schnell, L. and Schwab, L. (1990) Axonal regeneration in the rat spinal cord produced by an antibody against myelin associated neurite growth inhibitors. *Nature* **343,** 269–272.

28. Grill, R., Murai, K., Blasch, A., Gage, F. H., and Tuszynski, M. H. (1997) Cellular delivery of neurotrophin-3 promotes corticospinal axonal growth and partial functional recovery after spinal cord injury. *J. Neurosci.* **17,** 5560–5572.

29. Dushart, I. and Schwab, M. E. (1994) Secondary cell death and the inflammatory reaction after dorsal hemiscetion of the rat spinal cord. *Eur. J. Neurosci.* **2,** 654–662.

30. Hall, E. D. (1992) The neuroprotective pharmacology of methylprednisolone. *J. Neurosurg.* **76,** 13–22.

31. Johnson, E. M., Greenlund, L. J. S., Akins, P. T., and Hsu, C. Y. (1995) Neuronal apoptosis: current understanding of molecular mechanisms and potential role in ischemic brain injury. *J. Neurotrauma* **12,** 843–852.

32. Beattie, M. S., Shuman, S. L., and Bresnahan, J. C. (1998) Apoptosis and spinal cord injury. *Neuroscientist* **4,** 163–171.

33. Crowe, M. J., Bresnahan, J. C., Shuman, S. L., Masters, J. N., and Beattie, M. S. (1997) Apoptosis and delayed degeneration after spinal cord injury in rats and monkeys. *Nature Med.* **3,** 73–76.

34. Li, G. L., Brodin, G., Farooque, M., Funa, K., Holtz, A., Wang, W. L., and Olsson, Y. (1996) Apoptosis and expression of Bcl-2 after compression trauma to rat spinal cord. *J. Neuropathol. Exp. Neurol.* **55,** 280–289.

35. Liu, X. Z., Xu, X. M., Hu, R., et al. (1997) Neuronal and glial apoptosis after traumatic spinal cord injury. *J. Neurosci.* **17,** 5395–5406.

36. Emery, E., Aldana, P., Bunge, M. B., et al. (1998) Apoptosis after traumatic human spinal cord injury. *J. Neurosurg.* **89,** 911–920.

37. Shuman, S. L., Bresnahan, J. C., and Beattie, M. S. (1997) Apoptosis of microglia and oligodendrocytes after spinal cord injury in rats. *J. Neurosci. Res.* **50,** 798–808.

38. Blight AR. (1992) Macrophages and inflammatory damage in spinal cord injury. *J. Neurotrauma* **9 (Suppl 1),** S83–S92.

39. Jiang, N., Moyle, M., Soule, H. R., Rote, W. E., and Chopp, M. (1995) Neutrophil inhibitory factor is neuroprotective after focal ischemia in rats. *Ann. Neurol.* **38,** 935–942.

40. Popovich, P. G., Wei, P., and Stokes, B. T. (1997) Cellular inflammatory response after spinal cord injury in Sprague-Dawley and Lewis rats. *J. Comp. Neurol.* **377,** 443–464.

41. Streit, W. J. and Kincaid-Colton, C. A. (1995) The brain's immune system. *Sci. Am.* **Nov,** 54–61.

42. Gehrmann, J., Matsumoto, J., and Kreutzberg, G. W. (1995) Microglia: intrinsic immune effector cell of the brain. *Brain Res. Rev.* **20,** 269–287.

43. Hirschberg, D. L., Yoles, E., Belkin, M., and Schwartz, M. (1994) Inflammation after axonal injury has conflicting consequences for recovery of function: rescue of spared axons is impaired but regeneration supported. *J. Immunol.* **50,** 9–16.

44. Gomez-Pinilla, F., Vu, L., and Cotman, C. W. (1995) Regulation of astrocyte proliferation by FGF-2 and heparan sulfate in vivo. *J. Neurosci.* **15,** 2021–2029.

45. Teng, Y. D., Mocchetti, I., and Wrathall, J. R. (1998) Basic and acidic fibroblast growth factors protect spinal motor neurones in vivo after experimental spinal cord injury. *Eur. J. Neurosci.* **10,** 798–802.

46. Louis, J.-C., Magal, E., Takayama, S., and Varon, S. (1993) CNTF protection of oligodendrocytes against natural and tumor necrosis factor-induced death. *Science* **259,** 689–692.

47. Casaccia-Bonnefil, P., Carter, B. D., Dobrowsky, R. T., and Chao, M. V. (1996) Death of oligodendrocytes mediated by the interaction of nerve growth factor with its receptor p75. *Nature* **383,** 716–719.

48. Yoon, S. O., Casaccia-Bonnefil, P., Carter, B., and Chao, M. (1998) Competitive signaling between TrkA and p75 nerve growth factor receptors determines cell survival. *J. Neurosci.* **18,** 3273–3281.

49. Frade, J. M. and Barde, Y. A. (1998) Microglia-derived nerve growth factor causes cell death in the developing retina. *Neuron* **20,** 35–41.

50. McTigue, D. M., Horner, P. J., Stokes, B. T., and Gage, F. H. (1998) Neurotrophin-3 and brain-derived neurotrophic factor induce oligodendrocyte proliferation and myelination of regenerating axons in the contused adult rat spinal cord. *J. Neurosci.* **18,** 5354–5365.

51. Kobayashi, N. R., Fan, D. P., Giehl, K. M., Bedard, A. M., Wiegand, S. J., and Tetzlaff, W. (1997) BDNF and NT-4/5 prevent atrophy of rat rubrospinal neurons after cervical axotomy, stimulate GAP-43 and T alpha 1-tubulin mRNA expression, and promote axonal regeneration. *J. Neurosci.* **17,** 9583–9595.

52. Kuhn, P. L., Cantrall, J. L., Broude, E., and Bregman, B. S. (1998) GAP-43 re-expression in red nucleus neurons after thoracic spinal cord hemisection, transplantation and exogenous neurotrophin administration in adult rats. *Soc. Neurosci. Abst.* **24,** 70.

53. Deiner, P. S. and Bregman, B. S. (1994) Neurotrophic factors prevent the death of CNS neurons after spinal cord lesions in newborn rats. *Neuroreport* **5,** 1913–1917.

54. Ye, J. H. and Houle, J. D. (1997) Treatment of the chronically injured spinal cord with neurotrophic factors can promote axonal regeneration from supraspinal neurons. *Exp. Neurol.* **143,** 70–81.

55. Tetzlaff, W., Kobayashi, N. R., Giehl, K. M., Tsui, B. J., Cassar, S. L., and Bedard, A. M. (1994) Response of rubrospinal and corticospinal neurons to injury and neurotrophins. *Prog. Brain Res.* **103,** 271–286.

56. Sanner, C. A., Murray, M., and Goldberger, M. E. (1993) Removal of dorsal root afferents prevents retrograde death of axotomized Clarke nucleus neurons in the cat. *Exp. Neurol.* **123,** 81–90.

57. Sanner, C. A., Cunningham, T. J., and Goldberger, M. E. (1994) NMDA receptor blockade rescues Clarke's and red nucleus neurons after spinal hemisection. *J. Neurosci.* **14,** 6472–6480.

58. Shibayama, M., Hattori, S., Himes, B. T., Murray, M., and Tessler, A. (1998) Neurotrophin-3 prevents death of axotomized Clarke's nucleus neurons in adult rat. *J. Comp. Neurol.* **390,** 102–111.

58. Cotman, C. W. (1985) *Synaptic Plasticity,* Guilford Press, New York.

59. Liu, C. N. and Chambers, W. W. (1958) Intraspinal sprouting of dorsal root axons. *Arch. Neurol. Psychiatry* **79,** 46–61.

60. Goldberger, M. E. and Murray, M. (1988) Patterns of sprouting and implication for recovery of function, in *Advances in Neurology,* vol. 46 (Waxman, S. G., ed.), Raven, New York, pp. 361–385.

61. Goldberger, M. M., Murray, M., and Tessler, A. (1993) Sprouting and regeneration in the spinal cord: their roles in recovery of function after spinal injury, in *Neuroregeneration* (Gorio, A., ed.) Raven, New York, pp. 241–264.

62. Hermann, G. H., Holmes, G. M., Rogers, R. C., Bresnahan, J. C., and Beattie, M. S. (1998) Projections from nucleus raphe obscurus (NRO) to identified pudendal motoneurons in the male rat spinal cord. *J. Comp. Neurol.* **397,** 458–474.

63. Hermann, G. E., Holmes, G. M., Rogers, R. C., Beattie, M. S., and Bresnahan, J. C. (1996) Descending projections of the lateral paragigantocellular nucleus (LPGi): possible anatomical basis of role in lumbosacral reflexes. *Soc. Neurosci. Abstr.* **22,** 1835.

64. Holstege, G. and Tan, J. (1987) Supraspinal control of motoneurons innervating the striated muscles of the pelvic floor including urethral and anal sphincters in the cat. *Brain* **110,** 1323–1344.

65. Beattie, M. S., Leedy, M. G., and Bresnahan, J. C. (1993) Evidence for alterations of synaptic inputs to sacral spinal reflex circuits after spinal cord transection in the cat. *Exp. Neurol.* **123,** 35–50.

66. Li, Q., Beattie, M. S., and Bresnahan, J. C. (1995) Onuf's nucleus (ON) motoneurons (MNs) have more GABA-ergic synapses than other somatic MNs: a quantitative immunocytochemical study in the cat. *Soc. Neurosci. Abstr.* **21,** 1201.

67. Li, Q., Beattie, M. S., and Bresnahan, J. C. (1996) Ultrastructure of CGRP terminals in cat sacral spinal cord with evidence for sprouting in the sacral parasympathetic nucleus (SPN) after spinal hemisection. *Soc. Neurosci. Abstr.* **22,** 1843.

68. Krassioukov, A. V. and Weaver, L. C. (1996) Morphological changes in sympathetic preganglionic neurons after spinal cord injury in rats. *Neuroscience* **70,** 211–225.

69. Leedy, M. G., Beattie, M. S., and Bresnahan, J. C. (1987) Testosterone-induced plasticity of synaptic inputs to adult mammalian motoneurons. *Brain Res.* **424,** 386–390.

70. Matsumoto, A. Micevych, P., and Arnold, A. (1988) Androgen regulates synaptic input to motoneurons of adult rat spinal cord. *J. Neurosci.* **8,** 4168–4176.

71. Goshgarian, H. G., Yu, X.-J., and Rafols, J. (1989) Neuronal and glial changes in the rat phrenic nucleus occurring within hours after spinal cord injury. *J. Comp. Neurol.* **284,** 902–921.

72. Tai, Q. and Goshgarian, H. G. (1996) Ultrastructural quantitative analysis of glutamatergic and GABAergic synaptic terminals in the phrenic nucleus after spinal cord injury. *J. Comp. Neurol.* **372,** 343–355.

73. Christensen, M. D. and Hulsebosch, C. E. (1997) Chronic central pain after spinal cord injury. *J. Neurotrauma* **14,** 517–537.

74. Lindsey, A. E., LoVerso, R. L., Tovar, C. A., Beattie, M. S., Bresnahan, J. C. (1998) Mechanical allodynia and thermal hyperalgesia in rats with contsuion spinal cord injury (SCI). *J. Neurotrauma* **15,** 880 (Abstr.)

75. Wang, S. D., Goldberger, M. E., and Murray, M. (1991) Plasticity of spinal systems after unilateral lumbosacral dorsal rhizotomy in the adult rat. *J. Comp. Neurol.* **304,** 555–568.

76. Saruhashi, Y., Young, W., and Perkins, R. (1996) The recovery of 5-HT immunoreactivity in lumbosacral spinal cord and locomotor function after thoracic hemisection. *Exp. Neurol.* **139,** 203–213.

77. Thallmair, M., Metz, G. A. S., Z'Graggen, W. J., Raineteau, O., Kartje, G. L., and Schwab, M. E. (1998) Neurite growth inhibitors restrict plasticity and functional recovery following corticospinal tract lesions. *Nature Neurosci.* **1,** 124–131.

78. Thoenen, H. (1995) Neurotrophins and neuronal plasticity. *Science* **270,** 593–598.

79. Holmes, G. M., Bresnahan, J. C., Stephens, R. L., Rogers, R. C., and Beattie, M. S. (1998) Comparison of locomotor and pudendal reflex recovery in chronic spinally contused rats. *J. Neurotrauma* **15,** 874.

80. Bunge, M. B., Holets, V. R., Bates, M. L., Clarke, T. S., and Watson, B. D. (1994) Characterization of photochemically induced spinal cord injury in the rat by light and electron microscopy. *Exp. Neurol.* **127,** 76–93.

81. Bresnahan JC. (1978) An electron-microscopic analysis of axonal alterations following blunt contusion of the spinal cord of the rhesus monkey *(Macaca mulatta). J. Neurol. Sci.* **37,** 59–82.

82. Beattie, M. S., Lopate, G., and Bresnahan, J. C. (1990) Metamorphosis alters the response to spinal cord transection in *Xenopus laevis* frogs. *J. Neurobiol.* **21,** 1108–1122.

83. Michel, M. E. and Reier, P. J. (1979) Axonal-ependymal associations during early regeneration of the transected spinal cord in *Xenopus laevis* tadpoles. *J. Neurocytol.* **8,** 529–548.

84. Stensaas, L. J. (1983) Regeneration in the spinal cord of the newt Notopthalmus, in *Spinal Cord Reconstruction,* (Kao, C. C., Bunge, R. P. and Reier, P. J., eds.) Raven, New York, pp. 121–150.

85. Iwashita, Y., Kawaguchi, S., and Murata, M. (1994) Restoration of function by replacement of spinal cord segments in the rat. *Nature* **367,** 167–170.

86. Wang, X. M., Terman, J. R., and Martin, G. F. (1996) Evidence for growth of supraspinal axons through the lesion after transection of the thoracic spinal cord in the developing opossum, *Didelphis virginiana. J. Comp. Neurol.* **371,** 104–115.

87. Frisen, J., Johansson, C. B., Torok, C., and Lendahl, U. (1995) Rapid, widespread, and longlasting induction of nestin contributes to the generation of glial scar tissue after CNS injury. *J. Cell Biol.* **131,** 453–464.

88. Gage, F. H., Ray, J., and Fisher, L. J. (1995) Isolation, characterization, and use of stem cells from the CNS. *Annu. Rev Neurosci.* **18,** 159–192.

89. Weiss, J. S., Dunne, C., Hewson, J., et al. (1996) Multipotent CNS stem cells are present in the adult mammalian spinal cord and ventricular neuroaxis. *J. Neurosci.* **16,** 7599–7609.

90. Tzeng, S. F., Bresnahan, J. C., Beattie, M. S., and de Vellis, J. (1998) HLH Id gene family expression in neural cells of the rat spinal cord following contusion injury. *J. Neurochem.* **70,** S66 (abstract).

91. Kehl, L. J., Fairbanks, C. A., Laughlin, T. M., and Wilcox, G. L. (1997) Neurogenesis in postnatal rat spinal cord: a study in primary culture. *Science* **276,** 586–589.

92. Johansson, C. B., Momma, S., Clarke, D. L., Risling, M., Lendahl, U., and Frisen, J. (1999) Identification of a neural stem cell in the adult mammalian central nervous system. *Cell* **96,** 25–34.

93. D'Souza, S., Alinauskas, K., McCrea, E., Goodyear, C., and Antel, J. P. (1995) Differential susceptibility of human CNS-derived cell populations to TNF-dependent and independent immune-mediated injury. *J. Neurosci.* **15,** 7293–7300.

94. Shihabudden, L. S., Ray, J., and Gage, F. H. (1997) FGF-2 is sufficient to isolate progenitors found in the adult mammalian spinal cord. *Exp. Neurol.* **148,** 577–587.

95. Hill, C. E., Hermann, G. E., Beattie, M. S., and Bresnahan, J. C. (1998) Cortical and brainstem axons grow into lesion cavities after spinal cord contusion injury in the adult rat. *Soc. Neurosci. Abst.* **24,** 1729.

96. DePelipe, J. and Jones, E. G. (1991) *Cajal's Degeneration and Regeneration of the Nervous System,* Oxford University Press, New York.

97. Davies, S. J. A. and Silver, J. (1998) Adult axon regeneration in adult CNS white matter. *Trends Neurosci.* **15,** 515.

98. Gale, K., Kerasidis, H., and Wrathall, J. (1985) Spinal cord contusion in the rat: behavioral analysis of functional neurologic impairment. *Exp. Neurol.* **88,** 123–134.

99. Edgerton, V. R., Roy, R., Hodgeson, J., Prober, C., DeGuzman, C. P. and DeLeon, R., (1992) Potential of adult mammalian lumbosacral spinal cord to execute and acquire improved locomotion in the absence of supraspinal input. *J. Neurotrauma* **9(suppl 1),** S119–S128.

100. Wang, X. M., Basso, D. M., Terman, J. R., Bresnahan, J. C., and Martin, G. F. (1998) Adult opossums *(Didelphis virginiana)* demonstrate near normal locomotion after spinal cord transections as neonates. *Exp. Neurol.* **151,** 50–69.

101. Kunkel-Bagden, E., Dai, H. N., and Bregman, B. S. (1993) Methods to assess the development and recovery of locomotor function after spinal cord injury in rats. *Exp. Neurol.* **119,** 153–164.

102. Chancellor, M. B., Rivas, D. A., Huang, B., Kelly, G., and Salzman, S. K. (1994) Micturition patterns after spinal trauma as a measure of autonomic functional recovery. *J. Urol.* **151,** 250–254.

103. Pikov, V., Gillis, R. A., Jasmin, L., and Wrathall, J. R. (1998) Assesment of lower urinary tract functional deficit in rats with contusive spinal cord injury. *J. Neurotrauma* **15,** 375–386.

104. Holmes, G. M. and Sachs, B. D. (1992) Erectile function and bulbospongiosus EMG activity in estrogen-maintained castrated rats vary with behavioral context. *Horm. Behav.* **26,** 406–419.

105. Marson, L. and McKenna, K. E. (1996) CNS cell groups involved in the control of the ischiocavernosus and bulbospongiosus muscles: a transneuronal tracing study using pseudorabies virus. *J. Comp. Neurol.* **374,** 161–179.

106. Cosman, B. C., Stome, J. M., and Perkash, I (1991) Gastrointestinal complications of chronic spinal cord injury. *J. Am. Paraplegia Soc.* **14,** 175–181.

107. Longo, W. E., Ballantyne, G. H., and Modlin, I. M. (1989) The colon, anorectum, and the spinal cord patient. A review of the functional alterations of the denervated hindgut. *Dis. Colon Rectum* **32,** 261–267.

108. Holmes, G. M., Rogers, R. C., Bresnahan, J. C., and Beattie, M. S. (1998) External anal sphincter hyper-reflexia following spinal transection in the rat. *J. Neurotrauma* **15,** 451–457.

109. Braughler, J., Hall, E., Means, E., Waters, T., and Anderson, D. (1987) Evaluation of an intensive methylprednisolone-sodium succinate dosing regimen in experimental spinal cord injury. *J. Neurosurg.* **67,** 102–105.

110. Wirth, E. D. III, Fessler, R. G., Reier, P. J., et al. (1998) Feasibility and safety of neural tissue transplantation in patients with syringomyelia, *Soc. Neurosci. Abst.* **24,** 70.

111. Thompson, F. J., Uthman, B., Mott, S., et al. (1998) Neurophysiological assessment of neural tissue transplantation in patients with syringomyelia, *Soc. Neurosci. Abst.* **24,** 70.

112. Giovanini, M. A., Reier, P. J., Eskin, T. A., Wirth, E., and Anderson, D. K. (1997) Characteristics of human fetal spinal cord grafts in the adult rat spinal cord: influences of lesion and grafting conditions. *Exp. Neurol.* **148,** 523–543.

113. Harkema, S. J., Hurley, S. L., Patel, U. K., Requejo, P. K., Dobkin, B. H., and Edgerton, V. R. (1997) Human lumbosacral spinal cord interprets loading during stepping. *J. Neurophysiol.* **77,** 797–811.

114. Edgerton, V. R., Roy, R., DeLeon, R., Tillakaratne, N., and Hodgeson, J. (1997) Does motor learning occur in the spinal cord? *Neuroscientist* **3,** 287–294.

115. Jakema, L. B., Wei, P., Guan, Z., and Stokes, B. T. (1998) Brain-derived neurotrophic factor stimulates hindlimb stepping and sprouting of cholinergic fibers after spinal cord injury. *Exp. Neurol.* **154,** 170–184.

2

Calcium and Neuronal Death in Spinal Neurons

Gordon K.T. Chu, Charles H. Tator, and Michael Tymianski

1. INTRODUCTION

Spinal cord trauma can result in the death of neurons both immediately after injury and several hours afterwards. The cause of immediate neuronal death is thought to be mechanical impact, but delayed neuronal damage is believed to be caused by ischemia secondary to decreased spinal cord blood flow, hemorrhage, and edema *(198)*. The resultant neuronal death is thought to involve calcium entry similar to focal ischemic brain injuries. The predominant hypotheses on the mechanisms by which Ca^{2+}-dependent neurodegeneration is triggered are discussed in the context of current understanding of physiologic and pathologic Ca^{2+} signaling. Specifically, we address whether measurable relationships exist between Ca^{2+} influx and neurotoxicity, whether Ca^{2+} excess is a sufficient trigger for neurotoxicity, and whether Ca^{2+} neurotoxicity is a function of Ca^{2+} ion concentration, spatial distribution, and influx pathway. We focus primarily on the factors that relate cellular Ca^{2+} ions to the requirements for triggering neurotoxicity.

Calcium entry into neurons by glutamate activation of *N*-methyl-D-aspartate (NMDA) receptors resulting in neuronal death (the excitotoxicity hypothesis) are emphasized because studies have shown that glutamate may rise to toxic levels in the extracellular fluid after experimental central nervous system (CNS) trauma or ischemia *(15,56,57,68,108,119,120)*. Since Ca^{2+} ions regulate many processes that may lead to cytodestruction, such secondary processes are to be reviewed.

2. CALCIUM HOMEOSTASIS IN THE NEURON

Calcium ions are ubiquitous intracellular second messengers that regulate numerous cellular functions. Neurons must tightly control the free cytosolic Ca^{2+} ion concentration ($[Ca^{2+}]_i$) to allow efficient Ca^{2+}-dependent signaling to occur. To achieve an acceptable signal-to-noise ratio for Ca^{2+} signaling, resting $[Ca^{2+}]_i$), must remain at very low levels (around 100 nM, or 10^5 times lower than

From: *Neurobiology of Spinal Cord Injury*
Edited by: R. G. Kalb and S. M. Strittmatter © Humana Press Inc., Totowa, NJ

extracellular $[Ca^{2+}]$), so that relatively small or localized increases in $[Ca^{2+}]_i$ can be used to trigger physiologic effects. Because calcium is so vital to the survival of the cell, various mechanisms have evolved to maintain calcium homeostasis. These mechanisms can be divided into five categories: Ca^{2+} influx, Ca^{2+} buffering, internal Ca^{2+} storage, Ca^{2+} efflux, and intracellular Ca^{2+} diffusion.

2.1. Ca^{2+} Influx

Ca^{2+} influx occurs primarily through ion channels. These specialized pores in cell membranes are classified physiologically by their specific selectivities for certain ions (Ca^{2+}, K^+, Na^+, or Cl^-) and by their gating mechanism. Some ion channels are sensitive to membrane voltage, whereas others are associated with specific ligands. Voltage-gated ion channels may contain binding sites for certain ligands, and ligand-gated channels may exhibit certain forms of voltage dependence. Voltage-gated calcium channels are numerous and include N, L, P, Q, and T types *(121,131)*. Ligand-gated channels that permit calcium entry include NMDA and certain α-amino-3-hydroxy-5-methyl-4-isoxa zolepropionate (AMPA)/kainate receptors. The present review mainly focuses on calcium entry through NMDA receptors because it plays an important role in calcium-dependent neuronal death compared with calcium entry through other channels.

2.2. Ca^{2+} Buffering

Calcium buffering allows the neuron to regulate the spread of Ca^{2+} ions within the cell. As Ca^{2+} ions diffuse into the cell, they are rapidly buffered by a number of cytoplasmic proteins, such as calmodulin, calbindin, and parvalbumin *(7)*. Approximately 95–99% of Ca^{2+} ions entering the cell under physiologic conditions are buffered in this fashion *(138,221)*. The precise role of Ca^{2+} buffering substances remains poorly understood; however, they may act to keep $[Ca^{2+}]_i$ at high levels in localized areas within cells, to limit those high $[Ca^{2+}]_i$ levels to those specific areas, and to rapidly dissipate the Ca^{2+} gradients and thus limit the time-course of activation of Ca^{2+}-dependent processes *(23,90,140,160)*. Ca^{2+} buffers are also able to increase the apparent diffusion of Ca^{2+} ions within the cell. Thus, they may act as Ca^{2+} shuttles, carrying Ca^{2+} ions from their site of influx to and away from their site of action *(137,160,169,192,193)*. These effects are delicately balanced and depend on the distribution, type, and concentration of the Ca^{2+} buffer within the cell.

2.3. Ca^{2+} Storage

When Ca^{2+} loads exceed the buffering capacity of the neuron, Ca^{2+} ions may be sequestered into organelles such as the smooth endoplasmic reticulum, mitochondria, and synaptic vesicles *(17,33)*. These organelles can store large quantities of Ca^{2+} under a variety of conditions, using active and passive Ca^{2+} transport mechanisms similar to those found in the plasma membrane (see below). Although Ca^{2+} storage in organelles is an efficient mechanism for controlling cytoplasmic $[Ca^{2+}]_i$, this Ca^{2+} "lowering" system operates at a much slower time scale than cytoplasmic Ca^{2+} binding proteins. Therefore, it is incapable of modulating rapidly changing or highly localized changes in $[Ca^{2+}]_i$ *(215)*.

2.4. Ca^{2+} Efflux

Due to the large extracellular-to-intracellular Ca^{2+} ion concentration gradient and the electrical driving force propelling the positively charged Ca^{2+} ions toward the negatively charged inner plasma membrane, calcium efflux is an energy-dependent event in the cell. Neurons have at least two Ca^{2+} extrusion mechanisms, adenosine triphosphate (ATP)-driven Ca^{2+} pumps (Ca^{2+}-ATPases) and an Na^{2+}/Ca^{2+} exchange transport mechanism *(17,21,112,131,135,196)*. Ca^{2+} ATPases in the plasma membrane are modulated by calmodulin, a number of fatty acids, and protein kinases (protein kinases A and C). One ATP molecule is expended for each Ca^{2+} ion extruded. Ca^{2+} ATPases also exist in the membranes of the smooth endoplasmic reticulum, acting as a mechanism of intracellular Ca^{2+} sequestration. These are calmodulin independent and sequester two Ca^{2+} ions for each ATP molecule. The Na^{2+}/Ca^{2+} exchanger is triggered by a rise in $[Ca^{2+}]_i$, removing one Ca^{2+} ion for two to three Na^+ ions that enter. This process is dependent on the Na^+, K^+-ATPase.

2.5. Ca^{2+} Diffusion

Subcellular Ca^{2+} concentrations within the neuron can differ and are determined by the type and the subcellular distribution of Ca^{2+} entry sites (Ca^{2+} channels); they are also modulated by Ca^{2+} buffering and sequestration systems and by Ca^{2+} extrusion mechanisms. These modulators have marked effects on the ability of Ca^{2+} ions to diffuse within the cell. For example, Ca^{2+} ions bound to Ca^{2+} binding proteins possess markedly different diffusion characteristics from free Ca^{2+} ions *(90,137,140,159,160,192,193 195,221)*. Also, Ca^{2+} influx sites may be colocalized with Ca^{2+} binding sites within the cell (e.g., Ca^{2+}-dependent enzymes) *(93)*. These mechanisms serve to couple Ca^{2+} entry selectively with specific intracellular targets of Ca^{2+} ions. Thus any process that modifies cytoplasmic Ca^{2+} diffusion may disrupt this coupling process and prevent Ca^{2+} ions from reaching their sites of intracellular action. For example, synaptic transmitter release in neurons can be attenuated by introducing into cells exogenous Ca^{2+} chelating agents, which modify the intracellular diffusion of Ca^{2+} ions *(137,193)*. These agents act by disturbing the balance between Ca^{2+} influx and Ca^{2+} action at synaptic active zones where transmitter vesicles are released *(1,4)*.

The five mechanisms discussed above control the regulation of intracellular calcium in the neuron, and it is widely believed that when neurons experience Ca^{2+} increases that cannot be controlled by the Ca^{2+} regulating machinery, cell death ensues. How the Ca^{2+} ions trigger neurotoxicity, and by which molecular mechanisms, remains controversial.

3. EARLY WORK ON CALCIUM LONS AS MEDIATORS OF CYTOTOXICITY

Early on, Ca^{2+} ions were believed to play a role in cytotoxicity because of observations that livers damaged by toxins accumulated calcium, which suggested that calcium entry may be responsible for tissue damage *(127)*. Later experiments by Schanne et al. *(171)* revealed that adult hepatocytes in primary cultures were killed when exposed to various toxins (believed to affect plasma membrane integrity) in the presence, but not the absence, of extracellular Ca^{2+}. The authors believed that Ca^{2+} influx into cells was an absolute requirement for the expression of toxicity, and they termed this process the "final common pathway of cell death." Even prior to this, Zimmerman and colleagues *(222)* established a link between calcium and cytotoxicity; they observed in isolated heart preparations that perfusion with calcium-deficient solutions, followed by reperfusion with solutions containing calcium, resulted in rapid cessation of contractility followed by massive, widespread cell death (the calcium paradox). A Ca^{2+}-dependent toxicity was also then later shown by Duce and colleagues *(52)* in locust muscle cells stimulated by the excitatory neurotransmitter L-glutamate. These experiments demonstrate a link between a Ca^{2+}-mediated cytoxicity and neurotoxicity secondary to synaptic overactivtiy (see below).

However, the notion that cell death is absolutely dependent on Ca^{2+} ions was quickly challenged by other (for review, see ref. *25*). For example, several reports indicated that cytotoxicity in hepatocyte preparations can be produced in the absence of calcium *(58,189)*. More recently, the loss of hepatocyte viability during chemical hypoxia was shown to occur prior to any measured rise in intracellular Ca^{2+} concentration *(104)*, confirming that mechanisms other than those triggered by Ca^{2+} excess can also be cytotoxic. Similarly, cell death in the heart under some conditions can be triggered independently of variations in extracellular Ca^{2+} *(26)*, suggesting that in cardiac muscle, mechanisms other than those responsible for the calcium paradox may be operative.

Therefore all forms of cell death may not involve Ca^{2+} influx; nevertheless, considerable evidence indicates that in the adult mammalian nervous system, cellular Ca^{2+} overload is intimately related to traumatic/ischemic neuronal death. Early studies in tissue cultures showed that amputated axons degenerated only if Ca^{2+} ions were present in the culture medium *(172)*. In recent experiments in which Ca^{2+} accumulation was examined directly, neurodegeneration induced by neurotoxins such as capsaicin and glutamate was shown to be associated with increases in tissue Ca^{2+} *(85)*. Further investigations on the toxicity of excitatory amino acids (EAAs) in cultured neurons and brain slices confirmed an association between the observed toxicity and the presence of Ca^{2+} in the extracellular medium *(27,62,63,81)*. However, even this association was not absolute, as some subsequent studies produced evidence that was seemingly contradictory to the Ca^{2+} hypothesis *(35,152)*. Such contradictions illustrate that the rules governing the association between Ca^{2+} overload and neurotoxicity were poorly understood.

Animal studies also support the association between Ca^{2+} influx and damage to neural tissues. Within seconds to minutes after trauma to rat spinal cord, extracellular calcium ($[Ca^{2+}]_e$) decreases to <0.01 mM *(194,219,220)*. Experimental spinal cord injury also produces significant Ca^{2+} accumulation in white matter axons *(8–10)*, possibly as a consequence of white matter anoxia/ischemia *(196,199)*. Cerebral ischemia and epileptic seizures also appear to precipitate intracellular Ca^{2+} accumulation. This was demonstrated by studies using extracellular ion-selective microelectrodes, which showed a marked decrease in extracellular Ca^{2+} concentration following the induction of experimental cerebral ischemia *(66,75,139)*. Numerous later studies, using electron microscopy, autoradiographic techniques, atomic absorption spectroscopy, and free Ca^{2+} measurements with fluorescent Ca^{2+}-sensitive dyes, confirmed an increase in tissue and particularly cellular Ca^{2+} concentration following cerebral ischemia or seizures *(14,24,47,128,183,187,208,209,210)*.

The studies described above provide a foundation for what has been termed the *calcium hypothesis*, which states that neuronal Ca^{2+} overload leads to subsequent neurodegeneration. However, despite numerous subsequent overloads and neurotoxicity *(55,69,115)*, the mechanisms by which Ca^{2+} ions may trigger cell death remain poorly defined, as virtually every major Ca^{2+}-dependent intracellular process has, at some time, been implicated as the causative mechanism. For example, neurotoxic actions of Ca^{2+} overload have been ascribed to the overstimulation of enzymes such as phospholipases, plasmalogenase, calpains and other proteases, protein kinases, guanylate cyclase, nitric oxide synthetase, calcineurins, and endonucleases. Presumably, this leads to overproduction of toxic reaction products such as free radicals, to lethal alterations in cytoskeletal organization, or to activation of genetic signals leading to cell death *(59,203; see* Section 2 below).

4. CA²⁺ IONS, GLUTAMATE RECEPTORS, AND THE EXCITOTOXICITY HYPOTHESIS

Glutamate is the major excitatory neurotransmitter in the mammalian nervous system *(40,96,134)* and acts on several cell membrane receptors. The two groups of glutamate receptors are classified as ionotropic or metabotropic, based on pharmacologic, electrophysiologic, and biochemical studies (see refs. *40,96,134)*. The ionotropic receptors can be broadly subdivided into the NMDA receptors and the AMPA/kainate receptors, identified according to their selective agonists. These receptors are permeable to Na^+, K^+, and Ca^{2+} ions, although the permeabilities differ with each type of receptor. The metabotropic receptors mediate their actions through guanosine triphosphate (GTP) binding protein-dependent mechanisms to initiate phophoinositide hydrolysis, which forms inositol 1,4,5-triphosphate, and diacylglycerol, which can release Ca^{2+} ions from internal stores *(130,133,136,217)*. Recently, molecular approaches have demonstrated that many receptor subtypes exist within each receptor family, with different functional properties depending on their subunit composition and the subunit isoforms generated by alternative spicing *(134)*. The functional consequences of this diversity are incompletely understood and remain under intense investigation.

Rapid neuronal excitation, plasticity, and possibly processes involved in learning and memory *(49,134)* are associated with synaptic glutamate release *(34,134,217)*. The role of glutamate as a potential neurotoxin was noted prior to the realization that cellular Ca^{2+} accumulation may be critical in neurotoxicity (*see* section 3.). Systemic injections of L-glutamate and other related EAAs in neonatal mice by Lucas and Newhouse *(111)* and later Olney *(144,145)* destroyed the inner neural layers of the retina and produced lesions in brain areas lacking a blood-brain barrier. Attention was focused on excitatory synaptic transmission as the source of glutamate excess when Kass and Lipton *(91)* and Rothman *(163,164)* reported that the attenuation of synaptic transmission by magnesium ions and glutamate receptor antagonists reduced hypoxic/anoxic neuronal death in vitro. This concept was reinforced when glutamate receptor antagonists such as 2-amino-7-phosphonoheptanoic acid and MK-801 were found to diminish ischemic neuronal injury in vivo *(149,188)*. Thereafter, the term *excitotoxicity,* initially coined by Olney, came to indicate the process by which excessive synaptic release of EAAs resulted in neurotoxicity.

Several key experiments have been performed to dissect the mechanisms by which glutamate excess triggers toxicity. These have been comprehensively reviewed by others *(29,30,31,59,166,179,181)*. To date, virtually every glutamate receptor subtype has been implicated in mediating hypoxic/ischemic neuronal death *(31,64,71,95,147,211)*. However, there is general consensus that the NMDA glutamate receptor subtype plays a key role in mediating at least certain aspects of glutamate neurotoxicity, because of its high Ca^{2+} permeability compared with non-NMDA ionotropic glutamate receptors *(66,79,83,153,167)*. Since NMDA and other EAAs could produce calcium influx into cells *(16,113)*, studies of the pathophysiology of neurotoxicity began focusing on the ionic changes induced by glutamate. Rothman *(165)* initially reported that hippocampal neurons were susceptible to glutamate, kainate, and NMDA neurotoxicity in the absence of calcium ions and were resistant to the calcium ionophore A23187. Instead he noted that the absence of extracellular chloride was protective. He hypothesized that neurotoxicity of glutamate could be explained by a sequence involving glutamate-induced depolarization, followed by passive chloride influx. Cations would then be drawn into the cell, resulting in water entry and cell lysis. These experiments established the phenomenon of neuronal swelling as a consequence of excitotoxin application, but conflicted with the opinion that Ca^{2+} ions play a role in the neurodegeneration process *(27)*.

In 1987, Choi and colleagues *(28,32)* reported an extensive characterization of glutamate neurotoxicity in cultured cortical neurons. They observed that brief exposures to glutamate produced morphologic changes characterized by swelling, darkening (under phase optics), and increased granularity. In subsequent hours, this was followed by marked neurodegeneration. These effects were only seen in mature cultures, as reported previously by Rothman (*see* previous paragraph). Using ionic substitution experiments, Choi reported that glutamate neurotoxicity could be divided into two components distinguished on the basis of differences in time-course and ionic dependence. The first component, occurring within seconds, was characterized by cell swelling and depended on the presence of extracellular Na^+ and Cl^- but was reversible, and not necessarily neurotoxic. The second component was marked by neuronal disintegration, occurred within hours, was dependent on extracellular Ca^{2+}, and could be mimicked by A23187. Choi concluded that at lower glutamate exposures, the Ca^{2+} component predominated as the primary neurotoxic mechanism. These findings have stood the test of time and were subsequently reproduced by many investigators *(155)*.

Glutamate neurotoxicity was similarly tested in spinal cord neurons *(123)*. Spinal cord cell cultures from fetal mice were exposed for 5 min to varying concentrations of glutamate (10–1000 μM), and survival was assessed after 24 hr. It was determined that the half-maximal response (EC_{50}) for neuronal death was 100–200 μ*M* glutamate; in contrast, the glia were not injured. Cultures as young as 4 days in vitro were found to be susceptible, although greater cell death was experienced by older cultures. Ionic substitution experiments with these spinal cultures found that, similar to cortical cultures, neuronal swelling was dependent on extracellular sodium whereas neuronal degeneration was attenuated by removal of extracellular calcium. Neuronal death from glutamate exposure could also be reduced by the NMDA antagonists dextrophan and CGP 37849. However, there are some differences between cortical and spinal neurons. Regan *(122)* found that spinal neurons were less vulnerable to NMDA toxicity and more vulnerable to AMPA and kainate toxicity than cortical neurons. Terro et al. *(125)* found that motoneuron death in ventral spinal cord cultures caused by mild kainate exposure was associated with Ca^{2+} entry through Ca^{2+}-permeable AMPA/kainate receptors.

Although Ca^{2+}-independent mechanisms capable of killing neurons *(152)* exist undoubtedly, Ca^{2+} neurotoxicity has been felt to be a major factor in a number of physiologic preparations. However, this has been difficult to prove directly since both excitotoxicity and hypoxia-ischemia trigger numerous simultaneous changes in neurons. Thus, although many studies implicate Ca^{2+} influx in the neurotoxic process *(27,55,85,146)*, its contribution to toxicity following an insult has been difficult to separate from that of a multitude of non-Ca^{2+}-dependent processes set in motion at the same time (e.g., activation of Na^+, K^+, Cl^- conductances) *(165)*. The experiments that were most effective in overcoming this difficulty have employed Ca^{2+} concentration-response paradigms. One such experiment exposed cultured neurons to a fixed challenge with 250 μ*M* glutamate while varying, in different experiments, $[Ca^{2+}]_e$ from 1 to 10 m*M* *(205)*. There was a sigmoidal concentration-response relationship between $[Ca^{2+}]_e$ and neuronal death. Interestingly, despite this clear relationship, changes in $[Ca^{2+}]_e$ increases did not produce proportional changes in measured peak or average $[Ca^{2+}]_i$. demonstrating that factors other than the transmembrane Ca^{2+}

gradient affect $[Ca^{2+}]_i$ (these will be discussed in Section 7. below.). The strategy of varying the extracellular Ca^{2+} concentration and keeping all other study parameters unchanged isolates the neurotoxic effects of Ca^{2+} ions because the impact of any Ca^{2+}-influx-independent neurotoxic processes would probably remain unaltered. This approach has been used in cell cultures *(142,205)* and in brain slices *(62,110)* to test the Ca^{2+} dependence of EAA toxicity and of hypoxia-ischemia. The results have consistently demonstrated a marked dependence of neuronal survivability on extracellular Ca^{2+} concentrations during excitotoxic or hypoxic insults. These data have provided the most direct evidence to date that Ca^{2+} ions play a key role in the neurotoxicity process.

5. NEURODEGENERATION IS A CONSEQUENCE OF A DEREGULATION OF CA^{2+} HOMEOSTATIC MECHANISMS

Cellular energy failure from trauma or hypoxia-ischemia may result in Ca^{2+} overload through a number of mechanisms, including increased Ca^{2+} influx, decreased Ca^{2+} efflux, and altered internal Ca^{2+} buffering and sequestration. It has been hypothesized that this Ca^{2+} overload contributes further to cell death. Energy depletion causing depolarization may produce increased neurotransmitter release from synaptic terminals, and a decrease in transmitter reuptake. Depolarization-induced relief of magnesium block of NMDA channels may promote NMDA receptor activation, adding to excess Ca^{2+} influx and neurotoxicity *(31)*. Ca^{2+} influx through non-NMDA receptors and Ca^{2+} release from internal stores may also add to cellular Ca^{2+} overload and eventual neurotoxicity. Furthermore, after CNS trauma, Ca^{2+} ions may enter the cell through gaps in the cell membrane caused by the mechanical insult *(223,224)*.

Several studies have suggested a link between deregulation of Ca^{2+} homeostasis and neuronal death. Using isolated hippocampal neurons, Ogura et al. *(142)* assessed the dependency of glutamate-induced neurodegeneration on cytosolic Ca^{2+} concentrations. Neurons were loaded with the Ca^{2+} indicator fura-2 and briefly exposed to glutamate. They experienced a rise in $[Ca^{2+}]_i$ followed by an adaptive phase lasting several minutes during which $[Ca^{2+}]_i$ declined toward basal levels. The authors reported that the time required for neurons to recover from the acute $[Ca^{2+}]_i$ increase correlated with neuronal survival at 24 hr. More recently, we and others have studied the regulation of $[Ca^{2+}]_i$ during neurotoxic challenges with glutamate. We challenged cultured spinal neurons with 250 μM glutamate for 50 min. Upon glutamate application, $[Ca^{2+}]_i$ transiently increased to a peak value and then decayed to a steady-state plateau that lasted for the duration of the challenge. During recovery from the glutamate application $[Ca^{2+}]_i$ declined further, although it often failed to return to baseline, possibly owing to persistent neuronal depolarization *(190)*. Many neurons then underwent a delayed, sustained, and generally irreversible rise in $[Ca^{2+}]_i$ that often exceeded the dynamic range of the Ca^{2+} indicator. This phenomenon preceded neuronal staining with the vital dye trypan blue by about

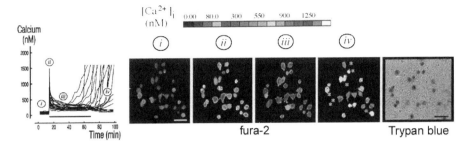

Fig. 1. Deregulation of [Ca²⁺]ᵢ homeostasis in cultured spinal neurons. The neurons were exposed to 250 μ*M* L-glutamate for a 50-min period, and [Ca²⁺]ᵢ was measured with the fluorescent indicator fura-2. At the end of the experiment, the same cells were stained with the viability indicator trypan blue. All neurons experienced a transient increase in [Ca²⁺]ᵢ from baseline (i) to a peak (ii), which was then reduced by homeostatic mechanisms to a lower plateau (iii). Neurons that underwent a deregulation in calcium homeostasis then underwent an uncontrolled secondary rise in [Ca²⁺]ᵢ (iv), which preceded cell death. All neurons that exhibited calcium deregulation also stained with trypan blue.

15–30 min, indicating that it must have preceded neuronal death. Our observations in spinal neurons were in agreement with other reports noting that glutamate-induced [Ca²⁺]ᵢ transients can trigger delayed Ca²⁺ overload and neurotoxicity in a variety of neuronal types following single *(45,155)* or repeated *(67)* EAA challenges. However, in contrast to the results of Ogura et al. noted just above, we have failed to correlate the decay rate of the initial [Ca²⁺]ᵢ transient with subsequent cell death (M. Tymianski, unpublished data).

Delayed Ca²⁺ overload in cultured neurons appears to be triggered by the initial EAA-induced [Ca²⁺]ᵢ increase, as it cannot be averted once the EAA insult has occurred *(155,205)*. Also, once initiated, the Ca²⁺ rise cannot be arrested in neurons by removing the agonist from the perfusion solution, or by blockers of Ca²⁺ channels or glutamate receptors *(44,67,115,206)*. In our hands, this secondary Ca²⁺ rise was not immediately reversible upon removal of extracellular Ca²⁺, but persisted for 2–3 min in 34 of 46 neurons *(205)*, suggesting that some of this [Ca²⁺]ᵢ rise is a consequence of Ca²⁺ release from intracellular sources (Fig. 1). Although the [Ca²⁺]ᵢ rise could eventually be reversed by removing extracellular Ca²⁺, cell death was not averted by this maneuver. In fact, cell death was independent of both internal and external Ca²⁺ concentrations once Ca²⁺ deregulation began *(155,205)*. Neurons were able to maintain their fura-2 fluorescence for up to 60 min following the onset of the delayed [Ca²⁺]ᵢ changes *(206)*. We called these secondary elevations in [Ca²⁺]ᵢ *Ca²⁺ deregulations*, because they represented a decompensation of neuronal Ca²⁺ homeostasis rather than nonspecific plasma membrane leakiness to ions.

The fact that cell death is independent of $[Ca^{2+}]_i$ by the time of the secondary Ca^{2+} increase conflicts with the hypothesis that the process of delayed Ca^{2+} accumulation is responsible for the ensuing cell death *(44,115,116)*, and suggests instead that Ca^{2+} deregulation occurs as a consequence of processes triggered by the initial excitotoxic insult. Therefore, Ca^{2+}-dependent neurotoxicity must be triggered very early, which limits what can be achieved in the treatment of excitotoxic neuronal injury. These results imply that the delay between the initial insult and the Ca^{2+} accumulation during which $[Ca^{2+}]_i$ and membrane integrity appear normal *(50,155,205,206)* does not necessarily represent a window of opportunity during which toxic cascades are dormant so that appropriate therapy could completely avert neurodegeneration. Instead, it appears that secondary neurotoxic cascades are already in operation during this time. Therefore, appropriate therapy may have to be instituted at the earliest possible time, to achieve maximal effectiveness. Future therapeutic strategies for ischemic brain damage may need to focus not only on the blockade of Ca^{2+} influx, but also on intracellular processes occurring downstream from the initial insult.

6. QUANTITATIVE STUDIES RELATING CA²⁺ IONS TO NEUROTOXICITY

Although numerous studies have implicated Ca^{2+} as a major factor in neurotoxicity, actual Ca^{2+} measurements did not correlate with neuronal death. Investigators using fluorescent Ca^{2+} indicators such as fura-2 for measuring free cytoplasmic Ca^{2+} ions found that $[Ca^{2+}]_i$ measurements correlated poorly with eventual survival outcome. Michaels and Rothman *(129)*, using cultured hippocampal neurons, examined the impact of a number of EAA receptor agonists on $[Ca^{2+}]_i$ and cell survival. A poor correlation was observed between the $[Ca^{2+}]_i$ rise and cell death, because certain insults could raise $[Ca^{2+}]_i$ but did not kill cells. For example, neurons depolarized with potassium produced $[Ca^{2+}]_i$ increases that were not neurotoxic, whereas glutamate applications produced both $[Ca^{2+}]_i$ increases and cell death. In another study with a model of in vitro anoxia *(51)*, hippocampal neurons challenged with NaCN experienced large $[Ca^{2+}]_i$ increases but little neurotoxicity. It was concluded that a general elevation in cytoplasmic calcium does not necessarily predict neurodegeneration. Other studies have also confirmed that $[Ca^{2+}]_i$ and neurotoxicity cannot be linked by simple linear correlations. Peak and average $[Ca^{2+}]_i$ measurements were plotted against the fractional cell death for spinal neurons exposed to a 50-min challenge with different agonists capable of raising $[Ca^{2+}]_i$ (250 μM glutamate, 100 μM NMDA, 100 μM kainate, or 50 mM K⁺). Neither parameter correlated with survival outcome *(205)*.

It must be noted that Ca^{2+} measurements with fluorescent indicators reflect only free Ca^{2+} ion concentrations, not total Ca^{2+} loads. A large range of $[Ca^{2+}]_e$, which produces a high probability of cell death, evokes a small change in measured $[Ca^{2+}]_i$, illustrating the insensitivity of this method when high Ca^{2+} loads are expected. The inability of measured $[Ca^{2+}]_i$ to reflect neurotoxicity suggests that intracellular Ca^{2+} homeostatic mechanisms such as Ca^{2+} buffering and sequestration must be operative and should be taken into account. One such mechanism involves the ability of at least some types of neurons to "recruit" Ca^{2+} buffering capacity when faced with high Ca^{2+} loads. For example, using dorsal root ganglion neurons, Thayer and Miller *(200)* found that as Ca^{2+} influx increases, the rise of $[Ca^{2+}]_i$ in the cytoplasm approaches an asymptotic "ceiling" such that higher Ca^{2+} loads are buffered with greater efficiency than lesser loads. Thus, the relationship between $[Ca^{2+}]_i$ and Ca^{2+} influx is not linear, but rather $[Ca^{2+}]_i$ follows a reverse exponential rise when plotted against absolute Ca^{2+} influx. Similar conclusions have been reported by other authors, e.g., in bovine chromaffin cells *(3)*.

These observations can be explained by at least two (not mutually exclusive) Ca^{2+} buffering processes. The first is that neurons to which these observations apply contain large quantities of a highly diffusible Ca^{2+} buffer having a relatively low affinity for Ca^{2+} ions (K_d for Ca^{2+} around 1 μM). Candidates for such a buffering system are the endogenous Ca^{2+} binding proteins calbindin and parvalbumin *(7,159)*. The other possibility is the presence of large quantities of a relatively immobile Ca^{2+} buffering or sequestration system. Candidates for this are mitochondria or the endoplasmic reticulum *(17,200,215)*. The impact of Ca^{2+} buffering mechanisms on Ca^{2+} homeostaies in the cell is briefly described in Section 2.2. above.

Another limitation of the fluorescent Ca^{2+} imaging approach is that as Ca^{2+} increases to micromolar levels, it may approach the saturation limits of the Ca^{2+} indicator, and fluorescence measurements may no longer reflect true $[Ca^{2+}]_i$. However, in the experiments described above *(51,129,205)* $[Ca^{2+}]_i$ rarely exceeded 1 μM. Therefore, saturation of fura-2 was unlikely to add a significant bias to the interpretation of this and Rothman's studies.

Due to the limitations of free Ca^{2+} measurements, some investigators turned to measurements of total Ca^{2+} fluxes using radiolabeled $^{45}Ca^{2+}$ in an attempt to establish quantitative relationships between Ca^{2+} and excitotoxic/anoxic neurotoxicity. Manev et al. *(115)* and Marcoux et al. *(116)* reported that cultured neurons exposed to glutamate or anoxia experienced delayed $^{45}Ca^{2+}$ accumulations, which probably represented the Ca^{2+} deregulation phenomenon described above. Abstracts from Choi and colleagues reported that in cortical neurons exposed to glutamate *(98)* or anoxia *(70)*, $^{45}Ca^{2+}$ measurements correlated precisely with cell death. Additional quantitative data on glutamate neurotoxicity were recently published by the same laboratory *(76)*. These demonstrated a very strong linear correlation between $^{45}Ca^{2+}$ accumulation and cell death when neurons were exposed to differing concentrations of glutamate, or to 500 μM glutamate for different exposure durations. Furthermore, agents known to attenuate acute glutamate neurotoxicity such as the NMDA antagonist D-aminophosphonovaleric acid (D-APV) and dextrorphan reduced neuronal $^{45}Ca^{2+}$ accumulation in a manner proportional to their protective effect. The tight linear correlation between $^{45}Ca^{2+}$ accumulation and glutamate-induced cell death has been reproduced. Schramm and Eimerl *(175)* reported a linear correlation coefficient of 0.85 ($p < 0.001$) when plotting $^{45}Ca^{2+}$ uptake against cell death due to NMDA and glutamate. This linear phenomenon applies not only to tissue cultures, but also to acute hippocampal brain slice preparations, in which anoxia-induced electrophysiologic cell damage correlated linearly with $^{45}Ca^{2+}$ accumulation (r = 0.927) *(110)*.

It is difficult to comment on the actual quantity of Ca^{2+} that produced damage in the above experiments and even more difficult to relate these quantities to physiologic phenomena. Lobner and Lipton *(110)* showed that 50% damage after 5 min of anoxia in hippocampal slices was due to about 4 nmol of Ca^{2+} per milligram dry weight. Schramm and Eimerl *(175)* report the figure of about 10 fmol $^{45}Ca^{2+}$ per cell, about three times the endogenous Ca^{2+} content, as producing 60% toxicity in their paradigms. This translates to millimolar Ca^{2+} concentrations in each cell (about 2.5 mM in a spherical, 20 µm-diameter neuron).

When glutamate, NMDA, or anoxia were used, $^{45}Ca^{2+}$ accumulation measurements appear to have resolved the difficulty in correlating Ca^{2+} quantity with neurotoxicity. However, certain inconsistencies remain. For example, if the relationship between Ca^{2+} content and cell death is simply a linear 1:1, then Ca^{2+} influx evoked by pathways other than through NMDA receptor channels should evoke cell death exactly in proportion to the degree of Ca^{2+} accumulation. However, this is not always the case. Hartley et al. *(76)* showed that when neurons were depolarized with high potassium, or challenged with non-NMDA receptor agonists such as kainate, $^{45}Ca^{2+}$ accumulation far exceeded the observed neurotoxicity. A similar dissociation between $^{45}Ca^{2+}$ accumulation and cortical neuronal death was observed in the laboratory of Marcoux; they showed that while NMDA-dependent Ca^{2+} influx and neurotoxicity were well correlated, anoxia-induced $^{45}Ca^{2+}$ accumulation—but not neurotoxicity—could be blocked by Ca^{2+} channel blockers, tetrodotoxin, and phenytoin *(117,214)*. These results are reminiscent of those described by Rothman using free Ca^{2+} measurements, which suggested that a general elevation in cytoplasmic calcium does not necessarily predict neurodegeneration *(51)*. It is apparent from the above that while linear relationships between Ca^{2+} accumulation and neurotoxicity may exist, additional factors may influence the slope of such relationships.

7. CA^{2+} NEUROTOXICITY: THE SOURCE-SPECIFICITY HYPOTHESIS

Given the above controversies, it was necessary to clarify further the role of Ca^{2+} ions as triggers for neurotoxicity. Recently, experiments were performed in our laboratory to examine formally whether the amount of Ca^{2+} loading of the cell is the single determinant of cell death *(124,205)*. If so, then neuronal death should depend solely on the magnitude and the time-course of the Ca^{2+} accumulation rather than on other factors such as the Ca^{2+} source. Under this hypothesis, equal $[Ca^{2+}]_i$ increases should produce equal cell death, regardless of the route of Ca^{2+} influx. To test this hypothesis, cultured spinal neurons were loaded with fura-2, and $[Ca^{2+}]_i$ influx was evoked through NMDA receptor channels, non-NMDA receptor channels, or through dihydropyridine-sensitive (L-type) Ca^{2+} channels. Agonists and antagonists of Ca^{2+} channels and glutamate receptors were applied to restrict the routes of Ca^{2+} influx appropriately. Experimental conditions were arranged so as to produce matching $[Ca^{2+}]_i$ increases through the different pathways and to relate them to the neurons' survival outcome.

The neurons were challenged for 40 min with glutamate, NMDA, kainate, or high K^+ in the presence of varying combinations of D-APV, nimodipine, or 6-cyano-7-nitroquinoxaline-2,3-dione (CNQX). Glutamate in the presence of NMDA receptor blockade (D-APV) elicited a $[Ca^{2+}]_i$ increase comparable to that evoked by glutamate alone. However, this $[Ca^{2+}]_i$ increase was not toxic and was mediated primarily by non-NMDA channels sensitive to CNQX. Thus, large $[Ca^{2+}]_i$ increases were not in themselves sufficient to cause neurotoxicity. Conversely, NMDA application produced high toxicity with a smaller $[Ca^{2+}]_i$ rise than that evoked by glutamate, whereas kainate caused an equivalent rise in $[Ca^{2+}]_i$ to NMDA, which was not neurotoxic even when kainate receptor desensitization was blocked by pretreating the cultures with concanavalin A *(126)*. Finally, depolarizing the neurons with high K^+ also produced a large, nonneurotoxic $[Ca^{2+}]_i$ increase. These results are qualitatively very comparable with those obtained using total $^{45}Ca^{2+}$ accumulation measurements in cortical neurons *(76)*, and they illustrate the striking protective effect of NMDA receptor blockade on acute Ca^{2+} neurotoxicity. Most importantly, however, they show quantitatively that this toxicity was not due to the ability of NMDA receptors to trigger greater initial $[Ca^{2+}]_i$ increases than other pathways, but rather to some other attribute specifically associated with NMDA-mediated Ca^{2+} influx.

However, these experiments assumed that the free cytoplasmic Ca^{2+} concentration measured by fluorescent calcium indicators correlated well with actual total Ca^{2+} influx. As suggested previously, this may not necessarily be true. This assumption was therefore tested by examining the relationship between $[Ca^{2+}]_i$ using calcium indicator dyes and total Ca^{2+} loading using $^{45}Ca^{2+}$ accumulation measured in the same neurons *(124)*. The free $[Ca^{2+}]_i$ changes evoked by NMDA or high extracellular K^+ were found to be highly correlated with $^{45}Ca^{2+}$ accumulation, thereby substantiating the above assumption. Furthermore, NMDA application produced a higher average $[Ca^{2+}]_i$ increase for the same $^{45}Ca^{2+}$ load compared with high K^+, implying a difference in calcium buffering depending on the route of influx. Using total calcium loading, we found that a threshold of 80% of maximum calcium load must be reached before any significant amount of neuronal cell death is experienced. However, similar to our previous experiment, this calcium-dependent neurotoxicity could only be achieved by influx through NMDA receptors because identical calcium loading achieved with high K^+ loading was unable to cause cell death. The possibility that sodium entry through NMDA receptors rather than calcium may be responsible for the cell death was also tested by performing the experiments in zero sodium or low sodium. In either case, cell death was not decreased.

This study further revealed a component of glutamate neurotoxicity that appears to be calcium independent. A comparison between glutamate- and NMDA-induced cell death revealed that for the same calcium load, glutamate application caused greater cell death. This additional cell death was not reduced by inhibiting metabotropic glutamate receptors, which may have mediated release of calcium from intracellular stores. Endogenous glutamate release was also blocked to eliminate the possibility that exogenous glutamate application increased endogenous glutamate release compared with exogenous NMDA application, and thereby increased neurotoxicity. However, inhibition of endogenous release of glutamate did not decrease cell death. It was also possible that glutamate increased intracellular Ca^{2+} in astrocytes and that this increase spread to the neurons via gap junctions between the cells. A gap junction blocker was utilized, but it had no effect on cell death. These experiments implicate an additional calcium-independent mechanism of neurotoxicity for glutamate compared with NMDA.

The above two studies demonstrate that, irrespective of either free $[Ca^{2+}]_i$ changes or total Ca^{2+} loading, calcium-dependent neurotoxicity is related to the route of calcium entry rather than the amount. Several possibilities could account for this source specificity of neurotoxicity. One possibility is that the rate-limiting processes that trigger early Ca^{2+}-dependent neurotoxicity (toxicity within minutes to hours) may be preferentially associated with NMDA receptors. It is plausible that the molecules involved in those processes are physically located in the submembrane region closely surrounding NMDA channels and are thereby preferentially activated by Ca^{2+} influx through this source. Under this hypothesis, Ca^{2+} influx through alternate sources such as voltage-gated Ca^{2+} channels would be insufficient to trigger cell death because these influx pathways are not linked with neurotoxic processes. By the time Ca^{2+} diffuses from these alternate sources, the resultant $[Ca^{2+}]_i$ in the vicinity of NMDA receptors and their associated trigger sites for Ca^{2+} neurotoxicity would not suffice to cause neurotoxicity.

Implicit in this assumption that NMDA receptors are preferentially associated with neurotoxic processes is the notion that Ca^{2+}-dependent processes in neurons are somehow compartmentalized along with specific Ca^{2+} influx pathways. This might occur by clustering of receptors, and their physical colocalization with effector mechanisms. Substantial evidence has been uncovered in recent years to support the existence of such processes. For example, NMDA and non-NMDA receptors can trigger distinct Ca^{2+} signaling pathways regulating different forms of gene expression in neurons *(6,105)*. Presumably, this occurs by activating spatially distinct sites of Ca^{2+} entry, resulting in the activation of different enzymes located at distinct sites in the cell. It is now accepted that certain Ca^{2+} fluxes in neurons are compartmentalized both spatially and temporally. For example, voltage-gated Ca^{2+} channels are strategically distributed throughout the plasma membrane so as to maximize Ca^{2+} influx in subcellular regions subserving relevant cellular functions *(87,106,161,162,216a)*. Similarly, NMDA and non-NMDA receptors are clustered throughout neurons on synapses, dendrites, and somata *(13,36,38,86,88,132)*, thereby optimizing Ca^{2+} fluxes in relevant areas. There is also confirmation that subcellular Ca^{2+} transients can have different temporal profiles in different compartments. For example, intense afferent stimulation can evoke long-lasting $[Ca^{2+}]_i$ gradients between hippocampal dendritic spines and dendritic shafts even in the absence of a physical diffusion barrier *(73,132)*. Several sets of experimental results have now been reported demonstrating that large and highly localized rises in $[Ca^{2+}]_i$ can occur in neurons so that Ca^{2+} ions can separately activate multiple processes within the same cell *(46,78,102)*.

Concurrently with evidence for compartmentalization of Ca^{2+} fluxes, studies have shown that activation of glutamate receptors, particularly of the NMDA subtype, is associated with initiation of a number of secondary processes known to occur in neuronal injury. These include the formation of free radicals *(150)*, initiation of the arachidonic acid cascade *(53,101,170)*, and activation of proteases *(185)*, lipases *(59)*, nitric oxide *(12, 42,43,54)*, and protein kinase C *(5)*. According to the source-specificity hypothesis, rate-limiting substrates or enzymes governing these reactions may be found in high concentrations near NMDA-gated channels. Recently, direct physical evidence has arisen to support this. For example, protein kinase C and calmodulin-dependent protein kinase II, which phosphorylate NMDA receptors, have been found along with NMDA receptors in postsynaptic densities of rat brain *(216b,216c)*.

Alternative explanations can be sought to explain the higher toxicity of NMDA receptors for a given Ca^{2+} load in spinal neurons. However, these still require a "subcellular" approach. For example, if $[Ca^{2+}]_i$ transients were buffered with different efficiencies when they are triggered by different pathways, then this requires cytoplasmic Ca^{2+} buffers to be preferentially compartmentalized with a given class of Ca^{2+} sources, rather than requiring that these sources be colocalized with Ca^{2+} neurotoxicity "trigger sites." A second hypothetical possibility is that elevated $[Ca^{2+}]_i$ triggers neurotoxicity by acting synergistically with some undiscovered (glutamate-triggered, Ca-independent) process associated specifically with NMDA receptors. Under this possibility, neurotoxic Ca^{2+} elevations may be produced by mechanisms not involving influx specifically through NMDA receptor channels as long as $[Ca^{2+}]_i$ still reaches "toxic" levels in the vicinity of NMDA receptors. Finally, it may be that NMDA receptors and non-NMDA receptors have significantly different cellular/neuritic distributions. By this hypothesis, Ca^{2+} influx through some sources would occur nearer to critical subcellular compartments necessary for cell survival.

It must be noted that specific details obtained in a given in vitro paradigm may not be generalizable to other systems. For example, cerebellar Purkinje cells are much more susceptible to Ca^{2+} neurotoxicity mediated by

AMPA/kainate receptors than by NMDA receptors *(20)*. However, the physical principles deduced from such experiments should remain valid.

8. SECONDARY PHENOMENA PRECIPITATED BY CA^{2+} EXCESS

Given the previous discussion on the source-specificity hypothesis of Ca^{2+} neurotoxicity, the question of which molecular mechanisms are activated by the Ca^{2+} ion remains. A number of specific processes (some of which have been listed above) are thought to contribute significantly to Ca^{2+}-related neuronal and axonal injury. These processes are reviewed in more detail here.

8.1. Formation of Free Radical Species

A free radical is any molecule, atom, or group of atoms with an unpaired electron in its outmost orbital, which accounts for its extreme reactivity. Free radical species of potential importance in cerebral ischemia/trauma include superoxide (O$_2\bullet$) and hydroxyl (OH$\bullet$) radicals. Free radicals are produced in small amounts by normal cellular processes, such as the mitochondrial electron transport system, reactions catalyzed by prostaglandin hydroperoxidase, and the auto-oxidation of small molecules, such as catecholamines, and by the microsomal cytochrome P-450 reductase system *(174)*. Free radicals may play physiologic roles, such as the modulation of a number of membrane receptors, including NMDA receptor function *(2)*.

Free radicals are formed in excess during traumatic and hypoxic injuries, because insufficient O$_2$ is available to accept electrons passed along the mitochondrial electron transport chain, leading to the reduction of other components of the system, such as flavin adenine dinucleotide and coenzyme Q. These molecules then auto-oxidize to produce free radical species. Also, during reperfusion, reactive oxygen radicals may form as byproducts of the reactions of arachidonic acid to produce prostaglandins and leukotrienes *(174)*. Arachidonic acid is particularly abundant during ischemic conditions, because it is released from membrane phospholipids during this time. This release is thought to be triggered, at least in part, by NMDA receptor activation and subsequent stimulation of Ca^{2+}-dependent phospholipase A2 *(27,59)*. Other injuries that contribute to the rise in [Ca^{2+}]$_i$ during ischemia may further contribute to the activation of the arachidonic acid cascade and free radical formation. A number of other Ca^{2+}-dependent lipases and phospholipases are also triggered during ischemia and contribute further to the rise in free fatty acid release and free radical formation.

Free radicals can react with and damage proteins, nucleic acids, lipids, and other classes of molecules, such as extracellular matrix glycosaminoglycans. Sulfur-containing amino acids and polyunsaturated fatty acids are particularly vulnerable. The latter are found at very high concentrations in the CNS as components of the plasma membrane. Thus, EAA release and free radical formation may act synergistically in lipid peroxidation and the production of ischemia-induced neuronal membrane damage *(118,150)*.

8.2. Formation of Nitric Oxide

Nitric oxide (NO), first identified as endothelium-derived relaxing factor *(60)*, is a short-lived, diffusible, highly reactive gas that has recently gained wide attention as a member of newly discovered class of messenger molecules. No serves a variety of functions in different tissues, including vascular endothelium, immune cells, neurons, smooth muscle, and cardiac muscle *(114)*. NO produced in one cell can diffuse and produce biologic effects in neighboring cells. In the brain, it is produced by both vascular endothelium and neurons. NO produced in vascular endothelium can have major effects on governing cerebral blood flow *(82)* and can act as a potent vasodilator *(213)*. In neurons, NO has been implicated in affecting cellular processes related to synaptic plasticity, including NMDA receptor function *(103)* and long-term potentiation *(11,141, 176)*.

NO production occurs at low levels during physiologic function and is governed by NO synthase *(18,19,41,80)*. Its synthesis is regulated by the Ca^{2+}-dependent regulatory enzyme calmodulin *(19)*, which is induced by the influx of Ca^{2+} ions through NMDA receptors *(54,191)*. NO production can affect intracellular Ca^{2+} pools in neurons and adjacent cells by modulating Ca^{2+} release from intracellular stores through a cyclic guanosine monophosphate kinase-mediated mechanism. NO by itself is not highly toxic but can lead to the formation of toxic species by reacting with oxygen radicals, particularly superoxide. NO reacts with superoxide to form the powerful oxidant peroxynitrite ($ONOO^-$), which directly oxidizes sulfhydryl groups, lipids, DNA, and proteins *(107)*. The extent of this reaction can be moderated by the presence of the endogenous free radical scavenger superoxide dismutase (SOD), which competes with NO for superoxide radicals. $ONOO^-$ can also lead to the production of hydroxyl radicals and reacts with certain metals to produce the highly toxic nitronium ion (NO^{2+}). The possible physiologic purpose of such processes is to maintain dynamic control of the neuronal membrane, cytoskeleton, and other vital cellular structures.

Excessive NO production has been postulated as a causative mechanism in neurotoxicity *(42,43,99)*, presumably via Ca^{2+} overload, leading to the ultimate overexpression of the mechanisms described above. Currently, it seems that NO production can be either protective or toxic, depending on a number of associated modulatory factors *(12,107)*. The ability of SOD to compete with NO for superoxide radicals, as noted just above, may explain the relative resistance of tissues containing high levels of SOD to NO and EAA toxicity *(37)*. This finding has been extended to experimental cerebral ischemia in vivo, using transgenic mice that overexpress SOD. These transgenics have been resistant to cold-and ischemia-induced cerebral injury *(22,218)*.

8.3. Calcium-Activated Proteases

The activation of EAA acid receptors and the subsequent sustained rise in $[Ca^{2+}]_i$ during cerebral ischemia induces a family of Ca^{2+}-dependent cysteine proteases, calpains *(185,186)*. Calpains are present in virtually all mammalian cells and appear to be largely associated with membranes in conjunction with a specific inhibitory protein, calpastatin. Proteases participate in the physiologic remodeling of the cellular cytoskeleton and membrane, membrane receptor cleavage, enzyme activation, and modulation of mitosis in dividing cells. Although the specific substrates for protease activity during neuronal ischemia remain largely uncertain, it seems that cytoskeletal proteins are a major target for Ca^{2+}-activated proteases during the cellular response to ischemic injury *(173)*. The Ca^{2+}-activated protease, calpain I, is present in neurons, including those known to be vulnerable to certain types of cerebral ischemic damage *(158)*. It is activated by low micromolar $[Ca^{2+}]_i$, a level known to be reached during neuronal hypoxia-ischemia and excitotoxicity *(204,205,207)*. Evidence suggests that calpain I-mediated cytoskeletal breakdown is a significant event in neuronal hypoxia-ischemia and other neurodegenerative disorders in vivo *(151,157,158,184,185,186)*.

8.4. Endonucleases, Apoptosis, and Necrosis

Ca^{2+} overload also activates endonucleases, a series of Ca^{2+}-dependent enzymes that degrade DNA and that may play a role in two well-recognized forms of cell death, necrosis and apoptosis *(143)*. Apoptosis is a form of programmed cell death that occurs during fetal development as well as during adult life *(154)*. The physiologic function of apoptosis in the CNS seems to be to remove those neurons whose processes have failed to find their targets during development or regeneration. Apoptosis is pathologically distinct from necrosis, involving compaction of the cell body, nuclear fragmentation, and the formation of surface blebs. Apoptosis can also be identified by a characteristic DNA fragmentation resulting from the cleavage of cell chromatin into oligonucleosome-length fragments seen as a "ladder" pattern on electrophoretic gels. Although the existence of apoptotic mechanism in neurons in vitro is controversial (cf. refs. *48* and *97*), it is now apparent that apoptosis may play a significant role in the process of delayed neuronal death *(92)* after transient global cerebral ischemic injury *(203)*. More recent studies have shown that apoptosis can also occur after spinal cord trauma, although most cells that die by this route are thought to be nonneuronal cells such as oligodendrocytes *(109)*. Whether apoptosis plays a major role in neuronal death after spinal cord injury has yet to be defined.

8.5. Mitochondrial Damage

Mitochondira serve as the energy generators of the cell. Considerable evidence indicates that they buffer Ca^{2+} ions during physiologic and pathologic states *(17,72, 74,131,200,201,215).* Work from several laboratories suggests that mitochondrial dysfunction is a common event in cell injury caused by toxins, ischemia, and mechanical trauma *(74,104,148,156,177,178,212).* Anoxia causes the transmitochondrial membrane potential to collapse, and then impairment of ATP production ensues. In addition to anoxia, the act of Ca^{2+} buffering by mitochondria also causes the transmembrane potential to decrease, resulting in intramitochondrial Ca^{2+} accumulation and a further collapse of ATP production. This process is accompanied by a concomitant release of H^+ ions from mitochondria. It is likely that extreme Ca^{2+} overload irreversibly damages mitochondria and that this event commits the cell to die.

8.6. Acidosis

CNS hypoxia-ischemia, trauma, and excitotoxicity produce acidosis. Ample evidence indicates that this occurs both at the tissue *(39,61,100,118,168,180)* and cellular levels *(77,84,94).* Several mechanisms of acidosis production are in effect during neuronal injury. First, a shift from aerobic to anaerobic metabolism due to cellular hypoxia results in lactate production and the release of two H^+ atoms for every two molecules of ATP produced. However, protons are also released during many other reactions, such as phospholipid hydrolysis. Particularly, Ca^{2+} influx causes a rapid intracellular acidification *(72,215)* through a number of mechanism, including a number of membrane exchangers ($Ca^{2+}/2H^+$ exchange at the cell and organelle membranes, Na^+/H^+ Na^+/H^+ exchange to restore Na^+ gradients), the displacement of bound H^+ by Ca^{2+} at negative groups intracellularly, and the release of H^+ from mitochondria during Ca^{2+} buffering as a consequence of $Ca^{2+}/2H^+$ exchange.

The mechanism by which acidosis *per se* produces neuronal damage is unclear. Some possibilities are that a rise in $[H^+]_i$ slows the cell's recovery from a deleterious rise in $[Ca^{2+}]_i$ *(215),* enhances free radical production *(182),* or accelerates DNA damage *(182).* However, some evidence suggests that acidosis could be beneficial by blocking NMDA receptors and thus reducing Ca^{2+} influx and the resultant neurotoxicity *(65,89,197,202).*

9. SUMMARY

Although the concept of Ca^{2+} ions as mediators of cytotoxicity has been accepted, there have been difficulties in correlating intracellular Ca^{2+} levels and neurotoxicity. It is now apparent that neurons succumbing to Ca^{2+} loading sustain a failure of Ca^{2+} homeostatic mechanisms due to processes set into motion very early following the initial insult. Recent work suggests that not only is total neuronal Ca^{2+} loading important but the route by which this loading occurs is also important. This influx of Ca^{2+} ions in neuronal cells can then lead to activation of various secondary processes resulting in cellular death. In the future, such processes may be better understood through examination of the subcellular interplay among membrane receptors, second messenger systems, and Ca^{2+}

buffering mechanisms. Newer techniques must be used to identify the molecules that are physically linked to the routes of Ca^{2+} influx in the hopes that such discoveries can reveal which downstream processes are responsible for neurotoxicity. Therapies may also be developed that attempt to uncouple the subcellular Ca^{2+} rises from their secondary neurotoxic process as well as limit the toxicity of these processes.

ACKNOWLEDGMENTS

G.K.T. Chu was supported by a Fellowship from the Joint Section on Neurotrauma and Critical Care of the American Association of Neurological Surgeons/Congress of Neurological Surgeons. Our work was supported by grants from the Medical Research Council of Canada, the Ontario Heart and Stroke Foundation, the Canadian Paraplegic Association, The Samuel Lunenfeld Research Foundation, and the Network on Neural Recovery and Regeneration of the Networks of Centres of Excellence of Canada.

REFERENCES

1. Adler, E. M., Augustine, G. J., Duffy, S. N., and Charlton, M. P. (1991) Alien intracellular calcium chelators attenuate neurotransmitter release at the squid giant synapse. *J. Neurosci.* **11,** 1496–1507.
2. Aizenman, E., Hartnett, K. A., and Reynolds, I. J. (1990) Oxygen free radicals regulate NMDA receptor function via a redox modulatory site. *Neuron* **5,** 841–846.
3. Artalejo, C. R., Perlman, R. L., and Fox, A. P. (1992) Omega-Conotoxin GVIA blocks a Ca^{2+} current in bovine chromaffin cells that is not of the "classic" N type. *Neuron* **1,** 85–95.
4. Augustine, G. J., Adler, E. M., and Charlton, M. P. (1991) The calcium signal for transmitter secretion from presynaptic nerve terminals. *Ann. NY Acad. Sci.* **635,** 365–381.
5. Baba, A., Etoh, S., and Iwata, H. (1991) Inhibition of NMDA-induced protein kinase C translocation by Zn^{2+} chelator: implication of intracellular Zn^{2+}. *Brain Res.* **557,** 103–108.
6. Bading, H., Ginty, D. D., and Greenberg, M. E. (1993) Regulation of gene expression in hippocampal neurons by distinct calcium signaling pathways. *Science* **260,** 181–186.
7. Baimbridge, K. G., Celio, M. R., and Rogers, J. H. (1992) Calcium binding proteins in the nervous system. *TINS* **15,** 303–308.
8. Balentine, J. D. (1988) Spinal cord trauma: in search of the meaning of granular axoplasm and vesicular myelin. *J. Neuropathol. Exp. Neurol.* **47,** 500–510.
9. Balentine, J. D., Paris, D. U., and Dean, D. L. (1982) Calcium-induced spongiform and necrotizing myelopathy. *Lab. Invest.* **47,** 286–295.
10. Balentine, J. D., Paris, D. U., and Greene, W. B. (1984) Ultrastructural pathology of nerve fibers in calcium-induced myelopathy. *J. Neuropathol. Exp. Neurol.* **43,** 500–510.

11. Barinaga, M. (1991) Is nitric oxide the "retrograde messenger?". *Science* **254,** 1296–1297.
12. Beckman, J. S. (1991) The double-edged role of nitric oxide in brain function and superoxide-mediated injury. *J. Dev. Physiol.* **15,** 53–59.
13. Benke, T. A. Jones, O. T., Collingridge, G. L., and Angelides, K. J. (1993) N-methyl-D-asparate receptors are clustered and immobilized on dendrites of living cortical neurons. *Proc. Natl. Acad. Sci. USA* **90,** 7819–7823.
14. Benveniste, H. and Diemer, N. H. (1988) Early postischemic ^{45}Ca accumulation in rat dentate hilus. *J. Cereb. Blood Flow Metab.* **8,** 713–719.
15. Benveniste, H, Drejer, J, Schousboe, A, and Diemer, N. H. (1984) Elevation of the extracellular concentrations of glutamate and aspartate in rat hippocampus during transient cerebral ischemia monitored by intracerebral microdialysis. *J. Neurochem.* **43,** 1369–1374.
16. Berdichevsly, E., Riveros, N., Sanches-Armass, S., and Orego, F. (1983) Kainate, N-methylaspartate and other excitatory amino acids increase calcium influx onto rat brain cortex cells in vitro. *Neurosci. Lett.* **36,** 75–80.
17. Blaustein, M. P. (1988) Calcium transport and buffering in neurons. *TINS* **11,** 438–443.
18. Bredt, D. S., Hwang, P. M., and Snyder, S. H. (1990) Localization of nitric oxide synthase indicating a neural role for nitric oxide. *Nature* **347,** 768–770.
19. Bredt, D. S. and Snyder, S. H. (1990) Isolation of nitric oxide synthetase, a calmodulin-requiring enzyme. *Proc. Natl. Acad. Sci. USA* **87,** 682–685.
20. Brorson, J. R., Manzolillo, P. A., and Miller, R. J. (1994) Ca^{2+} entry via AMPA/KA receptors and excitotoxicity in cultured cerebellar Purkinje cells. *J. Neurosci.* **14,** 187–197.
21. Carafoli, E. (1992) The Ca^{2+} pump of the plasma membrane. *J. Biol. Chem.* **267,** 2115–2118.
22. Chan, P. H., Yang, G. Y., Chen, S. F., Carlson, E., and Epstein, C. J. (1991) Cold-induced brain edema and infarction are reduced in transgenic mice overexpressing CuZn-superoxide-dismutase. *Ann. Neurol.* **29,** 482–486.
23. Chard, P. S., Bleakman, D., Christakos, S., Fullmer, C. S., and Miller, R. J. (1993) Calcium buffering properties of calbindin D28k and parvalbumin in rat sensory neurones. *J. Physiol.* **472,** 341–357.
24. Chen, S. T., Hsu, C. Y., Hogan, E. L., Juan, H. Y., Banik, N. L., and Balentine, J. D. (1987) Brain calcium content in ischemic infarction. *Neurology* **37,** 1227–1229.
25. Cheung, J. Y., Bonventre J. V., Malis, C. D., and Leaf, A. (1986) Calcium and ischemic injury. *N. Engl. J. Med.* **314,** 1670–1676.
26. Chizzonite, R. A. and Zak, R. (1981) Calcium-induced cell death: susceptibility of cardiac myocytes is age-dependent. *Science* **213,** 1508–1511.
27. Choi, D. W. (1985) Glutamate neurotoxicity in cortical cell culture is calcium dependent. *Neurosci. Lett.* **58,** 293–297.
28. Choi, D. W. (1987) Ionic dependence of glutamate neurotoxicity. *J. Neurosci.* **7,** 369–379.

29. Choi, D. W. (1988) Calcium-mediated neurotoxicity: relationship to specific channel types and role in ischemic damage. *TINS* **11,** 465–467.

30. Choi, D. W. (1988) Glutamate neurotoxicity and diseases of the nervous system. *Neuron* **1,** 623–634.

31. Choi, D. W. (1990) Cerebral hypoxia: some new approaches and unanswered questions. *J. Neurosci.* **10,** 2493–2501.

32. Choi, D. W., Maulucci-Gedde, M., and Kriegstein, A. R. (1987) Glutamate neurotoxicity in cortical cell culture. *J. Neurosci.* **7,** 357–368.

33. Clapham, D. E. (1995) Calcium signaling. *Cell* **80,** 259–268.

34. Collinridge, G. L. and Singer, W. (1990) Excitatory amino acid receptors and synaptic plasticity. *TIPS* **11,** 290–296.

35. Collins, F., Schmidt, M. F., Guthrie, P. B., and Kater, S. B. (1991) Sustained increase in intracellular calcium promotes neuronal survival. *J. Neurosci.* **11,** 2582–2587.

36. Connor, J. A., Wadman, W. J., Hockberger, P. E., and Wong, R. K. S. (1988) Sustained dendritic gradients of calcium induced by excitatory amino acids in CA1 hippocampal neurons. *Science* **240,** 649–653.

37. Coyle, J. T. and Puttfarcken, P. (1993) Oxidative stress, glutamate, and neurodegenerative disorders. *Science* **262,** 689–695.

38. Craig, A. M., Balckstone, C. D., Huganir, R. L., and Banker, G. (1993) The distribution of glutamate receptors in cultured rat hippocampal neurons: postsynaptic clustering of AMPA-selective subunits. *Neuron* **10,** 1055–1068.

39. Crockard, H. A., Gadian, D. G., Frackowiak, R. S. J., et al. (1987) Acute cerebral ischaemia: concurrent changes in cerebral blood flow, energy metabolites, pH, and lactate measured with hydrogen clearance and ^{31}P and ^{1}H nuclear magnetic resonance spectroscopy: II. Changes during ischaemia. *J. Cereb. Blood Flow Metab.* **7,** 394–402.

40. Curtis, D. R., Phillis, J. W., and Watkins, J. C. (1960) The chemical excitation of spinal neurons by certain acidic amino acids. *J. Physiol.* **150,** 656–682.

41. Dawson, T. M., Bredt, D. S., Fotuhi, M., Hwang, P. M., and Snyder, S. H. (1991) Nitric oxide synthase and neuronal NADPH diaphorase are identical in brain and peripheral tissues. *Proc. Natl. Acad. Sci. USA* **88,** 7797–7801.

42. Dawson, V. L., Dawson, T. M., Bartley, D. A., Uhl, G. R., and Snyder, S. H. (1993) Mechanism of nitric oxide-mediated neurotoxicity in primary cortical cultures. *J. Neurosci.* **13,** 2651–2661.

43. Dawson, V. L., Dawson, T. M., London, E. D., Bredt, D. S., and Snyder, S. H. (1991) Nitric oxide mediates glutamate neurotoxicity in primary cortical cultures. *Proc. Natl. Acad. Sci. USA* **88,** 6368–6371

44. de Erausquin, G. A., Manev, H., Guidotti, A., Costa, E., and Brooker, G. (1990) Gangliosides normalize distorted single cell intracellular free Ca^{2+} dynamics after toxic doses of glutamate in cerebellar granule cells. *Proc. Natl. Acad. Sci. USA* **87,** 8017–8021.

45. DeCoster, M. A., Koenig, M. L., Hunter, J. C., and Tortella, F. C. (1992) Calcium dynamics in neurons treated with toxic and non-toxic concentrations of glutamate. *Neuroreport* **3,** 773–776.

46. DeLisle, S. and Welsh, M. J. (1992) Inositol triphosphate is required for the propagation of calcium waves in *Xenopus* oocytes. *J. Biol. Chem.* **267,** 7963–7966.

47. Deshpande, J. K., Siesjo, B. K., and Wieloch, T. (1987) Calcium accumulation and neuronal damage in the rat hippocampus following cerebral ischemia. *J. Cereb. Blood Flow Metab.* **7,** 89–95.

48. Dessi, F., Charriault-Marlangue, C., Khrestchatisky, M., and Ben-Ari, Y. (1993) Glutamate-induced neuronal death is not a programmed cell death in cerebellar cultures. *J. Neurochem.* **60,** 1953–1955.

49. Dingledine, R., Boland, L. M., Chamberlin, N. L., et al. (1998) Amino acid receptors and uptake systems in the mammalian central nervous system. *CRC Crit. Rev. Neurobiol.* **4(1),** 1–97.

50. Dubinsky, J. M. (1993) Intracellular calcium levels during the period of delayed excitotoxicity. *J. Neurosci.* **13,** 623–631.

51. Dubinsky, J. M. and Rothman, S. M. (1991) Intracellular calcium concentration during "chemical hypoxia" and excitotoxic neuronal injury. *J. Neurosci.* **11,** 2545–2551.

52. Duce, I. R., Donaldson, P. L., and Usherwood, P. N. R. (1983) Investigations into the mechanism of excitant amino acid cytotoxicity using a well-characterized glutamatergic system. *Brain Res.* **263,** 77–87.

53. Dumuis, A., Sebben, M., Haynes, L., Pin, J. P., and Bockaert, J. (1988) NMDA receptors activate the arachidonic acid cascade system in striatal neurons. *Nature* **336,** 68–70.

54. East, S. J. and Garthwaite, J. (1991) NMDA receptor activation in rat hippocampus induces cyclic GMP formation through the L-arginine-nitric oxide pathway. *Neurosci. Lett.* **123,** 17–19.

55. Ellren, K. and Lehmann, A. (1989) Calcium dependency of N-methyl-D-aspartate toxicity in slices from the immature rat hippocampus. *Neuroscience* **32,** 371–379.

56. Faden, A. I., Demediuk, P., Panter, S. S., and Vink, P. (1989) The role of excitatory amino acids and NMDA receptors in traumatic brain injury. *Science* **244,** 798–800.

57. Faden, A. I. and Salzman, S. (1992) Pharmacological strategies in CNS trauma. *TIPS* **13,** 29–35.

58. Fariss, M. W., Pascoe, G. A., and Reed, D. J. (1985) Vitamin E reversal of the effect of extracellular calcium on induced toxicity in hepatocytes. *Science* **227,** 751–754.

59. Farooqui, A. A. and Horrocks, L. A. (1991) Excitatory amino acid receptors, neural membrane phospholipid metabolism and neurological disorders. *Brain Res. Rev.* **16,** 171–191.

60. Furchgott, R. F. and Zawadzki, J. V. (1980) The obligatory role of endothelial cells in the relaxation of arterial smooth muscle by acetylcholine. *Nature* **288,** 373–376.

61. Gadian, D. G., Frackowiak, R. S. J., Crockard, H. A., et al. (1987) Acute cerebral ischaemia: concurrent changes in cerebral blood flow, energy metabolites, pH, and lactate measured with hydrogen clearance and ^{31}P and ^{1}H nuclear magnetic resonance spectroscopy. I: Methodology. *J. Cereb. Blood Flow Metab.* **7,** 199–206.

62. Garthwaite, G. and Garthwaite, J. (1986) Neurotoxicity of excitatory amino acid receptor agonists in rat cerebellar slices: dependence on calcium concentration. *Neurosci. Lett.* **66,** 193–198.

63. Garthwaite, G., Hajos, F., and Garthwaite, J. (1986) Ionic requirements for neurotoxic effects of excitatory amino acid analogues in rat cerebellar slices. *Neuroscience* **18,** 437–447.

64. Garthwaite, J. and Garthwaite, G. (1983) The mechanism of kainic acid neurotoxicity. *Nature* **305,** 138–140.

65. Giffard, R. G., Monyer, H., Christine, C. W., Choi, and D. W. (1990) Acidosis reduces NMDA receptor activation, glutamate neurotoxicity, and oxygen-glucose deprivation neuronal injury in cortical cultures. *Brain Res.* **506,** 339–342.

66. Gilbertson, T. A., Scobey, R., and Wilson, M. (1991) Permeation of calcium ions through non-NMDA glutamate channels in retinal bipolar cells. *Science* **251,** 1613–1615.

67. Glaum, S. R., Scholz, W. K., and Miller, R. J. (1990) Acute- and long-term glutamate-mediated regulation of $[Ca^{2+}]_i$ in rat hippocampal pyramidal neurons in vitro. *J. Pharmacol. Exp. Ther.* **253,** 1293–1302.

68. Globus, M. Y.-T., Wester, P., Busto, R., and Dietrich, W. D. (1992) Ischemia induced extracellular release of serotonin plays a role in CA1 neuronal cell death in rats. *Stroke* **23,** 1595–1601.

69. Goldberg, M. P., Giffard, R. G., Kurth, M. C., and Choi, D. W. Role of extracellular calcium and magnesium in ischemic neuronal injury in vitro. *Neurology* **39(suppl),** 217 (abstract).

70. Goldberg, M. P., Kurth, M. C., Giffard, R. G., and Choi, D. W. (1989) ^{45}Calcium accumulation and intracellular calcium during in vitro "ischemia". *Soc. Neurosci. Abstr.* **15,** 803 (abstr).

71. Goldberg, M. P., Weiss, J. H., Pham, P. C., and Choi, D. W. (1987) N-methyl-D-aspartate receptors mediate hypoxic neuronal injury in cortical culture. *J. Pharmacol. Exp. Ther.* **243,** 784–791.

72. Gunter, T. E. and Pfeiffer, D. R. (1990) Mechanisms by which mitochondria transport calcium. *Am. J. Physiol.* **258,** C755–C786.

73. Guthrie, P. B., Segal, M., and Kater, S. B. (1991) Independent regulation of calcium revealed by imaging dendritic spines. *Nature* **354,** 76–80.

74. Halestrap, A. P., Griffiths, E. F., and Connern, C. P. (1993) Mitochondrial calcium handling and oxidative stress. *Biochem. Soc. Trans.* **21,** 353–358.

75. Harris, R. J., Symon, L., Branston, N. M., and Bayhan, M. (1981) Changes in extracellular calcium activity in cerebral ischaemia. *J. Cereb. Blood Flow Metab.* **1,** 203–209.

76. Hartley, D. M., Kurth, M. C., Bjerkness, L., Weiss, J. H., and Choi, D. W. (1993) Glutamate receptor-induced $^{45}Ca^{2+}$ accumulation in cortical cell culture correlates with subsequent neuronal degeneration. *J. Neurosci.* **13,** 1993–2000.

77. Hartley, Z. and Dubinsky, J. M. (1993) Changes in intracellular pH associated with glutamate excitotoxicity. *J. Neurosci.* **13,** 4690–4699.

78. Hernandez-Cruz, A., Sala, F., and Adams, P. R. (1990) Subcellular calcium transients visualized by confocal microscopy in a voltage-clamped vertebrate neuron. *Science* **247,** 858–862.

79. Hollman, M., Hartley, M., and Heinemann, S. (1991) Ca^{2+} permeability of KA-AMPA-gated glutamate receptor channels depends on subunit composition. *Science* **252,** 851–853.

80. Hope, B. T., Michael, G. J., Knigge, K. M., and Vincent, S. R. (1991) Neuronal NADPH diaphorase is a nitric oxide synthase. *Proc. Natl. Acad. Sci. USA* **88,** 2811–2814.

81. Hori, N., French-Mullen, J. M., and Carpenter, D. O. (1985) Kainic acid responses and toxicity show pronounced Ca^{2+} dependence. *Brain Res.* **358,** 380–384.

82. Iadecola, C., Pelligrino, D. A., Moscowitz, M. A., and Lassen, N. A. (1994) State of the art review: nitric oxide inhibition and cerebrovascular regulation. *J. Cereb. Blood Flow Metab.* **14,** 175–192.

83. Iino, M., Ozawa, S., and tsuzuki, K. (1990) Permeation of calcium through excitatory amino acid receptor channels in cultured rat hippocampal neurons. *J. Physiol.* **424,** 151–165.

84. Irwin, R. P. and Paul, S. M. (1992) Glutamate exposure rapidly decreases intracellular pH in rat hippocampal neurons in culture. *Soc. Neurosci. Abstr.* **19,** 58.1.

85. Jansco, G., Karcsu, S., Kiraly, E., et al. (1984) Neurotoxin induced nerve cell degeneration: possible involvement of calcium. *Brain Res.* **295,** 211–216.

86. Jones, K. A. and Baughman, R. W. (1991) Both NMDA and non-NMDA subtypes of glutamate receptors are concentrated at synapses on cerebral cortical neurons in culture. *Neuron* **7,** 593–603.

87. Jones, O. T., Kunze, D. L., and Angelides, K. J. (1989) Localization and mobility of w-conotoxin-sensitive Ca^{2+} channels in hippocampal CA1 neurons. *Science* **244,** 1189–1193.

88. Jones, O. T., McGurk, J. F., Bennett, M. V. L., et al. (1990) Distribution of NMDA receptors on hippocampal neurons. *Soc. Neurosci. Abstr.* **2,** 3 (abstract).

89. Kaku, D. A., Giffard, R. G., and Choi, D. W. (1993) Neuroprotective effects of glutamate antagonists and extracellular acidity. *Science* **260,** 1516–1518.

90. Kasai, H. and Peterson, O. H. (1994) Spatial dynamics of second messengers: IP3 and cAMP as long-range and associative messengers. *TINS* **17,** 95–101.

91. Kass I. S. and Lipton P. (1982) Mechanisms involved in irreversible anoxic damage to the in vitro rat hippocampal slice. *J. Physiol.* **332,** 459–472.

92. Kirino, T. (1982) Delayed neuronal death in the gerbil hippocampus following ischemia. *Brain Res.* **239,** 57–69.

93. Kitamura Y, Miyazaki, A., Yamanaka, Y., and Nomura, Y. (1993) Stimulatory effects of protein kinase C and calmodulin kinase II on N-methyl-D-aspartate receptor/channels in the postsynaptic density of rat brain. *J. Neurochem.* **61,** 100–109.

94. Koch, R. A. and Barish, M. E. (1994) Perturbation of intracellular calcium and hydrogen ion regulation in cultured mouse hippocampal neurons by reduction of the sodium ion concentration gradient. *J. Neurosci.* **14,** 2585–2593.

95. Koh, JY., Goldberg, M. P., Hartley, D. M., and Choi, D. W. (1990) Non-NMDA receptor mediated neurotoxicity in cortical culture. *J. Neurosci.* **10,** 696–705.

96. Krnjevic, K. (1974) Chemical nature of synaptic transmission in vertebrates. *Physiol. Res.* **54,** 418–540.

97. Kure, S., Tominaga, T., Yoshimoto, T., Tada, K., and Narisawa, K. (1991) Glutamate triggers internucleosomal DNA cleavage in neuronal cells. *Biochem. Biophys. Res. Commun.* **179,** 39–45.

98. Kurth, M. C., Weiss, J. H., and Choi, D. W. (1989) Relationship between glutamate-induced 45-calcium influx and resultant neuronal injury in cultured cortical neurons. *Neurology* **39(suppl),** 217 (abstract).

99. Lafon-Cazal, M., Pietri, S., Culcasi, M., and Bockaert, J. (1993) NMDA-dependent superoxide production and neurotoxicity. *Nature* **364,** 535–537.

100. LaManna, J. C. and McCracken, K. A. (1984) The use of neutral red as an intracellular pH indicator in rat brain cortex in vivo. *Anal. Biochem.* **142,** 117–125.

101. Lazarewicz, J. W., Wroblewski, J. T., and Costa, E. (1990) N-methyl-D-aspartate-sensitive glutamate receptors induce calcium-mediated arachidonic acid release in primary cultures of cerebellar granule cells. *J. Neurochem.* **55,** 1875–1881.

102. Lechleiter, J. D. and Clapham, D. E. (1992) Molecular mechanisms of intracellular calcium excitability in *X. laevis* oocytes. *Cell* **69,** 283–294.

103. Lei, S. Z., Pan, Z. H., Aggarwal, S. K., et al. (1992) Effect of nitric oxide production on the redox modulatory site of the NMDA receptor-channel complex. *Neuron* **8,** 1087–1099.

104. Lemasters, J. J., DiGiuseppi, J. D., Nieminen, A. L., and Herman, B. (1987) Blebbing free calcium and mitochondrial membrane potential preceding cell death in hepatocytes. *Nature* **325,** 78–81.

105. Lerea, L. S. and McNamara, J. O. (1993) Ionotropic glutamate receptor subtypes activate c-fos transcription by distinct calcium-requiring intracellular signaling pathways. *Neuron* **10,** 31–41.

106. Lipscombe, D., Madison, D., Poenie, M., Reuter, H., Tsien, R. Y., and Tsien, R. W. (1988) Spatial distribution of calcium channels and cytosolic transients in growth cones and cell bodies of sympathetic neurons. *Proc. Natl. Acad. Sci. USA* **85,** 2398–2402.

107. Lipton, S. A., Choi, Y. B., Pan, Z. H., et al. (1993) A redox-based mechanism for the neuroprotective and neurodestructive effects of nitric oxide and related nitroso-compounds. *Nature* **364,** 626–631.

108. Liu, D., Thagnipon, W., and McAdoo, D. J. (1991) Excitatory amino acids rise to toxic levels upon impact injury to the rat spinal cord. *Brain Res.* **547,** 344–348.

109. Liu, X. Z., Xu, X. M., Hu, R., et al. (1997) Neuronal and glial apoptosis after traumatic spinal cord injury. *J. Neurosci.* **17,** 5395–5406.

110. Lobner, D. and Lipton, P. (1993) Intracellular calcium levels and calcium fluxes in the CA1 region of the rat hippocampal slice during in vitro ischemia: relationship to electrophysiological cell damage. *J. Neurosci.* **13,** 4861–4871.

111. Lucas, D. R. and Newhouse, J. P. (1957) The toxic effect of sodium L-glutamate on the inner layers of the retina. *Arch. Ophthalmol.* **58,** 193–201.

112. Lwe, V. L., Tsien, R. Y., and Miner, C. (1982) Physiological $[Ca^{2+}]_i$ level and pump-leak turnover in intact red cells measured using an incorporated Ca chelator. *Nature* **298,** 478–481.

113. MacDermott, A. B., Mayer, M. L., Westbrook, G. L., Smith, S. J., and Barker, J. L. (1986) NMDA-receptor activation increases cytoplasmic calcium concentration in cultured spinal cord neurons. *Nature* **321,** 519–522.

114. Madison, D. (1993) Pass the nitric oxide. *Proc. Natl. Acad. Sci. USA* **90,** 4329–4331.

115. Manev, H., Favaron, Guidotti, A., and Costa, E. (1989) Delayed increase of Ca^{2+} influx elicited by glutamate: role in neuronal death. *Mol. Pharmacol.* **36,** 106–112.

116. Marcoux, F. W., Probert, A. W., and Weber, M. L. (1990) Hypoxic neuronal injury in tissue culture is associated with delayed calcium accumulation. *Stroke* **21(suppl III),** III-71-III74.

117. Marcoux, F. W., Probert, A. W., and Weber, M. L. (1989) Hypoxic neural injury in cell culture: calcium accumulation blockade and neuroprotection by NMDA antagonists but not calcium channel antagonists, in *Cerebrovascular Disease: Sixteenth Princeton Conference* (Ginsberg, M. G., Dietrich, W. D. eds.), Raven, New York, pp. 135–141.

118. Marmarou, A., Holdaway, R., Ward, J. D., et al. (1993) Traumatic brain tissue acidosis: experimental and clinical studies. *Acta Neurochir. Suppl. (Wien)* **57,** 160–164.

119. Matsumoto, K., Kamata, T., and Goto, N. (1993) The suppression by nizofenone of the release of glutamate and lactate as a mechanism of its neuroprotective effect: an in vivo brain microdialysis study. *J. Cereb. Blood Flow Metab.* **13(suppl 1),** S742 (abstract).

120. Matsumoto, K., Ueda, S., Hashimoto, T., and Kuriyama, K. (1991) Ischemic neuronal injury in the rat hippocampus following transient forebrain ischemia: evaluation using in vivo microdialysis. *Brain Res.* **543,** 236–242.

121. Nooney, J. M., Lamber, R. C., and Feltz, A. (1997) Identifying neuronal non-L Ca^{2+} channels—more than stamp collecting. *TIPS* **18,** 363–371.

122. Regan, R. F. (1996) The vulnerability of spinal cord neurons to excitotoxic injury: comparison with cortical neurons. *Neurosci. Lett.* **213,** 9–12.

123. Regan, R. F. and Choi, D. W. (1991) Glutamate neurotoxicity in spinal cord cell culture. *Neuroscience* **43,** 585–591.

124. Sattler, R., Charlton, M. P., Hafner, M., and Tymianski, M. (1998) Distinct influx pathways, not calcium load, determine neuronal vulnerability to calcium neurotoxicity. *J. Neurochem.* **71,** 2349–2364.

125. Terro, F., Yardin, C., Exclaire, F., Ayer-Lelievre, C., and Hugon, J. (1998) Mild kainate toxicity produces selective motoneuron death with marked activation of Ca^{2+}-permeable AMPA/kainate receptors. *Brain Res.* **809,** 319–324.

126. Mayer, M. L. and Vyklicky, L. (1989) Concanavalin A selectively reduces desensitization of mammalian neuronal quisqualate receptors. *Proc. Natl. Acad. Sci. USA* **86,** 1411–1415.

127. McLean, A. E. M., McLean, E., and Judah, J. D. (1965) Cellular necrosis in the liver induced and modified by drugs. *Int. Rev. Exp. Pathol.* **4,** 127–157.

128. Meyer, F. B. (1989) Calcium, neuronal hyperexcitability and ischemic injury. *Brain Res. Rev.* **14,** 227–243.

129. Michaels, R. L. and Rothman, S. M. (1990) Glutamate neurotoxicity in vitro: antagonist pharmacology and intracellular calcium concentrations. *J. Neurosci.* **10,** 283–292.

130. Milani, D., Malgaroli, A., Guidolin, D., et al. (1990) Ca^{2+} channels and intracellular Ca^{2+} stores in neuronal and neuroendocrine cells. *Cell Calcium* **11,** 191–199.

131. Miller, R. J. (1992) Neuronal Ca^{2+} getting it up and keeping it up. *TINS* **15,** 317–319.

132. Muller, W. and Connor, J. A. (1991) Dendritic spines as individual neuronal compartments for synaptic Ca^{2+} responses. *Nature,* **354,** 73–80.

133. Murphy, S. N. and Miller, R. J. (1988) A glutamate receptor regulates calcium mobilization in hippocampal neurons. *Proc. Natl. Acad. Sci. USA* **85,** 8737–8741.

134. Nakanishi, S. (1992) Molecular diversity of glutamate receptors and implications for brain function. *Science* **258,** 597–603.

135. Naschshen, D. A., Sanchez-Armass, S., and Weinstein, A. M. (1986) The regulation of cytosolic calcium in rat brain synaptosomes by sodium-dependent calcium efflux. *J. Physiol.* **381,** 17–28.

136. Neering, I. R., McBurney, R. N. (1984) Role for microsomal Ca storage in mammalian neurones? *Nature* **309,** 158–160.

137. Neher, E. (1986) Concentration profiles of intracellular calcium in the presence of a diffusible chelator, in *Calcium Electrogenesis and Neuronal Functioning.* Experimental Brain Research, Series 14 (Heinemann, U., Klee, M., and Neher, E., eds.), Springer-Verlag Berlin, pp. 80–96.

138. Neher, E. and Augustine, G. J. (1992) Calcium gradients and buffers on bovine chromaffin cells. *J. Physiol.* **450,** 273–301.

139. Nicholson, C., Bruggencate, G. T., Steinberg, R., and Stockle, H. (1977) Calcium modulation in brain extracellular microenvironment demonstrated with ion selective micropipette. *Proc. Natl. Acad. Sci. USA* **74,** 1287–1290.

140. Nowycky, M. C. and Pinter, M. J. (1993) Time courses of calcium and calcium-bound buffers following calcium influx in a model cell. *Biophys. J.* **64,** 77–91.

141. O'Dell, T. J., Hawkins, R. D., Kandel, E., and Arancio, O. (1991) Tests of the roles of two diffusible substances in long-term potentiation: evidence for nitric oxide as a possible early retrograde messenger. *Proc. Natl. Acad. Sci. USA* **88,** 11285–11289.

142. Ogura, A., Myamoto, M., and Kudo, Y. (1988) Neuronal death in vitro: parallelism between survivability of hippocampal neurones and sustained elevation of cytosolic calcium after exposure to glutamate receptor agonist. *Exp. Brain Res.* **73,** 447–458.

143. Ojcius, D. M., Zychlinsky, A., Zheng, L. M., and Young, D. (1991) Ionophore-induced apoptosis: role of DNA fragmentation and calcium fluxes. *Exp. Cell Res.* **197,** 43–49.

144. Olney, J. W. (1969) Brain lesion, obesity and other disturbances in mice treated with monosodium glutamate. *Science* **164,** 719–721.

145. Olney, J. W. (1978) Neurotoxicity of excitatory amino acids, in *Kainic Acid as a Tool in Neurobiology (McGeer, E. G., Olney, J. W., McGeer, P. L. eds.) Raven, New York,* pp. 95–121.

146. Olney, J. W., Price, M. T., Samson, L., and Labruyere, J. (1986) The role of specific ions in glutamate neurotoxicity. *Neurosci. Lett.* **65,** 65–71.

147. Opitz, T. and Reymann, K. G. (1991) Blockade of metabotropic glutamate receptors protects rat CA1 neurons from hypoxic injury. *Neuroreport* **2,** 455–457.

148. Ozawa, K., Seta, K., Araki, H., and Handa, H. (1966) The effect of ischemia on mitochondrial metabolism. *J. Biol. Chem.* **61,** 512–514.

149. Ozyurt, E., Graham, D. I., Woodruff, G. N., and McCulloch, J. (1988) Protective effect of the glutamate antagonist, MK-801 in focal cerebral ischemia in the cat. *J. Cereb. Blood Flow Metab.* **8,** 138–143.

150. Pellegrini-Giampietro, D. E., Cherici, G., Alesiani, M., Carla, V., and Moroni, F. (1990) Excitatory amino acid release and free radical formation may cooperate in the genesis of ischemia-induced neuronal damage. *J. Neurosci.* **10,** 1035–1041.

151. Potter, H., Nelson, R. B., Das, S., Siman, R., Kayyali, U. S., and Dressler, D. (1992) The involvement of proteases, protease inhibitors, and an acute phase response in Alzheimer's disease. *Ann. NY. Acad. Sci.* **674,** 161–173.

152. Price, M. T., Olney, J. W., Samson, L., and Labruyere, J. (1985) Calcium influx accompanies but does not cause excitotoxin-induced neuronal necrosis in retina. *Brain Res.* **14,** 369–376.

153. Pruss, R. M., Akeson, R. I., Racke, M. M., and Wilburn, J. L. (1991) Agonist-activated cobalt uptake identifies divalent cation-permeable kainate receptors on neurons and glial cells. *Neuron* **7,** 509–518.

154. Raff, M. C., Barres B. A., Burne, J. F., Coles, H. S., Ishizaki, Y., and Jacobson, M. D. (1993) Programmed cell death and the control of cell survival: lessons from the nervous system. *Science* **262,** 695–700.

155. Randall, R. D. and Thayer, S. A. (1992) Glutamate-induced calcium transient triggers delayed calcium overload and neurotoxicity in rat hippocampal neurons. *J. Neurosci.* **12,** 1882–1895.

156. Rehncrona, S., Mela, L., and Siesjo, B. K. (1979) Recovery of brain mitochondrial function in the rat after complete and incomplete cerebral ischemia. *Stroke* **10,** 437–446.

157. Roberts-Lewis, J. M., Marcy, V. R., Zhao, Y., Vaught, J. L., Siman, R., and Lewis, M. E. (1993) Aurintricarboxylic acid protects hippocampal neurons from NMDA- and ischemia-induced toxicity in vivo. *J. Neurochem.* **61,** 378–381.

158. Roberts-Lewis, J. M., Savage, M. J., Marcy, V. R., Pinsker, L. R., and Siman, R. (1994) Immunolocalization of calpain I-mediated spectrin degradation to vulnerable neurons in the ischemic brain. *J. Neurosci.* **14,** 3934–3944.

159. Roberts, W. M. (1993) Spatial calcium buffering in saccular hair cells. *Nature* **363,** 74–76.

160. Roberts, W. M. (1994) Localization of calcium signals by a mobile calcium buffer in frog saccular hair cells. *J. Neurosci.* **15,** 3246–3262.
161. Robitaille, R., Adler, E. M., and Charlton, M. P. (1990) Strategic location of calcium channels at transmitter release sites of frog neuromuscular synapses. *Neuron* **5,** 773–779.
162. Robitaille, R., Garcia, M. L., Kaczorowski, G. J., and Charlton, M. P. (1993) Functional co-localization of calcium and calcium-gated potassium channels in control of transmitter release. *Neuron* **11,** 645–655.
163. Rothman, S. M. (1983) Synaptic activity mediates death of hypoxic neurons. *Science* **220,** 536–537.
164. Rothman, S. M. (1984) Synaptic release of excitatory amino acid neurotransmitter mediates anoxic neuronal death. *J. Neurosci.* **4,** 1884–1891.
165. Rothman, S. M. (1985) The neurotoxicity of excitatory amino acids is produced by passive chloride influx. *J. Neurosci.* **5,** 1483–1489.
166. Rothman, S. M. and Olney, J. W. (1986) Glutamate and the pathophysiology of hypoxic-ischemic brain damage. *Ann. Neurol.* **19,** 105–111.
167. Rothman, S. M. and Olney, J. W. (1987) Excitotoxicity and the NMDA receptor. *TINS* **10,** 299–302.
168. Sako, K., Kobatake, K., Yamamoto, Y. L. and Kiksic, M. (1985) Correlation of local cerebral blood flow, glucose utilization and tissue pH following a middle cerebral artery occlusion in the rat. *Stroke* **16,** 828–934.
169. Sala, F. and Hernandez-Cruz, A. (1990) Calcium diffusion modeling in a spherical neuron: relevance of buffering properties. *Biophys. J.* **57,** 313–324.
170. Sanfeliu, C., Hunt, A., Patel, A. J. (1990) Exposure to N-methyl-D-aspartate increases release of arachidonic acid in primary cultures of rat hippocampal neurons and not in astrocytes. *Brain Res.* **526,** 241–248.
171. Schanne, F. A. X., Kane, A. B., Young, E. A., and Farber, J. L. (1979) Calcium dependence of toxic cell death: a final common pathway. *Science* **206,** 700–702.
172. Schlaepfer, W. W. and Bunge, R. P. (1973) Effects of calcium ion concentration on the degeneration of amputated axons in tissue culture. *J. Cell Biol.* **59,** 456–470.
173. Schlaepfer, W. W. and Zimmerman, U. J. (1985) Mechanisms underlying the neuronal response to ischemic injury: calcium-activated proteolysis of neurofilaments. *Prog. Brain. Res.* **63,** 185–196.
174. Schmidley, J. W. (1990) Free radicals in central nervous system ischemia. *Stroke* **21,** 1086–1090.
175. Schramm, M. and Eimerl, S. (1993) The quantity of Ca that enters a neuron to cause its death in glutamate toxicity. *Soc. Neurosci. Abstr.* **19,** 1501 (abstract).
176. Schuman, E. M. and Madison, D. V. (1991) A requirement for the intracellular messenger nitric oxide in long term potentiation. *Science* **254,** 1503–1506.
177. Schutz, H., Silverstein, P. R., Vapalahti, M., Bruce, D. A., Mela, L., and Langfitt, T. W. (1973) Brain mitochondrial function after ischemia and hypoxia: I. Ischemia induced by increased intracranial pressure. *Arch. Neurol.* **29,** 408–416.

178. Schutz, H., Silverstein, P. R., Vapalahti, M., Bruce, D. A., Mela, L., and Langfitt, T. W. (1973) Brain mitochondrial function after ischemia and hypoxia: II. Normotensive systemic hypoxemia. *Arch. Neurol.* **29,** 417–419.

179. Siesjo, B. K. (1988) Mechanisms of ischemic brain damage. *Crit. Care Med.* **16,** 954–963.

180. Siesjo, B. K. (1992) Pathophysiology and treatment of focal cerebral ischemia: part 1. Pathophysiology. *J. Neurosurg.* **77,** 169–184.

181. Siesjo, B. K. and Bengtsson, F. (1989) Calcium fluxes, calcium antagonists, and calcium-related pathology in brain ischemia, hypoglycemia, and spreading depression: a unifying hypothesis. *J. Cereb. Blood Flow Metab.* **9,** 127–140.

182. Siesjo, B. K., Katsura, K., and Tibor, K. (1996) Acidosis related brain damage, *in Advances in Neurology: Cellular and Molecular Mechanisms of Ischemic Brain Damage* (Siesjo, B. K. and Wieloch, T. eds.), Raven, New York, pp. 209–236.

183. Silver I. A. and Erecinska M. (1990) Intracellular and extracellular changes of [Ca^{2+}] in hypoxia and ischemia in rat brain in vivo. *J. Gen. Physiol.* **95,** 837–866

184. Siman, R. (1992) Proteolytic mechanism for the neurodegeneration of Alzheimer's disease. *Ann. NY Acad. Sci.* **674,** 193–202.

185. Siman, R., Noszek, C., and Kegerise, C. (1989) Calpain I activation is specifically related to excitatory amino acid induction of hippocampal damage. *J. Neurosci.* **9,** 1579–1590.

186. Siman, R. and Noszek, J. C. Excitatory amino acids activate calpain I and induce structural protein breakdown in vivo. *Neuron* **1,** 279–287.

187. Simon, R. P., Griffiths, T., Evans M. C., Swan, J. H., and Meldrum, B. S. (1984) Calcium overload in selectively vulnerable neurons of the hippocampus during and after ischemia: an electron microscopy study in the rat. *J. Cereb. Blood Flow Metab.* **4,** 350–361.

188. Simon, R. P., Swan, J. H., and Meldrum, B. S. (1984) Blockade of N-methyl-D-aspartate receptors may protect against ischemic damage in the brain. *Science* **226,** 850–852.

189. Smith, M. T., Thor, H., and Orrenius, S. (1981) Toxic injury to isolated hepatocytes is not dependent on extracellular calcium. *Science* **213,** 1257–1259.

190. Sombati, S. Coulter, D. A., and DeLorenzo, R. J. (1991) Neurotoxic activation of glutamate receptors induces an extended neuronal depolarization in cultured hippocampal neurons. *Brain Res.* **566,** 316–319.

191. Southam, E. S., East, S. J., and Garthwaite, J. (1991) Excitatory amino acid receptors coupled to the nitric oxide/cyclic GMP pathway in rat cerebellum during development. *J. Neurochem.* **56,** 2072–2081.

192. Speksnijder, J. E., Miller, A. L., Weisenseel, M. H., Chen, T. H., and Jaffe, L. F. (1989) Calcium buffer injections block fucoid egg development by facilitating calcium diffusion. *Proc. Natl. Acad. Sci. USA* **86,** 6607–6611.

193. Stern, M. D. (1992) Buffering of calcium in the vicinity of a channel pore. *Cell Calcium* **13,** 183–192.

194. Stokes, B. T., Fox, P., and Hallinden, G. (1983) Extracellular calcium activity in the injured spinal cord. *Exp. Neurol.* **80,** 561–572.

195. Strautman, A. F., Cork, R. J., and Robinson, K. R. (1990) The distribution of free calcium in transected spinal axons and its modulation by applied electrical fields. *J. Neurosci.* **10,** 3564–3575.

196. Stys, P. K., Ransom, B. R., Waxman, S. G., and Davis, P. K. (1990) Role of extracellular calcium in anoxic injury of mammalian central white matter. *Proc. Natl. Acad. Sci. USA* **87,** 4214–4216.

197. Taira, T., Smirnov, S., Voipio, J., and Kaila, K. (1993) Intrinsic proton modulation of excitatory transmission in rat hippocampal slices. *Neuroreport* **4,** 93–96.

198. Tator, C. H. (1991) Review of experimental spinal cord injury with emphasis on the local and systemic circulatory effects. *Neurochirurgie* **37,** 291–302.

199. Tator, C. H. and Fehlings, M. G. (1991) Review of the secondary injury theory of acute spinal cord trauma with emphasis on vascular mechanisms. *J. Neurosurg.* **75,** 15–26.

200. Thayer, S. A. and Miller, R. J. (1990) Regulation of the intracellular free calcium concentration in single rat dorsal root ganglion neurones in vitro. *J. Physiol.* **425,** 85–115.

201. Toescu, E. C., Gardner, J. M., and Petersen, O. H. (1993) Mitochondrial Ca^{2+} uptake at submicromolar $[Ca^{2+}]_i$ in permeabilised pancreatic acinar cells. *Biochem. Biophys. Res. Commun.* **192,** 854–859.

202. Tombaugh, G. C. and Sapolsky, R. M. (1990) Mechanistic distinctions between excitotoxic and acidotic hippocampal damage in an in vitro model of ischemia. *J. Cereb. Blood Flow Metab.* **10,** 527–535.

203. Tominaga, T., Kure, S., and Yoshimoto, T. (1993) Temporal profile of DNA degradation in injured rat brain. *J. Cereb. Blood Flow Metab.* **13(suppl 1),** S460 (abstract).

204. Tymianski, M. (1996) Cytosolic calcium concentrations and cell death in vitro, in *Advances in Neurology Vol. 71: Cellular and Molecular Mechanisms of Ischemic Brain Damage* (Siesjo, B. K., Wieloch, T., eds.), Raven, New York, pp. 85–105.

205. Tymianski, M., Charlton, M. P., Carlen, P. L., and Tator, C. H. (1993) Source-specificity of early calcium neurotoxicity in cultured embryonic spinal neurons. *J. Neurosci.* **13,** 2085–2104.

206. Tymianski, M., Charlton, M. P., Carlen, P. L., and Tator, C. H. (1993) Secondary Ca^{2+} overload indicates early neuronal injury which precedes staining with viability indicators. *Brain Res.* **607,** 319–323.

207. Tymianski, M., Charlton, M. P., Carlen, P. L., and Tator, C. H. (1994) Properties of neuroprotective cell-permeant Ca^{2+} chelators: effects on $[Ca^{2+}]_i$ and glutamate neurotoxicity in vitro. *J. Neurophysiol.* **267,** 1973–1992.

208. Uematsu, D., Araki, N., Greenberg, J. H., and Reivitch, M. (1990) Alterations in cytosolic free calcium in the cat cortex during bicuculline-induced epilepsy. *Brain Res. Bull.* **24,** 285–288.

209. Uematsu, D., Greenberg, J. H., Hickey, W. F., and Reivich, M. (1989) Nimodipine attenuates both increase in cytosolic free calcium and histologic damage following focal cerebral ischemia and reperfusion in cats. *Stroke* **20,** 1531–1537.

210. Uematsu, D., Greenberg, J. H., and Karp, A. (1988) In vivo measurement of cytosolic free calcium during cerebral ischemia and reperfusion. *Ann. Neurol.* **24,** 420–428.

211. Urca, G. and Urca, R. (1990) Neurotoxic effects of excitatory amino acids in the mouse spinal cord: quisqualate and kainate but not N-methyl-D-aspartate induce permanent neural damage. *Brain Res.* **529,** 7–15.

212. Vlessi, A. A., Widener, L. L., and Bartos, D. (1990) Effect of peroxide, sodium, and calcium on brain mitochondrial respiration in vitro: potential role in cerebral ischemia and reperfusion. *J. Neurochem.* **54,** 1412–1418.

213. Wang, Q., Paulson, O. B., and Lassen, N. A. (1992) Effect of nitric oxide blockade by NG-nitro-L-alginin on cerebral blood flow response to changes in carbon dioxide tension. *J. Cereb. Blood Flow Metab.* **12,** 947–953.

214. Weber, M. L., Probert, A. W., Boxer, P. A., and Marcoux, F. W. (1988) The effect of ion channel modulators on hypoxia-induced calcium accumulation and injury in cortical neuronal cultures. *Soc. Neurosci. Abstr.* **14,** 1117 (abstract).

215. Werth, J. L. and Thayer, S. A. (1994) Mitochondria buffer physiological calcium loads in cultured rat dorsal root ganglion neurons. *J. Neurosci.* **14,** 348–356.

216a. Westenbroek, R. E., Ahlijanian M. K., and Catterall, W. A. (1990) Clustering of L-type channels at the base of major dendrites in hippocampal neurons. *Nature* **347,** 281–284.

216b. Putney, J. W. (1998) Calcium signalin: Up, down, up, down. . . . what's the point? *Science* **279,** 191–192.

216c. Ghosh, A. and Greenberg, M. E. (1995) Calcium signalling in neurons: Molecular mechanisms and cellular consequences. *Science* **268,** 239–247.

217. Worley, P. F., Baraban, J. M., and Synder, S. H. (1987) Beyond receptors: multiple second-messenger systems in brain. *Ann. Neurol.* **21,** 217–229.

218. Yang, G., Chan, P. H., Chen, J., et al. (1994) Human copper-zinc superoxide dismutase transgenic mice are highly resistant to reperfusion injury after focal cerebral ischemia. *Stroke* **25,** 165–170.

219. Young, W. and Flamm, E. S. (1982) Effect of high dose corticosteroid therapy on blood flow, evoked potentials, and extracellular calcium in experimental spinal injury. *J. Neurosurg.* **57,** 667–673.

220. Young, W., Yen, V., and Blight A. (1982) Extracellular calcium activity in experimental spinal cord contusion. *Brain Res.* **253,** 115–123.

221. Zhou, Z. and Neher, E. (1993) Mobile and immobile calcium buffers in bovine adrenal chromaffin cells. *J. Physiol.* **469,** 245–273.

222. Zimmerman, A. N. E., Daems, W., Hulsmann, W. C., Snijder, J., Wisse, E., and Durrer, D. (1967) Morphological changes of heart muscle caused by successive perfusion with calcium-free and calcium-containing solutions (calcium paradox). *Cardiovasc. Res.* **1,** 201–209.

223. Ziv, N. E. and Spira, M. E. (1993) Spatiotemporal distribution of Ca^{2+} following axotomy and throughout the recovery process of cultured aplysia neurons. *Eur. J. Neurosci.* **5,** 657–668.

224. Ziv, N. E. and Spira, M. E. (1995) Axotomy induces a transient and localized elevation of the free intracellular calcium concentration to the millimolar range. *J. Neurophysiol.* **74,** 2625–2637.

3

The Spinal Cat

Serge Rossignol, Marc Bélanger, Connie Chau,
Nathalie Giroux, Edna Brustein, Laurent Bouyer,
Claude-André Grenier, Trevor Drew,
Hughes Barbeau, and Tomas A. Reader

1. INTRODUCTION

A number of reviews have summarized important insights on the role played by various nervous system structures in the control of locomotion *(1–8)*. These reviews have also highlighted the remarkable locomotor capacities of the spinal cord after a complete spinal transection, which removes all the ascending and descending pathways normally exerting important control over spinal cord functions. The purpose of this chapter is to focus specifically on the locomotor capabilities of the spinal cat, not so much to show that "spinal" locomotion resembles "normal" locomotion but rather to illustrate the extent to which the spinal cord can express and adapt its locomotor functions in the absence of these regulatory mechanisms. Does this spinal behavior represent the contribution of the spinal cord to normal locomotion? Probably not, because in all pathologic conditions, the central nervous system utilizes whatever circuitry is available to optimize its functions. It is possible that some mechanisms are less important in the normal cat but become essential for locomotion after spinalization, such as some sensory afferents. Thus, a better understanding of the "physiopathology" of locomotion after spinal cord injury in animal models is important both in highlighting some of the principles that may help understand normal locomotion and in increasing our understanding of some of the mechanisms of recovery of a motor function following a spinal trauma. Such knowledge is important for improving the design of various types of therapeutic approaches in spinal-cord-injured patients *(9,10)*.

Although this chapter mainly reviews work in recent decades, it is important to remember that pioneering work in this area dates from the beginning of the century. Indeed, Sherrington *(11–13)* described the locomotion of spinal dogs and cats, 1 or 2 days after a spinal section. In the high spinal cat, he showed that stimulation of the perineum, or faradization of the foot or of an afferent nerve, could elicit stepping, as could stimulation of the cut end of the bulb or the spinal cord. Sherrington *(13)* stated that "The burden of shunting from flexion to exten-

From: *Neurobiology of Spinal Cord Injury*
Edited by: R. G. Kalb and S. M. Strittmatter © Humana Press Inc., Totowa, NJ

Table 1
Chronic Complete Spinal Section in Cats

	Level of Transection	References
Kittens and infants	T12–L1 and L3	15, 18, 19, 21, 23, 85, 86
Adults	Cervical	87, 88
Adults	T3–T4	24, 25, 30, 89
Adults	T6–T10	24, 30, 89
Adults	T13–L1	12, 23, 25, 26, 28, 31, 32, 36, 52, 56, 58, 78, 80, 81, 86, 89–93

sion and vice versa is thrown greatly upon mechanisms wholly intrinsic to the cord." This early view on the generation of locomotion by spinal cord mechanisms was even more strongly held by T. G. Brown *(14),* as extensively reviewed before *(1,6).*

Table 1 lists some key references on locomotor functions in cats after complete chronic transections of the spinal cord at various levels and ages. Shurrager and Dykman *(15)* reported that kittens spinalized at the age of 2 days to 12 weeks could display walking movements over ground that were maintained after a second transection above the first lesion, thus ruling out all possibilities of recovery due to regrowth of descending axons. Much insistence was put on daily training as well as on the fact that young animals performed much better than older animals. An important study reported that kittens spinalized within a few days after birth (before they had expressed any locomotor pattern) could eventually walk and gallop with their hindlimbs, with correct foot placement and weight support of the hindquarters *(16,17).*

A more complete account of the locomotor performances found in chronic spinal kittens *(1,18,19)* showed that the kinematics and the electromyograms (EMGs) of spinal kittens were very similar indeed to those of normal cats, and also that the spinal kittens had the ability to adapt locomotion to various treadmill speeds as well as to differential treadmill speeds. The latter capability highlighted the importance of afferent feedback for adapting the locomotor pattern to external conditions. Moreover, these spinal kittens adapted their stepping to perturbations that were applied during the swing and stance phase and generated specific compensatory responses in the various phases (see ref. *20* for a review).

The age of spinalization and the amount of training on the treadmill were shown to have important effects on the recovery of locomotion pattern *(21–23).* Spinal animals of 2 weeks of age walked much better than those spinalized at 12 weeks of age. However, training in 12-week-old cats had an observable effect on weight bearing during locomotion. The EMG pattern appeared normal in many respects, except that clonus was frequently observed. Defects in the locomotor pattern such as uncoupling between the knee and ankle as well as an absence of yield in the early part of stance were reported.

Early studies also reported locomotor capabilities in adult cats *(12,24–27)*. Ten Cate *(28)* noted that spinal cats could generate hindlimb movements only when the body was pulled forward by the forelimb movements. A very important paper appeared *(29)* showing that acutely spinalized adult cats could walk with the hindlimbs on a treadmill after an iv injection of α_2-noradrenergic agonist, clonidine. A report on the locomotion of adult chronic spinal cats *(30)* placed with all four limbs on a treadmill concluded that two-thirds of the cats had clear rhythmic movements of the hindlimbs but that, because of the loss of fore-hindlimb coordination, the movements were erratic. They also reported that these cats were also generally incapable of weight support, that they had a paw drag during swing, and that the knee and ankle movements were uncoupled. Their locomotion did not improve with repeated trials (training). This rather negative report obscured the fact that adult spinal cats could indeed walk with their hindlimbs on a treadmill, much like the spinal kittens, as will be seen later.

In the following paragraphs we summarize various aspects of our work on adult spinal cats. We first describe the characteristics of spinal locomotion compared with that of the same animal recorded previously in intact conditions. We show some of the adaptive capacities of this spinal locomotion (speeds, slopes), how it can be modulated by various neurotransmitters, and finally how recent evidence suggests that controlling mechanisms of locomotion may undergo plastic changes after either central or peripheral lesions. For more details, some of the key original papers should be consulted *(31–34)*.

2. GENERAL METHODOLOGY

2.1. General Protocol

Adult cats were first trained for 1–4 weeks to walk at constant speeds (0.2–0.7 m/s) for periods of about 20 min in an enclosed transparent Plexiglass box over a motor-driven treadmill belt. Thereafter, EMG recording wire electrodes were implanted and, in some cases, an intrathecal cannula was also inserted at the same time. After recovery from these implantations, locomotor movements and associated EMGs were obtained to establish baseline values before spinalization. Following spinalization, the same type of data were obtained starting 3 days after the transection, at a time when the cat had recovered from the surgery. Cats were kept for 3 months up to more than a year. At the end of the study, the spinal cords were removed for histology and, in some cases, for autoradiographic studies of receptors.

2.2. Surgical Procedures

All surgeries were performed in aseptic conditions and under general anesthesia (either halothane after intubation or, in some cases, pentobarbital 35 mg/kg ip) with appropriate preoperative medication (acepromazine maleate

[Atravet™, 0.1 mg/kg sc], atropine [0.05 mg/kg sc], and penicillin G [40,000 IU/kg, im]).

2.2.1. Implantation of Chronic EMG Electrodes

One to four multipin head connectors (TRW Electronic Components Group, Elk Grove Village, IL) were used. Fifteen Teflon-insulated stainless steel wires (AS633; Cooner Wire, Chatsworth, CA) were soldered to each connector fixed to the skull using acrylic cement and led subcutaneously to various muscles.

2.2.2. Nerve Cuffs

In some cats bipolar cuff electrodes (~1 cm length) *(35)* were used to stimulate the superficial peroneal nerve (~6 mm between electrode leads). The leads were soldered to the EMG connectors.

2.2.3. Intrathecal Cannula

An intrathecal cannula (Teflon 24 LW tubing) connected to an adaptor was cemented on the skull together with the EMG connectors. The other end of the cannula was inserted into the intrathecal space through an opening in the atlanto-occipital ligament down to L4–L5, as measured from external landmarks.

2.2.4. Spinal Cord Lesion

A laminectomy was performed at the T13 vertebra, the dura was carefully removed, and xylocaine (2%) was applied topically before the cord was transected to prevent brisk movements of the hindlimbs when performing the lesion. The spinal cord was completely severed, and the ventral surface of the spinal canal could be clearly visualized. Absorbable hemostat (Surgicel™) was inserted at the lesion site. In some of the cats reported on here, a large lesion of the ventral and ventrolateral tracts was made by a bilateral approach through the vertebral pedicles.

2.3. Postoperative Care

Cats were placed in an incubator until they regained consciousness before being returned to their cages with ample food and water. For analgesia, buprenorphine (0.005–0.01 mg/kg sc) was also given every 6 hr in the first few postoperative days. All spinal cats were placed in individual cages. The cages were specially lined with a foam mattress, in addition to the usual absorbent tissues, to reduce the risk of skin ulcers. Cats were attended to at least twice daily for manual bladder expression, general inspection, and cleaning of the hindquarters. All procedures followed a protocol approved by the local Ethics Committee, and the well-being of the cats was always ensured (see also ref. *36* for more general care of spinal cats used by others).

2.4. Recording and Analysis Procedures of Locomotor performance

Recording of locomotion in the intact state was done while the animal walked freely at different speeds imposed by the treadmill belt. For intact locomotion, the cats were placed on the treadmill belt enclosed by the Plexiglass box, and their free walking was recorded at different speeds. For spinal locomotion, the forelimbs were placed on a platform while the hindlimbs walked on the moving treadmill belt. Early after spinalization, the experimenter supported the weight of the hindquarters of the cat and provided equilibrium; later on, cats were capable of walking with complete weight support of the hindquarter, and the tail was held only to secure lateral stability.

Reflective markers were placed on the skin overlying the iliac crest, the femoral head, the knee joint, the lateral malleolus, the metatarsophalangeal joint (MTP), and the tip of the fourth toe. Video images of the side view of the cat were captured using a digital camera (Panasonic 5100, shutter speed 1/1000 s) and recorded on a video tape (Panasonic AG 7300 recorder). Calibration markers (10-cm distance) were placed either on the treadmill frame, or on the trunk of the cat, to reduce parallax errors.

The EMG signals were differentially amplified (bandwidth of 100 Hz to 3 KHz) and recorded on a 14-channel tape recorder with an approriate frequency response. The EMG recordings were synchronized to the recorded video images by means of a digital Society for Motion Picture and Television Engineers (SMPTE) time code recorded on the EMG tape and the audio channel of the VHS tape, and also inserted into the video images themselves.

The recorded EMG data during locomotion were played back on an electrostatic polygraph (Model ES 2000, Gould Instruments, Valley View, OH), and representative records of the animal's performance in various states and conditions were selected for analysis. The EMG signals were digitized at 1 kHz, and the onset and offset of bursts of activity were detected automatically and then verified manually and corrected when necessary. The duration and amplitude of the muscle bursts were measured using custom-designed software. The mean amplitude was calculated as the integral of the rectified EMG divided by the burst duration. The EMGs were often displayed as averages after rectification and normalization and synchronized to kinematic events such as toe off or touch down (as in Fig. 1C,F)

Kinematic analysis of the hindlimbs was performed with a two-dimensional PEAK Performance (Englewood, CA) system. Selected video images were digitized, and the x and y coordinates of different joint markers were obtained at 60 fields/s. The x-y coordinates of the markers were used to derive various measurements. Stick diagrams of one step cycle are reconstructions of the actual hindlimb movement during the stance and swing phases (Fig. 1A,D). Normalized average joint angular plots (running average of three) show changes in the angular excursion during one normalized step cycle (Fig. 1B,E) repeated for two normalized step cycles.

A complete step cycle consists of a stance and a swing phase. The stance phase begins as soon as the foot contacts the supporting surface, in this case the treadmill belt, and terminates when the foot starts its forward movement. The swing phase begins at the onset of forward movement (not necessarily paw lift

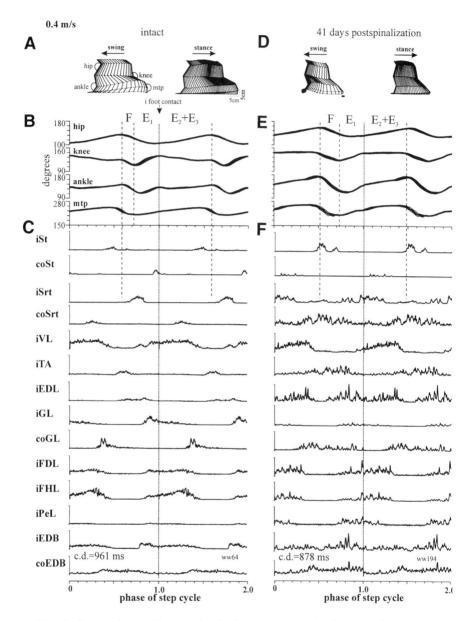

Fig. 1. Comparison of locomotion in the same cat under intact and spinal conditions at 0.4 m/s. **(A)** Stick figures reconstructed from a video sequence of one normal step cycle, displaying the swing and stance phases separately, with arrows pointing to the direction of motion of the leg. The orientations of joint measurements are given. Note that to prevent overlap of the stick figures, each stick figure is displaced by an amount equal to the displacement of the foot on the horizontal axis. **(B)** Angular excursion of the four joints averaged over 10 cycles. Flexion always

because of paw drag) and terminates as the foot strikes the treadmill belt again. The normalized average angular plots, with the swing and stance phases, are subdivided into different components (F and E1 for swing and E2 and E3 for stance) based on the terminology introduced by Philippson (37). Foot drag, which was seen after spinalization, was defined as a period at the beginning of swing during which the dorsum of the toes drag on the treadmill belt during the forward movement of the foot.

2.5. Histology and Autoradiography

At the end of the experiments, animals were killed with an overdose of sodium pentobarbital, and the spinal cord was removed for histologic analysis (Klüver-Barrera method) to ensure the completeness of the spinal transection. Coronal sections 10 μm thick were made, including the area of the spinal lesion. Detailed methods for assessing the extent of the partial spinal lesions can be found in ref. 38.

For autoradiographic studies, the spinal cords were removed, divided into 4- to 6-mm-thick blocks preserving the spinal level, and rapidly frozen by immersion in *N*-methylbutane cooled to –40°C with liquid nitrogen. The spinal cord blocks were then stored at –80°C. Series of 20-μm-thick sections of segments C6–C8, T11–T12, directly above the level of the transection, and L4–L6 were obtained with a cryostat, mounted on gelatin-chrome alum-coated slides, and stored at –80°C (39).

3. KINEMATICS OF SPINAL LOCOMOTION

3.1. Recovery of Locomotion

To induce locomotion during the first 24–48 hr following a complete spinal cord transection, the hindquarters of the cat needed to be supported by the experimenter, and strong cutaneous stimulation had to be applied to the perineal

corresponds to downward deflections of the angular traces. The first dotted line from the left represents the onset of knee flexion and determines the onset of the flexion (F) phase. The second dotted line is aligned with the onset of knee extension during swing (E1, or first extension phase according to Philippson [37]). The vertical line that extends throughout the figure represents the foot contact of the paw facing the camera (i, ipsilateral). All the angles and the averaged EMG data below were synchronized on this foot contact. After foot contact, a small dip can be seen in the angular trace of the ankle, which strictly corresponds to the second extension phase (E2) and then the third extension phase (E3). (**C**) Averaged rectified EMG traces of 10 cycles synchronized on the ipsilateral foot contact. Abbreviations for muscles are given in the text. c.d., cycle duration. (**D–F**) Same format as A, B, and C but for a cat spinalized at T13, 41 days previously. Note that the EMG averages are more "spikey" because the EMGs in spinal cats are often much more clonic and tend to discharge in small bursts within the main burst.

or abdominal region. Although the limbs were extended, very small but coordinated steps were observed even at this time. Over the next several days, hindlimb adductor tonus, which was minimal in the first 48 hr, increased and then remained strong thereafter, being apparent mainly during locomotion so that we usually put a separator between the hindlimb to prevent scissoring. At d 5, the cats were capable of walking with very small joint excursions with the limbs generally in an extensed position. The cat dragged the paw for most of the swing phase and at foot contact landed either on the tip of the toes, or on the dorsum of the paw. These features were typical of the locomotion at this stage. None of the cats were capable of independent weight support. By d 10, the cats increased their stance length mainly by augmenting the limb extension at the end of stance. However, some cats also showed a slight increase in the forward paw placement. At this stage, the cats began to support the weight of the hindquarters and place the plantar surface of the paw on the treadmill belt. Although decreased, the paw drag could still persist for up to approximately half of the swing phase in some cases.

Over the next 3–4 weeks, the paw was placed slightly further ahead of the hip at paw contact, thereby increasing the stance length. Full weight support was maintained, and locomotion could be elicited by movement of the belt without using perineal stimulation. It is important to note that, although cats differed slightly for the time needed to recover, a similar sequence of progress was observed in all cats.

3.2. Recovered Locomotion

The recovered locomotion retains many of the general features observed before transection. There are, however, several pre- and postspinalization differences. The first difference concerns the paw drag, which can be seen in the early part of the swing phase (compare the trajectory of the foot in Fig. 1A and D). This drag could cover a considerable portion of the swing length in some cats. During the swing phase, the paw dragged on the belt for 9.0–20.3% of the step cycle duration, which is equivalent to 26–58% of the swing duration.

The second difference is that the cats took smaller steps following spinalization. In several spinal cats that had a good locomotor recovery, the step cycle duration was decreased by 5–27% (from 961 to 878 ms in the specific example of Fig. 1) The consistent reduction in the stride length was due primarily to a decrease in the forward placement of the paw. In some cats, the step cycle length was also reduced, because the paw did not reach as far back at the end of the stance phase. It is possible that one of the contributing factors to this reduction in stride length was friction caused by the paw being dragged along the treadmill belt for nearly half the swing length. Because the stride length was reduced, the cadence, or number of steps/s, had to increase to maintain the same speed compared with before spinalization.

At the slow walking speed used in this study (0.4 m/s), there were two double support periods per cycle. The double supports were similar, either before or after spinalization. When expressed as a percentage of the cycle duration, the double-support phases represented either a similar proportion of the cycle or they were slightly reduced. The time interval between the ipsilateral and contralateral contacts (I-C interval) shows that stepping remained symmetrical following spinalization, with the contralateral step occurring approximately 50% out of phase with the ipsilateral one *(32)*.

The angular excursion of the hip was generally similar before and after spinalization, although in some cats it was somewhat more flexed or extended throughout the cycle. At the knee and ankle, flexion was some times more pronounced throughout the step cycle in some cats, giving the impression that they performed exaggerated stepping movements (compare the ankle in Fig. 1B and E). Before spinalization, the MTP joints started to plantarflex at the end of the stance and reached maximal plantarflexion early in the swing. Owever, the MTP joints rarely plantarflexed beyond 180°. After spinalization, some cats plantarflexed the MTP joint well beyond 180° as the paw dragged on the treadmill belt at the onset of swing.

Despite these differences, the adult spinal cats can generate a hindlimb locomotor pattern that retains the essentials of the kinematic characteristics documented before the transection.

4. ELECTROMYOGRAPHIC ACTIVITY IN SPINAL LOCOMOTION

Figure 1C and F compares the average rectified EMGs before and after spinalization. In Figures 1, 2, 4, and 5, all gains for each channel of the EMGs have been kept the same to facilitate comparisons even though the gain may not be ideal for the display of a particular condition. It is quite apparent, in this case particularly, that the envelope of EMG activity after spinalization is much more "spiky", which reflects the clonic characteristics of the EMG discharges. This spikiness remains even after averaging 10 cycles.

There was some alteration in the timing of the flexor muscles after spinalization. Whereas before spinalization the hip flexor sartorius (Srt) was clearly activated after the knee flexor semitendinosus (St), following spinalization, this delay was much reduced and, in some cats, the St and Srt were activated at the same time. Normally the St discharges prior to the knee flexion and before foot lift-off (see first dotted line in Fig. 1A). In several spinal cats, St activity was retarded and occurred during the actual knee flexion; the St onset was delayed by a range of 60–140 ms relative to the onset of the cycle. Interestingly, the start of the St activity was approximately the same when correlated with the onset of the actual lift of the paw after the period of drag. Another difference seen in all cats was an earlier onset of the tibialis anterior (TA) (Fig. 1F). The TA was activated between 5 and 20% before the beginning of a swing, and it remained active for a longer period. At the hip, iliopsoas (Ip) activity started at about the same time after spinalization in some cats or was delayed with respect to the onset of swing. Similarly, onset of the Srt activity could also be delayed (-0.5% vs. 9.8%) with respect to the beginning of the swing phase. It is postulated that this alteration in flexor coupling (i.e., earlier start of the ankle flexor and more or less synchronous knee and hip flexors) may cause the paw drag observed during the early part of swing.

Changes in the activity of extensor muscles were not as obvious. Ankle and knee extensors could have a variable amplitude depending on the tendency of the spinal cat to lean on one or the other side during locomotion. Smaller muscles such as the flexor digitorum longus (FDL) and the flexor hallucis longus (FHL) also consistently discharged in the extension phase after spinalization as before spinalization. The extensor digitorum brevis (EDB) had a burst of activity in E1, and its activity was continued during E2 + E3. Some muscles, such as Srt, which normally discharge only during swing, often discharged during stance in several cats; this is quite obvious in the example shown Figure 1F.

5. ADAPTATION OF SPINAL LOCOMOTION

5.1. Speed

The ability of the spinal cats to adapt their locomotion to various conditions such as changes in speed was evaluated. Walking speed can be increased by decreasing the cycle duration and increasing the number of steps per unit time (cadence) and/or by increasing step length. Comparing the stick figures in Figure 1 (A,D) and 2 (A,D), it is quite clear that in both the intact and spinal states there is an increase in step length with increasing speed from 0.4 to 0.6 m/s. The cycle duration is also shorter in both the intact (961 ms at 0.4 m/s and 725 ms at 0.6 m/s) and spinal state (from 878 to 707 ms). Thus in both cases the animal covers more ground in less time.

Some EMG changes are seen with increasing speeds. Figure 2C shows a more prominent second burst in St. (This muscle often discharges a more variable burst, which coincides with the onset of extensor activity in E.1) As suggested by others *(40),* this activity of flexor muscles at the knee may be needed to offset the increased angular acceleration. The extensors ipsilateral gastrocnemius lateralis (iGL) and contralateral gastrocnemius lateralis (coGL) are slightly augmented. In the spinal state (Fig. 2F), small changes are also seen in GLs and vastus lateralis (VL) in this case. Figure 3 illustrates a more systematic study of changes in extensor muscle characteristics as a function of treadmill speed. In the normal cat, there is a clear decrease in burst duration of extensors *(6,41,42)* such as the VL and an increase of amplitude with increasing speed. In the spinal cat the decrease in duration is also seen, but the increase in amplitude is not always clear in all cats.

5.2. Slopes

Cats were tested on 10° slopes (uphill or downhill) in both intact and spinal conditions. In the intact cat, walking uphill on a 10° slope induced a clear

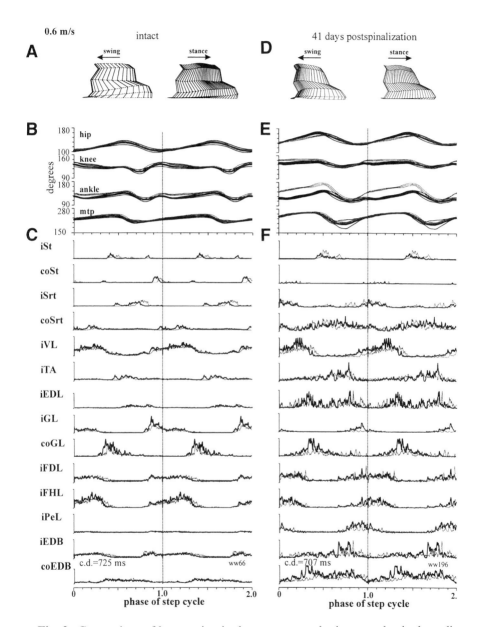

Fig. 2. Comparison of locomotion in the same cat under intact and spinal conditions at 0.4 m/s (gray or strippled lines) or at 0.6 m/s (solid black lines). **(A–F)** Same format as in Figure 1 but for a walking speed of 0.6 m/s.

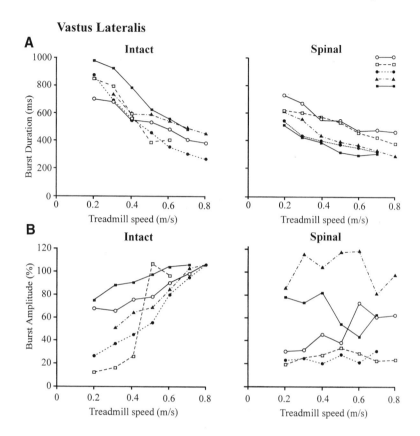

Fig. 3. Changes in the duration and amplitude of the vastus lateralis with speed. **(A)** Burst durations of 10–25 cycles were measured at different treadmill speeds from 0.2 to 0.8 m/s. in four cats before spinalization (Intact) and at different times after spinalization. Filled circles, 51 d; filled squares, 16 d; open circles, 27 d; filled triangles, 25 d. **(B)** Normalized amplitude of the EMG bursts as a function of speed. The EMG bursts were integrated, and the surfaces were expressed as a percentage of the maximal amplitude burst obtained before spinalization.

increase of the amplitude of some extensors such as GLs *(42),* with a prolongation of stance length (compare Fig 1A with Fig. 4A and Fig. 1D with Fig. 4D). After spinalization, the changes in extensors are minimal if present for at least 10° slopes.

In downhill slopes (Fig.5), the mechanics of the movements differ, and the hip-to-toe length changes from the onset of stance to the onset of swing. The foot drag is minimal if present in such a situation. Some changes were seen in the mode of discharge of the Srt, which now has a clear activity during stance.

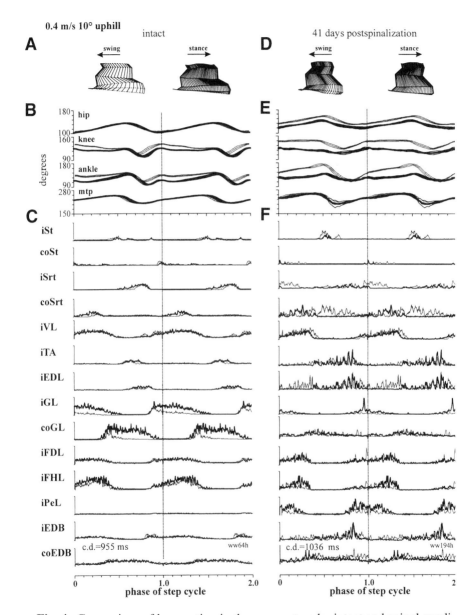

Fig. 4. Comparison of locomotion in the same cat under intact and spinal conditions at 0.4 m/s without slope (gray lines) and with an uphill slope of 10° (black lines). (**A–F**) Same format as in Figure 1 but for a walking speed of 0.4 m/s and an uphill slope of 10°.

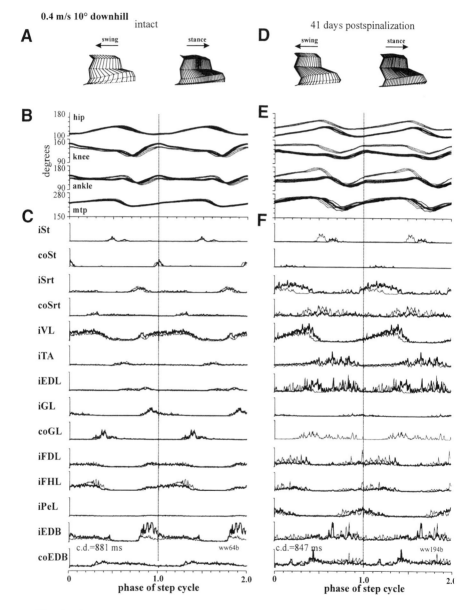

Fig. 5. Comparison of locomotion in the same cat under intact and spinal conditions at 0.4 m/s without slope (gray lines) and with an downhill slope of 10° (black lines). **(A–F)** Same format as in Figure 1 but for a walking speed of 0.4 m/s and a downhill slope of 10°.

For instance, in Figure 5C, the ipsilateral, Srt shows its usual discharge in swing but also an ongoing activity prolonged for about one-half of the stance. (Again, the gain of the amplifier is not optimal for the display in the intact cat). After spinalization, it appears that the St is even more delayed as if mechanical factors (gravity?) participating in the control of the swing of the knee were exaggerated in the downhill slope. The Srt now has a very clear burst in extension.

In conclusion, whereas the changes related to speed are clear in the intact state and in the spinal state, probably due to an afferent feedback regulation, the clear adaptive changes to slopes seen in the intact cat are only partially manifested in the spinal cat. Whereas compensation in the intact cat involves a sophisticated readjustment of posture of the whole body during walking on slopes through the brainstem and cortical mechanisms, the changes in the spinal cat are probably related to local changes of the loading characteristics on the hindquarters.

5.3. Perturbations

5.3.1. Mechanically Induced Reflexes

When the dorsum of the foot of a cat is contacted during the swing phase, there is an elaborate well-organized short latency flexion response of all the joints, as previously reported *(43,44)*. Such responses are illustrated, in Figure 6, for one cat, first in the intact state (Fig. 6A) and then at two different spinal periods, early (Fig. 6B) and late (Fig. 6C). In Figure 6A, note how after contact (tap), the knee rapidly flexes, withdrawing the foot upward and backward. Thereafter the limb is brought above and in front of the obstacle. In Figure 6B, taken at an early stage of locomotor recovery during which the limb is not normally brought forward as much as normal, there is a withdrawal response that lifts the foot above the obstacle, but this is not followed by a subsequent placing of the foot in front of the obstacle. In Figure 6C, the corrective response evoked at a later stage, when the steps are better developed, resembles much more that seen in the intact state. These responses are well integrated into the locomotor cycle in both preparations, and the disturbance lasts at the most for the duration of one cycle and does not disrupt the ongoing walk.

5.3.2. Electrically Induced Reflexes

Similarly, when an electrical stimulation is delivered to the skin of the dorsum of the paw during swing, in either intact or spinal cats, there are a clear stimulus-evoked facilitatory responses in several flexor muscles, in both the intact and spinal cat. During stance, stimulation of the superficial nerve (through an implanted cuff) generates an initial inhibition of the extensors, often followed by a rebound excitatory response. Therefore, stimuli delivered to the paw given in different phases of the step cycle evoke phase-dependent responses well adapted to compensate for the perturbation in the various phases *(6,20,43,44)*, even after spinalization.

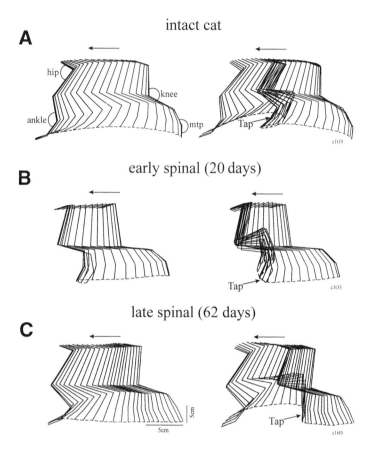

intact cat

early spinal (20 days)

late spinal (62 days)

Fig. 6. Mechanical perturbation of the swing in intact and spinal conditions. (**A**) On the left, unperturbed swing. To the right, the foot hits the tapper, and there is a marked flexion of the knee, which retracts the foot backward and upward before being placed in front of the obstacle. (**B**) The same cat, 20 d after spinalization, still had small step cycles. Taps were effective to evoke a marked knee flexion, but the foot was not placed forward in front of the obstacle. (**C**) At 62 d, when the cat performed longer step cycles, the response to tap consisted of a knee flexion followed by positioning of the foot in front of the obstacle.

6. PHARMACOLOGY OF SPINAL LOCOMOTION

6.1. Initiation of Locomotion and Training in Early-Spinal Cat

A number of previous studies (reviewed in ref. *6* have shown that the nora-drenergic system may be implicated in the generation of the complex locomotor rhythms recorded in hindlimb nerves of acutely spinalized and paralyzed cats *(45–51)*. As mentioned before, after an iv injection of clonidine, an α_2-noradrenergic agonist, acutely spinalized cats can also walk with the hindlimbs on a treadmill *(29)*. This important finding was pursued in adult spinal cats with either ip *(52–54)* or intrathecal it *(33,34)* injections of various noradrenergic drugs acting on different noradrenergic receptors.

Noradrenaline it could initiate locomotion in cats spinalized within a week (see also ref. *55* for acute spinal cats), but the effects were short lasting, as would be expected, given the efficient inactivating mechanisms for this neuro-transmitter. With an it injection of a single bolus of the α_2-agonist clonidine (100–200 μg/100 μL) a paraplegic cat that is unable to make more than a few uncoordinated stepping movements can, within minutes after the injection, step with the hindlimbs on the treadmill belt (Fig. 7). Moreover, this ability to walk on the treadmill is maintained for 4–6 hr. Other α_2-agonists, such as tizanidine and oxymetazoline, were also found to induce locomotion in such early-spinal cats. It was found that tizanidine had fewer side effects than clonidine (i.e., nau-sea) and that oxymetazoline could have effects lasting more than 2 days. The effects were largely blocked by the α_2-antagonist yohimbine, thus indicating pharmacologic specificity of the effects. The α_1-agonist methoxamine could, in some cases, induce a few steps, but the effects were never as long lasting as those obtained with α_2-agonists *(34)*.

Since clonidine allows walking for a time window of 4–6 hr, cats were trained intensively after injections of clonidine given every day (ip or it) for up to 11 days *(33,56)*. Each day, cats were trained in bouts of 15–20 min until the effects of clonidine wore off. As shown before *(52)*, in the first few days after spinalization, the induced locomotor pattern consisted of a rhythmic alternation of the hindlimbs with the hips being generally in a more extended position than normal, thus evoking a pattern normally seen only later on when cats regain the ability to walk spontaneously without drugs. Although analyses of the locomo-tion in control periods (before giving clonidine again) in the first 5–6 d showed no obvious carry-over effects of the training on the previous day, the effects of clonidine were normally enhanced compared with those of the previous day, inducing larger amplitude and better organized movements at all joints. In the four cats investigated in this manner, locomotion with weight support and plan-tar foot placement was achieved between d 6 and 11. This was the earliest period for the recovery of spontaneous locomotion that we had obtained in spinal cats (compare with 14–27 d in ref. *32*), suggesting a possible role for early daily training with drugs on the time course of locomotor recovery after a spinal lesion. These findings further suggest some degree of plasticity in the spinal neural network controlling locomotion, as well as the potential of combining pharmacologic treatments and locomotor training (*see* Section 6.2.).

6.2. Modulation of Locomotion and Reflexes by Various Drugs

6.2.1. The Noradrenergic System

Spinal cats normally recover the ability to walk spontaneously, i.e., without drugs, with the hindlimbs on the treadmill within a few weeks *(6,31,32,57)*. In

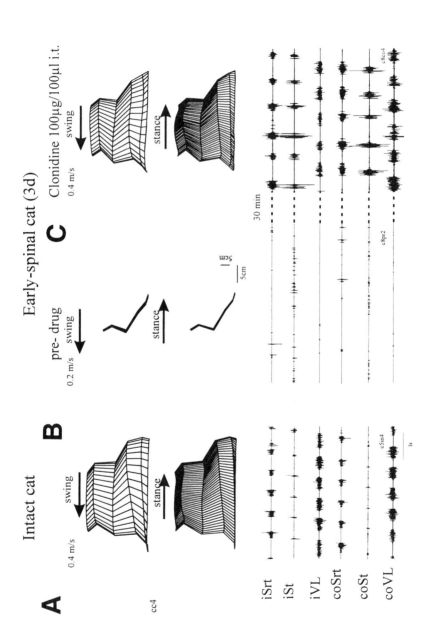

Fig. 7. *(p. 74)* Initiation of locomotion by an it injection of clonidine. **(A)** Stick figures of the swing and stance of one step cycle at 0.4 m/s. The arrows indicate the direction of movement of the leg. EMGs of six consecutive step cycles. **(B)** Same cat, 3 d after spinalization with the treadmill at 0.2 m/s. The cat had no locomotor movement. The calibrations apply to all stick figures. **(C)** Same cat, 30 min after an injection of a bolus (100 μl) of it clonidine. The cat stepped quite well despite the initial paw drag during swing. (Note that the MTP joint reverses.)

this state of recovered locomotion, clonidine has the ability to lengthen the step cycle significantly and also to regularize it when it is unstable. The duration of muscle bursts was generally increased, while the mean EMG amplitude tended to increase or remain the same in flexors and decrease in extensors. Clonidine often exacerbated the paw drag. With higher doses, the cats also tended to sag at the end of stance. The α_1-agonist methoxamine as well as noradrenaline itself had little effect on the already established pattern of locomotion *(34)*.

In the spinal cat, a decrease of excitability could be observed by stimulating cutaneous nerves directly with implanted cuffs or by recording the responses to tapping of the dorsum of the paw during the swing phase of locomotion, as detailed before. The normal brisk responses disappeared or were replaced by sluggish late responses, the paw often remaining in contact with the tapper. When the paw of a spinal cat is dipped in water, a brisk fast paw shake reaction is elicited. These paw shake responses are virtually abolished by clonidine *(34,58)*. The reduced cutaneous excitability may partly explain the enhanced paw drag and may suggest that spinal cats normally use this initial paw drag as a stimulus to lift the foot during swing (see Section 7.2.).

6.2.2. Serotonergic System

Because we did not find that Serotonin (5-HT) agonists could initiate locomotion in the early days after spinalization *(59)* (contrary to the neonatal rat; see Chap. 4), we investigated this system less. However, in already walking late spinal cats, 5-HT agonists such as quipazine and 5-Methoxy-*N, N*-dimethyl-tryptamine (5-MeODMT), acting broadly on different subclasses of receptors, exerted a powerful excitatory effect; the amplitude of EMG bursts, particularly in the extensor muscles, was much increased, and the locomotor pattern became much more vigorous *(52,59)*. St, which acts as a knee flexor as well as a hip extensor, had a larger amplitude during flexion and was also recruited during midstance, potentially contributing to the extensor thrust in this part of the stance. Back muscles, which normally do not discharge at low speed (below 0.6 m/s) *(60)*, now had 2 bursts per cycle.

A combination of a 5-HT agonist and an α_2-noradrenergic agonist resulted in additive effects, so that the cat walked with longer step cycles (noradrenergic effect mainly) and the EMG bursts were increased (serotoninergic effect).

6.2.3. The Glutamatergic System

Although NMDA can induce locomotion in several species such as the lamprey *(61–63)*, the neonatal rat *(see* Chap. 4 and refs.*64–66)*, and the decerebrate paralyzed cat *(67)*, it did not initiate treadmill locomotion during the first few days after spinalization in cats *(68)*. Instead, there was a marked increase in general excitability of the spinal cat as manifested by tremor, fast paw shake, and fanning of the toes. However, in the same cats, clonidine could induce locomotion as usual a few hours after the administration of NMDA. We have recently observed that, in early spinal cats (around 5 d), which are just starting to express some rhythmic hindlimb movements, NMDA dramatically improved locomotion and that this effect lasted for several days.

When injected in already walking spinal cat (late spinal), NMDA could evoke a somewhat brisker locomotion. On the other hand, the NMDA blocker AP5 could abolish locomotion for more than 30–60 mins. About 15 mins after the onset of the block, NMDA could reinstate locomotion, suggesting that activation of NMDA receptors might be essential in maintaining the spinal locomotor pattern. *(68)*.

6.3. Autoradiographic Studies

In an attempt to understand better the effects of different agonists and antagonists on the spinal cat, we initiated a study of various receptors in the spinal cord at different times post lesion. The α_1-, α_2-noradrenergic and serotonin$_{1A}$ (5-HT$_{1A}$) receptors were examined in the spinal cord of control cats as well as in animals spinalized, (at T13) some weeks or months previously *(39)*. The highest levels of α_1-noradrenergic receptors, labeled with [^{3}H]prazosin, were found in laminae II, IX, and X in control cats., whereas α_2-noradrenergic receptors, labeled with [^{3}H] idazoxan, were found mainly in laminae II, III, and X with moderate densities in lamina IX. Following the spinal transection, both types of receptors remained unchanged above the lesion. At 15 and 30 d post lesion, binding significantly increased in laminae II, III, IV, and X for α_2, and in laminae I, II, III, and IX for α_1 receptors in lumbar segments. At longer survival times, binding densities returned to near control values. The 5-HT$_{1A}$ receptors, labeled with [^{3}H]8-hydroxydipropylaminotetralin ([^{3}H]8-OH-DPAT), were found mainly in laminae I–IV and X. After transection, binding increased significantly only in laminae II, III, and X of lumbar segments at 15 and 30 d. At a later stage, binding levels returned to control values.

This transient upregulation of various monoaminergic receptors, observed in the lumbar region in the first month after spinal transection, suggests that important modifications occur during the period in which cats normally recover functions such as locomotion of the hindlimbs. We still have to find out whether these increased binding sites are functional receptors, whether we can modify these receptor changes, and whether pharmacologic manipulations would change the time-course of recovery of locomotor function.

6.4. Pharmacology of Locomotion in Intact and Partial Spinal Cats

The fact that receptors change after spinalization raised the question of the effects of various drugs in other conditions, such as the normal condition or after a partial spinal lesion.

Agonists such as clonidine injected intrathecally had fewer effects in intact cats than any other preparation, at least in the same dose range *(69,70)*. These reduced and short-lasting effects made it difficult to ascertain that the drugs had effects (as seen in the spinal cat), at least during undemanding treadmill locomotion. The short duration of the effects also suggests that the intact cats have efficient means of compensating for the neurotransmitter imbalance with various other systems. It may also reflect that the state of the receptors is such that the effects are not as prominent as after lesions. On the other hand, a noradrenergic blocker such as yohimbine given alone can have important effects in the normal cat, whereas it has little effect on its own in the spinal cat *(71)*. Similarly, although the NMDA blocker AP5 has some effects on the locomotion of the normal cat (reduced weight support), the normal cat can compensate, presumably through a compensatory modulation by other neurotransmitters, whereas the spinal cat stops walking with AP5.

After large ventral-ventrolateral lesions of the spinal cord, cats gradually recovered voluntary quadrupedal locomotion on the treadmill *(38,57)*. The recovery of locomotion was rather fast in cats with smaller lesions. However, with large lesions, the cats only used their forelimbs for the first few weeks, after which there was a recovery of voluntary quadrupedal locomotion. However, this locomotion was irregular and is characterized by frequent stumble and falls, weak support of the hindquarters, and abnormal coupling between the fore- and hindlimbs. Sometimes the forelimbs and hindlimbs walked at two different frequencies with periodic rephasing of the two girdles.

We studied the effects of noradrenergic and serotoninergic drugs in two cats with the largest lesions *(69,72)*. In the first 10 d post lesion, when the cat could barely walk or stand, noradrenaline was seemingly beneficial. The cat could sustain its weight and perform several consecutive step cycles, even though the fore-hindlimb coupling remained somewhat labile. Paradoxically, clonidine at this stage proved to have quite a negative effect and more or less abolished whatever locomotor capacities the cat had. During the plateau period, noradrenaline was beneficial, as was the α_1-agonist methoxamine. Quipazine and 5MeODMT had a beneficial effect but not the 5-HT$_{1A}$ agonist 8-OH-DPAT, which induced a paw drag not present before. The best overall result was obtained with the combination of methoxamine and quipazine; this combination significantly improved the overall quality of locomotion.

The combined results on the different effects of the same drugs on the intact, partial, and complete spinal cat raise several issues. The effects of the drugs used depend on the preparation and thus presumably on the state of the receptors and presumably on other cellular mechanisms. Although we have a good idea of the time-course of changes of these receptors in spinal cats, we still have to find out the evolution profile of receptors in partial spinal cats and determine whether they could explain the results reported above.

What seems to become more and more evident is that spinal locomotion does not depend on the activation of noradrenergic receptors since spinal cats can walk without this neurotransmitter and that blocking the receptor does not prevent spinal locomotion. On the other hand, blocking NMDA receptors in spinal cats totally abolishes locomotion. It is possible then that the basic mechanisms for locomotor rhythmicity in the spinal cat depend on NMDA mechanisms and that, normally, the noradrenergic and serotoninergic systems would modulate these fundamental mechanisms. Indeed, a number of pieces of evidence suggest such a potentiation of NMDA by 5-HT (*see* Chap. 4 as well as refs. *64, 66, 73,* and *74*). If this hypothesis is borne out, it could suggest new pharmacologic approaches in spinal-cord-injured patients.

7. SPINAL PLASTICITY AND LEARNING

The experiments on locomotor training with or without clonidine *(9,31,33,56,75)* reported above suggest some degree of activity-dependent spinal plasticity. Other work by Edgerton's group has addressed more directly the issue of plasticity by comparing two types of specific training such as standing and walking *(76–81)*. The approach we have adopted was to study the locomotor adaptations after peripheral nerve lesions of muscle or cutaneous nerves. The idea here was first to document the locomotor compensation in the intact state and then find out whether this compensation was carried over after a spinal section. We have also studied the ability of cats already spinalized to adapt to such peripheral nerve lesions.

7.1. Muscle Nerve Lesions

We investigated the locomotor performance of cats before and after a unilateral denervation of the ankle flexors TA and extensor digitourm longus (EDL) both in cats with intact spinal cord and after spinalization *(82)*. The effects of inactivation of the ankle flexors were studied in three cats with intact spinal cord over periods of 4–7 weeks. Cats adapted their locomotor performance very rapidly, within a few days, so that the locomotor behavior appeared to be practically unchanged after the neurectomies. However, kinematic analyzes of video recordings often revealed small, albeit consistent, increases in knee and/or hip flexion. These changes were accompanied by a moderate increase in the amplitude of knee and hip flexor muscle activity. Cats maintained a regular and symmetrical walking pattern over the treadmill for several minutes. Two of these cats were then spinalized at T13 and afterward, studied for about a month. Whereas normally cats regain a regular and symmetrical locomotor pattern after spinalization (*see* Section 3.2.), the cats that were neurectomized *before* spinalization expressed a disorganized and asymmetrical locomotor pattern characterized by a predominance of knee flexion and the absence of plantar foot contact of the denervated limb.

In a complementary experiment, a cat was first spinalized and allowed to recover regular symmetrical locomotor performance before being also submitted to the same unilateral ankle flexor inactivation; the cat was then studied for about 50 d. After denervation, the cat maintained a well-organized symmetrical gait although there was almost no ankle flexion on the denervated side. There was no exaggerated knee hyperflexion nor gait asymmetry as seen in the two cats spinalized only after they had adapted to the denervation of ankle flexors.

It is concluded that, after muscle denervation, locomotor adaptation is achieved through changes occurring at different levels. Since cats spinalized after adaptation to the neurectomy expressed an asymmetrical locomotor pattern dominated by hyperflexion, we propose that the spinal circuitry has been modified during the adaptive process, presumably through the action of corrective supraspinal inputs. Indeed, spinal cats do not normally display such abnormal hyperflexions and neither did the one cat denervated *after* spinalization. On the other hand, since the modified locomotor pattern in the spinal state is not functional and contains only some aspects of the compensatory response seen before spinalization, it is suggested that the complete functional adaptation observed in intact cats after peripheral nerve lesions may be dependent on changes occurring at both the spinal and supraspinal levels.

More recent collaborative work with Pearson *(83)* has shown that after a unilateral section of the gastrocnemius lateralis nerve in spinal cats that have just recovered locomotion, initially a marked yield of the ankle during stance is induced as well as an important increase of activity in the synergist, medial gastrocnemius (MG). After a week, there is a quasi-complete disappearance of the yield, and the amplitude of the MG remains increased. This again suggests that the spinal cat has some adaptive capacities, so that the locomotor pattern can be modified to accommodate changes in the state of the neuromuscular apparatus.

7.2. Cutaneous Nerve Lesions

Normal cats with a bilateral neurectomy of all major nerves innervating the skin of the hindpaws had very few deficits of treadmill locomotion (Fig. 8A), although there was an increase in the amplitude of the knee flexor St together with a minor shortening of the step cycle. However, for several weeks after such a denervation, they were incapable of walking on a horizontal ladder *(84)*. Once the cats had recovered the ability to walk on the ladder, they were spinalized at T13 and followed for several weeks. Whereas normally spinal cats recover locomotion with plantar foot placement and weight support of the hindquarters, these spinal and neurectomized cats were incapable of placing the foot properly. Instead, they dragged their feet back and forth on the treadmill, without ever regaining the capacity to place the foot normally (Fig. 8B). This suggests that cutaneous information is essential for the expression of spinal locomotion.

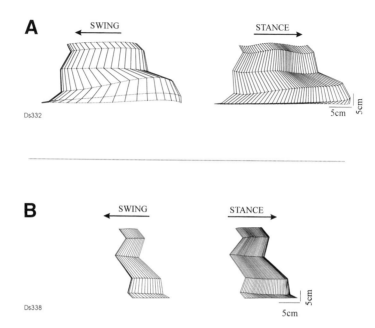

Fig. 8. Effect of a cutaneous neurectomy of the hindlimb paw on locomotion before and after spinalization. **(A)** At 41 d after cutting the following five nerves at ankle level: saphenous, sural, superficial peroneal, tibial, and cutaneous branch of deep peroneal. Note that the locomotion on the treadmill is essentially normal. **(B)** At 48 d after spinalization, the same cat not only makes very small steps but the feet are dragged across the belt during the whole swing phase and remain on the dorsal surface (toes buckled inward) during the whole stance period. This abnormal foot placement during locomotion does not disappear with training.

In one spinal cat, one nerve was left intact (i.e., the cutaneous branch of the deep peroneal nerve), and this animal was eventually capable of walking with almost a proper foot contact. When the receptive field of this nerve was inactivated through local anesthesia or neurectomy, the ability to place the foot was lost as for the other cats. Similarly, one cat was first spinalized and then trained to recover locomotion. Thereafter, cutaneous nerves were sectioned sequentially leaving enough time between the neurectomies to retrain the animal to walk. Whereas section of the cutaneous peroneal branch of the deep peroneal, the saphenous, and sural nerves produced few deficits, section of the superficial peroneal nerve induced buckling of the toes so that the cat dragged its foot during several weeks of training. However, this spinal cat eventually relearned to place the foot properly. When the last cutaneous nerve was cut (tibial nerve), the cat permanently lost the ability to place the foot, as did the others.

These experiments suggest first that cutaneous inputs are crucial for the expression of spinal locomotion and second that, even after spinalization, the spinal cord has some potential for optimizing locomotion when the sensory information is reduced. This is good evidence of spinal learning.

8. CONCLUSIONS

Collectively, the work on spinal locomotion has first shown that it is an innate pattern, that it has several characteristics of the normal locomotor pattern, that it is sensitive to training, that it is adaptable to perturbations, to speed, slopes, and tilts, that it can be initiated as well as modulated by different neurochemical systems, and finally that it can adapt to changes in the neuromuscular apparatus or sensory information. Much more work is now needed to understand other physiopathologic mechanisms such as changes in receptor functions and in the balance between the receptors of various neurotransmitter systems and the role of sensory information in the spinal state. It is hoped that part of this knowledge will serve to target therapeutic approaches for rehabilitation strategies in spinal cord-injured people.

ACKNOWLEDGMENTS

We acknowledge the support of various granting agencies, MRC, NCE, FCAR, and APA. A very special thanks to Janyne Provencher for her assistance in experiments as well as preparing figures. We also acknowledge the expert help of France Lebel, Claude Gagner, Philippe Drapeau, Gilles Messier, and Daniel Cyr. Special thanks to R. W. for inspiration.

REFERENCES

1. Grillner, S. (1981) Control of locomotion in bipeds, tetrapods, and fish, in *Handbook of physiology. The nervous system II* (Brookhart, J. M. and Mountcastle, V. B., eds.), American Physiological Society, Bethesda, MD, pp. 1179–1236.
2. Armstrong, D. M. (1986) Supraspinal contributions to the initiation and control of locomotion in the cat. *Prog. Neurobiol.* **26,** 273–361.
3. Pearson, K. G. (1993) Common principles of motor control in vertebrates and invertebrates. *Annu. Rev. Neurosci.* **16,** 265–297.
4. Grillner, S. and Dubuc, R. (1988) Control of locomotion in vertebrates: spinal and supraspinal mechanisms, in *Functional recovery in neurological disease* (Waxman, S. G., ed.), Raven, New York, pp. 425–453.
5. Rossignol, S. and Dubuc, R. (1994) Spinal pattern generation. *Curr. Opin. Neurobiol.* **4,** 894–902.
6. Rossignol, S. (1996) Neural control of stereotypic limb movements, in *Handbook of physiology, Section 12. Exercise: regulation and integration of multiple systems* (Rowell, L. B. and Sheperd, J. T. eds.), American Physiological Society, Oxford, pp. 173–216.
7. Shik, M. L. and Orlovsky, G. N. (1976) Neurophysiology of locomotor automatism. *Physiol Rev.* **56,** 465–500.

8. Gelfand, I. M., Orlovsky, G. N., and Shik, M. L. (1988) Locomotion and scratching in tetrapods, in *Neural control of rhythmic movements in vertebrates* (Cohen, A. H., Rossignol, S., and Grillner, S. eds.), John Wiley & Sons, New York, pp. 167–199.

9. Barbeau, H. and Rossignol, S. (1994) Enhancement of locomotor recovery following spinal cord injury. *Curr. Opin. Neurol.* **7,** 517–524.

10. Rossignol, S. and Barbeau, H. (1995) New approaches to locomotor rehabilitation in spinal cord injury. *Ann. Neurol.* **37,** 555–556.

11. Sherrington, C. S. (1899) On the spinal animal. *Med. Chir. Trans.* **82,** 449–486.

12. Sherrington, C. S. (1910) Flexion-reflex of the limb, crossed extension-reflex, and reflex stepping and standing. *J. Physiol.* **40,** 28–121.

13. Sherrington, C. S. (1910) Remarks on the reflex mechanism of the step.*Brain.* **33,** 1–25.

14. Brown, T. G. (1911) The intrinsic factors in the act of progression in the mammal. *Proc R. Soc Lond. [Biol.]* **84,** 308–319.

15. Shurrager, P. S. and Dykman, R. A. (1951) Walking spinal carnivores. *J. Comp. Physiol. Psychol.* **44,** 252–262.

16. Grillner, S. (1973) Locomotion in the spinal cat, in *Control of posture and locomotion,* vol. 7 in *Advances in Behavioral Biology* (Stein, R. B., Pearson, K. G., Smith, R. S., et al., Plenum, eds.), New York, pp. 515–535.

17. Forssberg, H., Grillner, S., and Sjostrom, A. (1974) Tactile placing reactions in chronic spinal kittens. *Acta Physiol. Scand.* **92,** 114–120.

18. Forssberg, H., Grillner, S., and Halbertsma, J. (1980) The locomotion of the low spinal cat. I. Coordination within a hindlimb. *Acta Physiol. Scand.* **108,** 269–281.

19. Forssberg, H., Grillner, S., Halbertsma, J., et al. (1980) The locomotion of the low spinal cat: II. Interlimb coordination. *Acta Physiol. Scand.* **108,** 283–295.

20. Rossignol, S., Lund, J. P., and Drew, T. (1988) The role of sensory inputs in regulating patterns of rhythmical movements in higher vertebrates. A comparison between locomotion, respiration and mastication, in *Neural control of rhythmic movements in vertebrates* (Cohen, A., Rossignol, S., and Grillner, S., eds.), John Wiley & Sons, New York, pp. 201–283.

21. Smith, J. L., Smith, L. A., Zernicke, R. F., et al. (1982) Locomotion in exercised and non-exercised cats cordotomized at two or twelve weeks of age. *Exp. Neurol.* **76,** 393–413.

22. Bregman, B. S. and Goldberger, M. E. (1983) Infant lesion effect: I. Development of motor behavior following neonatal spinal cord damage in cats. *Dev. Brain Res.* **9,** 103–117.

23. Robinson, G. A. and Goldberger, M. E. (1986) The development and recovery of motor function in spinal cats. I. The infant lesion effect. *Exp. Brain Res.* **62,** 373–386.

24. McCouch, G. P. (1947) Reflex development in the chronically spinal cat and dog. *J. Neurophysiol.* **10,** 425–428.

25. Kozak, W. and Westerman, R. (1966) Basic patterns of plastic change in the mammalian nervous system. *Symp. Soc. Exp. Biol.* **20,** 509–544.

26. Afelt, Z. (1970) Reflex activity in chronic spinal cats. *Acta Neurobiol. Exp.* **30,** 129–144.
27. Afelt, Z. (1974) Functional significance of ventral descending tracts of the spinal cord in the cat. *Acta Neurobiol. Exp.* **34,** 393–407.
28. Ten Cate, J. (1962) Innervation of locomotor movements by the lumbosacral cord in birds and mammals. *J. Exp. Biol.* **39,** 239–242.
29. Forssberg, H. and Grillner, S. (1973) The locomotion of the acute spinal cat injected with clonidine i.v. *Brain Res.* **50,** 184–186.
30. Eidelberg, E., Story, J. L., Meyer, B. L., et al. (1980) Stepping by chronic spinal cats. *Exp. Brain Res.* **40,** 241–246.
31. Barbeau, H. and Rossignol, S. (1987) Recovery of locomotion after chronic spinalization in the adult cat. *Brain Res.* **412,** 84–95.
32. Bélanger, M., Drew, T., Provencher, J., et al. (1996) A comparison of treadmill locomotion in adult cats before and after spinal transection. *J. Neurophysiol.* **76,** 471–491.
33. Chau, C., Barbeau, H., and Rossignol, S. (1998) Early locomotor training with clonidine in spinal cats. *J. Neurophysiol.* **59,** 392–409.
34. Chau, C., Barbeau, H., and Rossignol, S. (1998) Effects of intrathecal α_1- and α_2 noradrenergic agonists and norepinephrine on locomotion in chronic spinal cats. *J. Neurophysiol.* **79,** 2941–2963.
35. Julien, C. and Rossignol, S. (1982) Electroneurographic recordings with polymer cuff electrodes in paralyzed cats. *J. Neurosci. Methods* **5,** 267–272.
36. Roy, R. R., Hodgson, J. A., Lauretz, S. D., et al. (1992) Chronic spinal cord-injured cats: surgical procedures and management. *Lab. Anim. Scie.* **42,** 335–343.
37. Philippson, M. (1905) L'autonomie et la centralisation dans le système nerveux des animaux. *Trav. Lab. Physiol. Inst. Solvay (Bruxelles)* **7,** 1–208.
38. Brustein, E. and Rossignol, S. (1998) Recovery of locomotion after ventral and ventrolateral spinal lesions in the cat. I. Deficits and adaptive mechanisms. *J. Neurophysiol.* **80,** 1245–1267.
39. Giroux, N., Rossignol, S., and Reader, T. A. (1999) Autoradiographic study of α_1-, α_2-Noradrenergic and Serotonin $_{1A}$ receptors in the spinal cord of normal and chronically transected cats. *J. Comp. Neurol.* **406,** 402–414.
40. Wisleder, D., Zernicke, R. F., and Smith, J. L. (1990) Speed-related changes in hindlimb intersegmental dynamics during the swing phase of cat locomotion. *Exp. Brain Res.* **79,** 651–660.
41. Halbertsma, J. M. (1983) The stride cycle of the cat: the modelling of locomotion by computerized analysis of automatic recordings. *Acta Physiol scand. Suppl.* **521,** 1–75.
42. Pierotti, D. J., Roy, R. R., Gregor, R. J., et al. (1989) Electromyographic activity of cat hindlimb flexors and extensors during locomotion at varying speeds and inclines. *Brain Res.* **481,** 57–66.
43. Forssberg, H., Grillner, S., and Rossignol, S. (1975) Phase dependent reflex reversal during walking in chronic spinal cats. *Brain Res.* **85,** 103–107.
44. Forssberg, H., Grillner, S., and Rossignol, S. (1977) Phasic gain control of reflexes from the dorsum of the paw during spinal locomotion. *Brain Res.* **132,** 121–139.

45. Jankowska, E., Jukes, M. G., Lund, S., et al. (1967) The effects of DOPA on the spinal cord. 6. Half centre organization of interneurones transmitting effects from the flexor reflex afferents. *Acta Physiol scand.* **70,** 389–402.
46. Jankowska, E., Jukes, M. G., Lund, S., et al. (1967) The effect of DOPA on the spinal cord. 5. Reciprocal organization of pathways transmitting excitatory action to alpha motoneurones of flexors and extensors. *Acta Physiol scand.* **70,** 369–388.
47. Grillner, S. and Zangger, P. (1979) On the central generation of locomotion in the low spinal cat. *Exp. Brain Res.* **34,** 241–261.
48. Pearson, K. G. and Rossignol, S. (1991) Fictive motor patterns in chronic spinal cats. *J. Neurophysiol.* **66,** 1874–1887.
49. Viala, D. and Valin, A. (1972) Reflexe à longue latence et activités à caractère locomoteur chez le chat spinal aigu sous DOPA. *J. Physiol. (Paris)* **65,** 518A
50. Baev, K. V. (1977) Rhythmic discharges in hindlimb motor nerves of the decerebrate, immobolized cat induced by intravenous injection of DOPA. *Neurophysiology* **9,** 165–167.
51. Chandler, S. H., Baker, L. L., and Goldberg, L. J. (1984) Characterization of synaptic potentials in hindlimb extensor motoneurons during L-Dopa-induced fictive locomotion in acute and chronic spinal cats. *Brain Res.* **303,** 91–100.
52. Barbeau, H. and Rossignol, S. (1991) Initiation and modulation of the locomotor pattern in the adult chronic spinal cat by noradrenergic, serotonergic and dopaminergic drugs. *Brain Res.* **546,** 250–260.
53. Rossignol, S., Barbeau, H., and Julien, C. (1986) Locomotion of the adult chronic spinal cat and its modification by monoaminergic agonists and antagonists, in *Development and plasticity of the mammalian spinal cord,* (Fidia Research Series III, Goldberger, M., Gorio, A., and Murray, M. eds.), Liviana Press, Padova, pp. 323–345.
54. Rossignol, S., Chau, C., and Barbeau, H. (1995) Pharmacology of locomotion in chronic spinal cats, in *Alpha and gamma motor systems* (Taylor, A., Gladden, M. H., and Durbaba, R., eds.), Plenum, New York, pp. 449–455.
55. Kiehn, O., Hultborn, H., and Conway, B. A. (1992) Spinal locomotor activity in acutely spinalized cats induced by intrathecal application of noradrenaline. *Neurosci. Lett.* **143,** 243–246.
56. Barbeau, H., Chau, C., and Rossignol, S. (1993) Noradrenergic agonists and locomotor training affect locomotor recovery after cord transection in adult cats. *Brain Res. Bull.* **30,** 387–393.
57. Rossignol, S., Chau, C., Brustein, E., et al. (1996) Locomotor capacities after complete and partial lesions of the spinal cord. *Acta Neurobiol Exp.* **56,** 449–463.
58. Barbeau, H., Julien, C., and Rossignol, S. (1987) The effects of clonidine and yohimbine on locomotion and cutaneous reflexes in the adult chronic spinal cat. *Brain Res.* **437,** 83–96.
59. Barbeau, H. and Rossignol, S. (1990) The effects of serotonergic drugs on the locomotor pattern and on cutaneous reflexes of the adult chronic spinal cat. *Brain Res.* **514,** 55–67.
60. Zomlefer, M. R., Provencher, J., Blanchette, G., et al. (1984) Electromyographic study of lumbar back muscles during locomotion in acute high decerebrate and in low spinal cats. *Brain Res.* **290,** 249–260.

61. Grillner, S., McClellan, A., Sigvardt, K., et al. (1981) Activation of NMDA-receptors elicits "fictive locomotion" in lamprey spinal cord in vitro.*Acta Physiol scand.* **113,** 549–551.

62. Brodin, L., Grillner, S., and Rovainen, C. M. (1985) N-methyl-D-aspartate (NMDA), kainate and quisqualate receptors and the generation of fictive locomotion in the lamprey spinal cord. *Brain Res.* **325,** 302–306.

63. Sigvardt, K. A., Grillner, S., Wallen, P., et al. (1985) Activation of NMDA receptors elicits fictive locomotion and bistable properties in the lamprey spinal cord. *Brain Res.* **336,** 390–395.

64. Cazalets, J. R., Grillner, P., Menard, I., et al. (1990) Two types of motor rhythm induced by NMDA and amines in an in vitro spinal cord preparation of neonatal rat. *Neurosci. Lett.* **111,** 116–121.

65. Cowley, K. C. and Schmidt, B. J. (1994) A comparison of motor patterns induced by N-methyl-D-aspartate, acetylcholine and serotonin in the in vitro neonatal rat spinal cord. *Neurosci. Lett.* **171,** 147–150.

66. Maclean, J., Cowley, K. C., and Schmidt, B. J. (1998) NMDA receptor-mediated oscillatory activity in the neonatal rat spinal cord is serotonin dependent. *J. Neurophysiol.* **79,** 2804–2808.

67. Douglas, J. R., Noga, B. R., Dai, X., et al. (1993) The effects of intrathecal administration of excitatory amino acid agonists and antagonists on the initiation of locomotion in the adult cat. *J. Neurosci.* **13,** 990–1000.

68. Chau C, Provencher J, Lebel F, et al. (1994) Effects of intrathecal injection of NMDA receptor agonist and antagonist on locomotion of adult chronic spinal cats. *Soc. Neurosci. Abstr.* **20,** 573.

69. Giroux N, Brustein E, Chau C, et al. (1998) Differential effects of the noradrenergic agonist clonidine on the locomotion of intact, partially and completely spinalized adult cats, in *Neuronal mechanisms for generating locomotor activity* (Kiehn, O., Harris-Warrick, R. M., Jordan, L. M., et al., eds.), The New York Academy of Sciences, New York, NY, pp. 517–520.

70. Rossignol, S., Chau, C., Brustein, E., et al. (1998) Pharmacological activation and modulation of the central pattern generator for locomotion in the cat, in *Neuronal mechanisms for generating locomotor activity* (Kiehn, O., Harris-Warrick, R. M., Jordan, L. M., et al., eds.), pp. 346–359.

71. Giroux N, Lebel F, Provencher J, et al. (1998) The effects of intrathecal administration of the noradrenergic antagonist yohimbine on locomotion in adult intact cats. *Soc. Neurosci. Abstr.* **24,** 917.

72. Brustein, E. and Rossignol, S. (1999) Recovery of locomotion after ventral and ventrolateral spinal lesions in the cat. II. The effects of noradrenergic and serotoninergic drugs. *J. Neurophysiol.* **81,** 1513–1530.

73. Wikstrom, M., Hill, R., Hellgren, J., et al. (1995) The action of 5-HT on calcium-dependent potassium channels and on the spinal locomotor network in lamprey is mediated by 5-HT 1A-like receptors. *Brain Res.* **678,** 191–199.

74. Cazalets, J. R., Sqalli-Houssaini, Y., and Clarac, F. (1992) Activation of the central pattern generators for locomotion by serotonin and excitatory amino acids in neonatal rat. *J. Physiol.* **455,** 187–204.

75. Edgerton, V. R., Johnson, D. J., Smith, L. A., et al. (1983) Effects of treadmill exercises on hindlimb muscles of the spinal cat, in *Spinal cord reconstruction* (Kao, C. C., Bunge, R. P., and Reier, P. J. eds.), Raven, New York, pp. 435–444.

76. De Leon, R., Hodgson, J. A., Roy, R. R., et al. (1998) Locomotor capacity attributable to step training versus spontaneous recovery after spinalization in adult cats. *J. Neurophysiol.* **79,** 1329–1340.

77. De Leon, R., Hodgson, J. A., Roy, R. R., et al. (1998) Full weight-bearing hindlimb standing following stand training in the adult spinal cat. *J. Neurophysiol.* **80,** 83–91.

78. Edgerton, V. R., de Guzman, C. P., Gregor, R. J., et al. (1991) Trainability of the spinal cord to generate hindlimb stepping patterns in adult spinalized cats, in *Neurobiological basis of human locomotion* (Shimamura, M., Grillner, S., and Edgerton, V. R. eds.), Japan Scientific Societies Press, Tokyo, pp. 411–423.

79. Hodgson, J. A., Roy, R. R., De Leon, R., et al. (1994) Can the mammalian lumbar spinal cord learn a motor task? *Med. Sci. Sports Exer.* **26,** 1491–1497.

80. Lovely, R. G., Gregor, R. J., Roy, R. R., et al. (1986) Effects of training on the recovery of full-weight-bearing stepping in the adult spinal cat. *Exp. Neurol.* **92,** 421–435.

81. Lovely, R. G., Gregor, R. J., Roy, R. R., et al. (1990) Weight-bearing hindlimb stepping in treadmill-exercised adult spinal cat. *Brain Res.* **514,** 206–218.

82. Carrier, L., Brustein, L., and Rossignol, S. (1997) Locomotion of the hindlimbs after neurectomy of ankle flexors in intact and spinal cats: model for the study of locomotor plasticity. *J. Neurophysiol.* **77,** 1979–1993.

83. Rossignol, S., Bouyer, L. J. G., Whelan, P. J., et al. (1997) Chronic spinal cats can recover locomotor function following transection of an extensor nerve. *Soc. Neurosci. Abstr.* **23,** 761.

84. Bouyer, L. J. G., and Rossignol, S. (1998) The contribution of cutaneous inputs to locomotion in the intact and the spinal cat. *Ann. N.Y. Acad. Sci.* **860,** 508–512.

85. Freeman, L. W. (1952) Return of function after complete transection of the spinal cord of the rat, cat and dog. *Ann. Surg.* **136,** 193–205.

86. Goldberger, M. E. (1986) Autonomous spinal motor function and the infant lesion effect, in *Development and plasticity of the mammalian spinal cord,* Fidia Research Series. (Goldberger, M. E., Gorio, A., and Murray, M., eds.), Liviana Press, Padova, pp. 363–380.

87. Miller, S. and Van der Meche, F. G. A. (1976) Coordinated stepping of all four limbs in the high spinal cat. *Brain Res.* **109,** 395–398.

88. Zangger, P. (1981) The effect of 4-aminopyridine on the spinal locomotor rhythm induced by L-Dopa. *Brain Res.* **215,** 211–223.

89. Ranson, S. W. and Hinsey, J. C. (1930) Reflexes in the hind limbs of cats after transection of the spinal cord at various levels. *Am. J. Physiol.* **94,** 471–495.

90. Baker, L. L., Chandler, S. H., and Goldberg, L. J. (1984) L-Dopa induced locomotorlike activity in ankle flexor and extensor nerves of chronic and acute spinal cats. *Exp. Neurol.* **86,** 515–526.

91. Robinson, G. A. and Goldberger, M. E. (1986) The development and recovery of motor function in spinal cats. II. Pharmacological enhancement of recovery. *Exp. Brain Res.* **62,** 387–400.

92. Giuliani, C. A. and Smith, J. L. (1987) Stepping behaviors in chronic spinal cats with one hindlimb deafferented. *J. Neurosci.* **7,** 2537–2546.
93. Rossignol, S., Bélanger, M., Barbeau, H., et al. (1989) Assessment of locomotor functions in the adult chronic spinal cat. Conference proceedings: criteria for assessing recovery of function: behavioral methods. APA, Springfield, NJ, pp. 10–11.

4

Organization of the Spinal Locomotor Network in Neonatal Rat

Jean-René Cazalets

1. INTRODUCTION

Among the various models used for studying the generation of rhythmic motor behaviors *(1)*, the locomotor system in all its various forms (walking, flight, swimming, etc.), has been the favorite. From studies on mammalian spinal cord at the beginning of the century *(2,3)*, it has been postulated that the central nervous system has the endogenous capability of generating rhythmic motor activities (for an extensive review, see ref. *4*). The generation of these complex motor patterns does not require sensory feedback, as shown in many in vitro preparations of isolated nervous systems or in decerebrated/spinal curarized animals in which a fictive motor pattern can be elicited. Moreover, in the last 30 years, convergent information from many invertebrate and vertebrate preparations has unambiguously confirmed the idea that defined neuronal networks organize motor behaviors and has led to the concept of a central pattern generator (CPG) *(5)*.

This concept has provided a solid basis for fruitful analysis, since it became possible to study the functioning of motor networks at the cellular level and under restrictive conditions and to identify their components. This has been achieved in many invertebrate preparations *(5)* and in lower vertebrates, in which the cellular organization of the networks responsible for swimming is beginning to be well understood *(6,7)*. When compared with undulatory swimming, walking in tetrapods exhibits specificity, however, and the circuitry has not yet been analyzed. The recent use of an in vitro isolated spinal cord preparation from newborn rat has provided significant progress in this field, and the general outline of the network is beginning to be drawn. In addition to their importance for the field of spinal cord physiology, these recent findings may also have a deep impact on research in locomotor recovery after spinal cord injury. In fact, establishing where the locomotor networks are precisely located in the spinal cord and defining their anatomic contours will help in functional repair of the spinal cord whatever the strategy used for this purpose. The present

From: *Neurobiology of Spinal Cord Injury*
Edited by: R. G. Kalb and S. M. Strittmatter © Humana Press Inc., Totowa, NJ

paper reviews some of these data, which were obtained by various groups, and focuses on the localization and the organization of the lumbar spinal locomotor networks. A comparison with data obtained in other species is made, in an attempt to establish some general insight on locomotor networks in vertebrates.

2. SPINAL CORD AND RHYTHMIC MOTOR BEHAVIORS

2.1. The Spinal Cord Generates Many Different Motor Outputs

The spinal cord is the privileged structure for generating rhythmic motor activities. First, concerning respiration, although most rhythmic activity is generated in the brainstem (for review, see ref., 8), some important respiratory mechanisms take place at the spinal level, and the existence of a respiratory spinal network has been postulated (9,10). Second, it has been demonstrated that in quadrupeds distinct networks for fore- and hindlimbs exist, although the forelimb network has been much less studied. In the curarized spinal rabbit it was found that fore- and hindlimb networks can be activated separately (11), demonstrating that several independent networks are implicated. It has been also shown that coordinated locomotor movements between fore- and hindlimbs occur in high spinal cats (12) or in isolated spinal cord from newborn rat (13).

Another important source of rhythmicity concerns the trunk muscle activity. If this is obvious for swimming animals in which locomotion is sustained by undulatory movements of the body (6), such an activity is nevertheless found in all classes of vertebrates. The long back muscles play an important role in aquatic tetrapods and reptiles (14,15). In mammals comparable activities are observed. During walking in humans (16) and cat (17), the lumbar back muscles show a bilateral contraction, with two bursts of activity per step cycle. During development in rat, close relationships between the back and hindlimb muscle activity emerge at around postnatal d 11 (18). Another important finding is that in high spinal paralyzed cats, trunk muscles are rhythmically active during fictive locomotion, which demonstrates a central origin for long back muscle activity (19). All these networks, however, do not normally operate in isolation, and they are interacting with each other according to behavioral needs. For example, as previously underlined, fore- and hindlimb movements are centrally coordinated at the spinal level (12). Similarly, a coupling between respiratory and hindlimb activity, whose efficiency is dependent on pCO_2, has been demonstrated in both the rabbit (20) and cat (21).

From the brief overview provided here, it is clear that for the "networker," the spinal cord consists of a mixture of various areas, each of them with its own specificity. A consequence of all these findings is that virtually all parts are susceptible to rhythmic activation and that these various networks will be active according to their particular and specific properties during motor tasks. This topographic specialization of the cord is also attested to by the location of specialized motoneuronal populations along the cord. For example, in the cat, the phrenic motoneurons, which innervate the diaphragm, are restricted to segments

C4–C6 *(8);* those innervating the hindlimbs extend from L4 to S2 *(22).* This functional segregation, which appears at the motoneuronal level, will also be reflected at the interneuronal level. The problem, however, is to know the extent to which an interneuronal population involved in one type of motor behavior will be restricted to one portion of the cord. This raises the problem of localizing a network; some general comments on this topic should be made. First, one should precisely define the type of activity studied, for which the prerequisite is a correct evaluation at the behavioral level in the intact animal. This is necessary to be sure that the activity observed under restricted conditions, when trying to dissect the neuronal network that underlies it, is actually related to the behavior considered. Second, since it is known that the various networks interact with each other (see above), one should be careful to avoid cross-interactions. Third, how should we define the network? Should we consider the neurons that at a given time will be directly involved in the generation of the considered motor pattern or, alternately, the neurons that are also indirectly concerned by the ongoing activity? In other words, should we consider the minimal neuronal assembly required for generating a motor pattern identical to the one recorded in intact animal, or should we include all relevant neuronal populations that can generate a rhythmic pattern? All the various points raised here underline the difficulties that can be encountered in this approach and that explain some discrepancies that may occur.

2.2. Characteristics of the Locomotor Behavior: In Vivo Versus In Vitro

In vitro preparations are indisputably useful models for investigating the cellular basis of behavior. An important point, however, is to know to what extent the activity recorded under isolated conditions reflects the one observed in the intact animal. It is critical to establish a comparison between these two levels. A paradox in the newborn rat preparation is that at birth, rat pups do not perform spontaneous locomotor movements. In fact, various studies have shown that the locomotor networks are not expressed during the first postnatal week, because of postural constraints *(23–25).* When body weight is suppressed, as during swimming *(23,26,27)* or air-stepping suspended *(25,28)* experiments, locomotor movements can be performed at day 0 (d 0), demonstrating that locomotor networks are functional at this very early age. The swimming and air-stepping patterns consist in an alternating gait in which the phase relationships between the right and left legs and the ipsilateral legs are out of phase, which is similar to what is recorded in vitro. The motor period is also a critical criterion to be determined for characterizing the locomotor activity. In intact animals, we have shown that it decreases from 1 s at birth to 0.2 s at adulthood *(27)* and that it ranges from 1 to 0.5 s between d 0 and 3. This is similar to what is recorded in vitro, since the motor period recorded in the presence of various concentrations of excitatory amino acids (from 4 to 0.5 s) *(29,30),* fits the one recorded in vivo (see also ref. *31).*

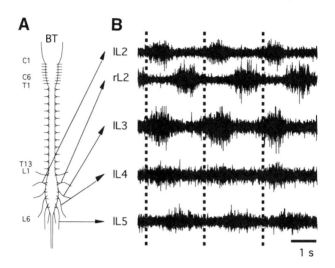

Fig. 1. The brainstem/spinal cord preparation of neonatal rat. **(A)** Schematic view of the isolated preparation. BT, brainstem; C, cervical; T, thoracic; L, lumbar. **(B)** Extracellular recordings of lumbar ventral roots during bath application of a mixture of NMDA (2×10^{-5} M) and 5-HT (10^{-4} M). Out-of-phase motor bursts are recorded from the right and left side (here the L2 ventral roots, top two traces). The bursts of action potentials recorded from the L2–L3 ventral roots (flexor phase) are out of phase with the one recorded from the L5 ventral roots (extensor phase).

The detailed pattern of intralimb muscular activities has only been studied in the adult animal *(32–35),* and 2-week-old rats *(36),* since this is much more difficult in very young pups. Electromyographic activities, performed in a semiisolated preparation in which the legs were left attached *(37),* have shown, however, that a complex pattern of intraleg muscle relationships can be generated in vitro at birth closely resembling the one recorded in vivo.

The locomotor repertoire in vivo or in vitro does not include galloping, which first appears under L-DOPA activation at d 10 in intact animals *(28).* To date, use of an in vitro preparation has not determined whether the absence of gallop in the newborn rat is caused by some immaturity in spinal circuitry, in the descending pathways, or in both.

3. ISOLATED BRAIN STEM SPINAL CORD PREPARATION OF NEWBORN RAT

Some difficulties inherent in the use of intact animals can be overcome using isolated preparations. They provide an accurate control of the extracellular medium and since there is no blood-brain barrier, the effects of transmitters and of their agonists or antagonists can be directly studied by bath application.

Moreover, stable intracellular recordings can easily be performed on identified neurons (*see* Figs. 3 and 4).

The in vitro brainstem-spinal cord preparation of newborn rat (Fig. 1A) can survive several hours (8–10 hr) due to its small size and the absence of myelin. A completely isolated brainstem-spinal cord (Fig. 1A) or a semi-isolated spinal cord in which the hind limbs are kept *(29,37)* can also be used.

The best known element of the spinal neuronal circuitry is indeed the motoneuron. The location of the motoneurons innervating most of the muscle groups of the hindlimbs has been determined in the rat, by retrograde labeling of horseradish peroxidase *(38)*. Motoneuron columns in rats extend from L1 to L6, as presented in Figure 6. The enlargement reaches its maximum at L4–L5, which contain most of the hindlimb motoneurons. As a consequence, the L4 and L5 ventral roots are much bigger than the others (*see* Fig. 6). This distribution of motoneuronal populations as well as the number of spinal segments differs, however, from one species to another. In the cat, for example, the motoneurons are also distributed over six segments but with a shift (from L4 to S2), and there is an additional L7 segment *(22)*. In the chick, they are present from the last thoracic segment to L8 *(39)*. Therefore, when comparing data from one species to another, one should be careful that the homology between segments is respected.

3.1. Generation of Locomotor-like Activity

The activity recorded in the ventral roots in response to the bath application of neurotransmitter consists in right and left alternating bursts of action potentials (Fig. 1B), whose period depends on the concentration of bath-applied transmitter. In the isolated spinal cord of neonatal rat, it was found initially that locomotor-like activity was induced by activating N-methyl-D-aspartate (NMDA) receptors *(40,41)*. Following this report we found that excitatory amino acid action also initiates locomotion via the activation of kainate receptors *(29)*. At the same time, we demonstrated that serotonin (5-HT) alone is also able to turn on the locomotor network and to activate it in a dose-dependent manner *(29)*. We subsequently observed that applying a mixture of 5-HT and an NMDA receptor agonist was a powerful means of eliciting long-lasting (several hours) locomotor-like activity *(42,43)*. This methodology has now been adopted by most of the groups working with the same preparation *(44–47)*. Other neurotransmitters such as acetylcholine and dopamine *(37)* have also been demonstrated to activate the locomotor network. We found that the ipsilateral L2 and L5 ventral roots exhibit out-of-phase activity, suggesting that both flexor and extensor units are active during one cycle (Fig. 1B). Kiehn and Kjaerulff *(37)* subsequently identified the burst in L5 as the extensor phase of the cycle, and they also demonstrated that complex motor patterns are generated in vitro. Most of the neurochemicals used, with the exception of acetylcholine, *(46,48)* and including the excitatory amino acids *(29,30,40,49),* elicit a motor pattern that is compatible with locomotor activity in terms of period and phase relationship. It should be noted that for a given transmitter concentration, the resulting motor period will depend on other state parameters such as temperature *(42)* or external K^+ concentration *(43)*.

4. LOCALIZATION OF THE LOCOMOTOR NETWORK

Recently, a series of paper has addressed the problem of localizing the locomotor network, using the in vitro preparation of isolated spinal cord. These studies are reviewed here, and we will see that they have contributed complementary information, although some points are in part controversial. We first approach the localization over the transverse plane and then along the rostrocaudal axis. As a preliminary, we review the various techniques used.

4.1. In Search of Lost Networks: The Use of Various Strategies

Whatever the preparation used, numerous approaches have been developed for localizing active zones involved in ongoing behavior. First, anatomic methods have been based on the labeling of cells, which, when they are active, may take up extracellular probes such as deoxyglucose *(11)* or sulphorhodamine *(44,50)* or express cellular endogenous markers *(51,52)*. Other attempts have been made to visualize active cells on-line, by using calcium-sensitive dye in isolated spinal cord *(53)*. The most widely used procedures, however, are lesion experiments in which part of the nervous system is cut or removed *(37,47,54–57)*. To overcome the disadvantage of the irreversible interruption of conduction and the possible traumatic action of lesioning, alternative methods were developed. In the chick embryo, for example, cobalt gel around the cord permits the temporary and local interruption of synaptic transmission *(58)*. In the cat, cooling various parts of the spinal cord reversibly blocks action potential propagation *(55)*. Finally, taking advantage of an in vitro system, compartmentalization using Vaseline of the walls of various areas of the spinal cord has been performed, to activate separate parts of the nervous system independently *(31,59)*.

4.2. Localization Along the Transverse Axis

Cells located in the transverse plane and active during locomotion have been identified by both anatomic and electrophysiologic methods. Sulphorhodamine, a fluorescent tracer that is taken up by neurons in an activity-dependent manner, labeled cells mainly in the intermediate gray area and around the central canal, in a totally isolated cord of the neonatal rat *(44)*. Viala et al. *(60)* also found that in the spinal curarized rabbit, cells from lumbar segments were preferentially labeled by 2-deoxy(C14)glucose in the intermediate gray matter during L-DOPA-induced fictive locomotion. A comparable labeling pattern was also found in the cat during fictive scratching *(51)* or locomotion *(52)*, using c-fos immunoreactivity.

In their elegant study, Kjaerulff and Kiehn *(61)* identified the crucial elements for locomotion in the transverse plane, by performing section experiments. Horizontal sections along the entire spinal cord length permitted testing of locomotor capabilities of the ventral part. When they removed the dorsal part of the spinal cord, the locomotor-like activity did not change, providing that the section did not pass the level of the central canal. When the lesion was performed at the anterior rim of the central canal, the rhythmicity persisted only in the more rostral segments *(61)*, probably because they have major rhythmic capabilities. These authors also found that the extreme lateral part of the gray matter did not contain key elements for the in vitro functioning of the locomotor network.

In parallel, various attempts have been made to identify physiologically neurons involved in the generation of the locomotor activity. The main purpose of these studies is to provide evidence that cells with oscillatory properties may be the substrate for locomotor rhythmicity. Hochman et al. *(62)* first recorded neurons around the central canal that exhibited NMDA-mediated voltage oscillations. Subsequently, it was reported that 80% of the interneurons recorded in lamina VII were rhythmically active during drug-induced locomotor activity *(63)*, although the possibility was raised that a proportion of these neurons was activated nonspecifically under these conditions *(64)*. Kiehn et al. *(65)* confirmed that the large majority of neurons recorded in the medial intermediate gray area and around the central canal displayed rhythmicity in relationship with drug-induced locomotor-like activity. If in most of the cells (88%) the activity was synaptically driven, a small proportion (12%) exhibited intrinsic bursting capabilities. A common finding, underlined in all these studies *(62,63,65)*, is that cells exhibiting active intrinsic properties that could be involved in locomotor processes only represent a small fraction (about 10%) of the cells tested. As noted by Kiehn et al. *(65)*, their role in locomotor rhythm generation has not yet been determined, so that the important point that remains to be established is the relationship between cells active in that zone and the motoneurons. Nevertheless, these different studies strongly indicate that during locomotion, important processes are taking place around and close to the central canal. Figure 2 summarizes all the experimental data that have been collected on the transverse organization of locomotor structures.

4.3. Localization Along the Longitudinal Axis

The conclusions drawn from our initial study *(31)* were based on results obtained with a partitioned spinal cord. Vaseline walls were built at various levels along the thoracolumbar cord, to determine the minimal portion of spinal cord required for generating a locomotor-like activity. Figure 3 presents the result of such an experiment. When the transmitters were applied on the rostral segments (i.e., T13–L2) locomotor-like activity was recorded in the upper ventral roots (L2), and the intracellularly recorded motoneuron from the L4 segments received a synaptic drive in relationship to the activity recorded in the L2 ventral roots (Fig. 3A). In contrast, when the transmitters were bath applied on the caudal pool, no locomotor-like activity was recorded in any part, and only

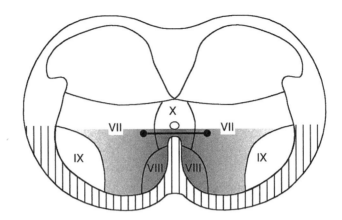

Fig. 2. Location of locomotor areas in the transverse plane. Schematic representation of a transverse section of the neonatal rat lumbar cord. The gray zone in layer VII represents the area in which most of the locomotor processes take place (*see* ref. *61*). The gradient illustrates the fact that oscillating neurons are found around the central canal. Pathways mediating reciprocal inhibition (dots) pass through the ventral commissure. The hatched area contains fibers mediating the rostrocaudal alternation.

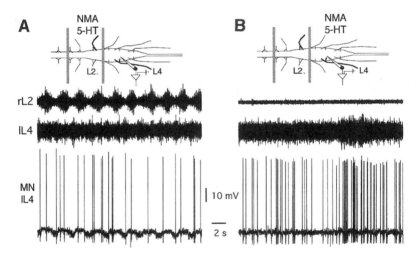

Fig. 3. Locomotor-like activity is generated in the rostral segments in an intact spinal cord. Vaseline walls (vertcal bars at top in A and B) built at the T12 and the L2 levels delimited two compartments. (**A**) Bath application of NMA/5-HT to the upper lumbar cord induced locomotor-like activity. The membrane potential of the motoneuron (MN) recorded in the caudal pool (MN from the left L4 segment) was phasically modulated in relationship with the activity recorded in L2. (**B**) After a washout with normal saline, NMA/5-HT was bath-applied to the caudal pool. It only induced tonic firing in the L4 ventral root and in the intracellularly recorded motoneuron. In A and B, a positive current (2 nA) was injected through the microelectrode to elicit spiking activity in the intracellularly recorded motoneuron.

tonic firing occurred (Fig. 3B) in the L4 ventral root and the tested motoneuron. Due to the Vaseline wall thickness, there is some uncertainty regarding segmental borders, and it is likely that, in fact, the network we studied is even slightly more rostral, i.e., it is retained in part of T13 and probably does not occupy the whole L2 segment. This was already mentioned in our previous study, in which we frequently used the term T13–L2 network *(31)*. In agreement with our results is the finding by Cowley and Schmidt *(46)* that after sectioning at the junction between T13 and L1, 5-HT-induced locomotor-like activity permanently disappeared in lumbar segments.

These authors also performed compartmentalized bath application of the spinal cord divided into three different pools. They found that 5-HT in the caudal bath (from L3 to the sacral level) failed to induce any rhythmic activity. Conversely, bath application of 5-HT to the more rostral bath (C1–T9) produced rhythmic activity in the cervicothoracic region. More surprisingly, they reported that the simultaneous bath application of 5-HT on the cervicothoracic and low lumbar region produced rhythmic activity in cervical, thoracic, and lumbar regions. Although they did not mention it for the two experiments performed under these conditions, there was no doubt that 5-HT-in the middle pool (T11–L2) would have induced a stable and regular locomotor-like activity. They concluded that a 5-HT-sensitive circuitry involved in the generation of locomotor-like pattern is distributed throughout the entire cervicothoracic spinal cord.

These findings *(46)* are, however, contradicted by Kremer and Lev-Tov *(47)*, who observed that the left-right alternating rhythm did not persist in the rostral segments (T6) when sectioning in T11, while the rhythm remained intact below the section. Therefore, as in Kremer and Lev-Tov's study, Cowley and Schmidt *(46)* did not provide the features of these cervicothoracic activities and thus questions arise about their locomotor nature. This is of particular importance since they are recorded in ventral roots, which do not innervate limb muscles. As emphasized earlier (*see* Section 2.1.), rhythmic activities that occur in relation to locomotor activity in postural long back muscles are centrally generated *(19)*. Alternatively, these rhythms could originate in forelimb networks *(11,12)* or spinal respiratory networks *(66,67)*, and interactions through propriospinal connections might occur.

Several studies with section experiments have led to the conclusion that, when isolated, the caudal lumbar segments express rhythmic activities *(46,47,61)*. According to Cowley and Schmidt *(46)*, irregular and nonalternating activity that they considered as nonlocomotor activity can be recorded in caudal segments with NMA or acetylcholine after sectioning between L3 and L4. They also found that when the T13–L1 segment was sectioned, whatever the portion of the spinal cord they observed, i.e., L1–L3 or L1–L6, all 5-HT-induced locomotor activity disappeared in the lumbar roots. Kremer and Lev-Tov *(47)* observed that a section in the mid-L3 segment blocked the activity in the detached caudal cord without affecting rhythmicity in the rostral lumbar cord at all. Increasing the concentration of NMDA restored irregular bursts of weak amplitude at a period 3 times longer than in the rostral segments under the same conditions *(47)*.

Other experiments, performed in the presence of inhibitory acid antagonists to reveal some "latent" CPGs *(47)* by removing inhibition, have shown that single isolated segments or hemisegments can generate a rhythmic activity. Bicuculline at high concentrations (i.e., above 10^{-5} M and up to 5×10^{-5} M) alone *(46,47,68)* or in combination with various transmitters *(46,47)* or with strychnine *(69)* elicits slow synchronous bursts of activity *(47,59,69)*. These characteristic rhythms elicited by antagonists of inhibitory amino acids occur, however, at a much longer period (15–30 s) than the control locomotor activity (1–2 s), and in an intact spinal cord there is a perfect synchrony between all ventral roots (ipsi- and contralateral). It is probable that under these conditions, many processes other than locomotor ones are released, in addition to nonspecific effects. In light of pharmacologic data, this may reflect abnormal functioning of the nervous system. Both strychnine *(70,71)* and bicuculline *(72)* block several types of K^+ and Ca^{2+} voltage-gated channels. Therefore, the role of bicuculline and strychnine remains ambiguous at the very high doses at which they were generally used in most of these studies (see detailed discussion of this problem in ref. *73*).

4.4. Inhibitory Transverse Coupling

In our previous study in newborn rats, we found that after a longitudinal section in the sagittal plane from the caudal end of the cord to the L2 segment, the alternation between the right and left side at the L3–L5 level persisted, despite the lack of physical connections *(31)*. Thus, we concluded that the cross-inhibitory connections occurred in the rostral segments. This was also observed by Mc Lean et al. *(63)* and Magnusson et al. *(64)*, who performed a midsagittal section up to L1–T13. Other authors have subsequently found that reciprocal left and right pathways are not only present in L1–L2. Kiehn and Kjaerulff *(61)*, performing the reverse experiment, i.e., a longitudinal section in the midline, from rostral (T12) to caudal (L2), found that cross-inhibition in the caudal segments was sufficient by itself to maintain a loose out-of-phase relationship in the entire lumbar cord. Kremer and Lev-Tov *(47)* amplified this view finding that only cross-connections established at a single segmental level (from T12 to L4) were sufficient to preserve the right and left alternation, although it was more unstable *(47)*.

Although the right and left alternation was found to occur at all levels between T13 and L4 *(37,46,47)*, it was clear from these studies that removing the cross-pathways by midsagittal section at the T13–L2 level induced more important disturbances than sectioning in other segments, suggesting that rostral coupling is of greater functional importance than caudal coupling. The pathways mediating the contralateral inhibition have been identified, since transverse coupling fibers appear to run exclusively in the most ventral surface of the cord *(37)*.

4.5. Synaptic Drive Onto Motoneurons

The pattern of motoneuron action potential firing drives the muscle contraction underlying coordinated movement. Motoneurons integrate the synaptic inputs from numerous sources, perhaps most importantly those arising from the premotoneuronal locomotor network. The synaptic input arising in the locomotor network and impinging on the motoneurons has been called locomotor drive. Here we address the question: what are the constituents of this locomotor drive (nature and properties of the postsynaptic potentials, transmitters involved, type of receptors activated, etc.)?

By using the spinal cord, partitioned with Vaseline walls (Fig. 4; *see* also Fig. 3), one can activate the rhythmogenic compartment (located in L1–L2) independently from its output motoneuronal target (located in L3–L5). Therefore, it has become feasible to study the characteristics of the locomotor drive, by modifying the synaptic functioning in the caudal compartment using various drugs, without affecting the locomotor-like activity induced by bath application to the rostral pool (T13–L2 segments; Fig. 4A) *(59)*. Under this experimental setup, locomotor-like activity was induced by bath-applying NMA/5-HT to segments L1–L2 and monitoring by extracellular (top traces, Fig. 4) or intracellular (bottom trace, Fig. 4) recordings. The membrane potential of the intracellularly recorded motoneurons was phasically modulated in relationship with the ongoing locomotor activity recorded in the upper segments. One cycle of activity was composed of a hyperpolarizing phase and a depolarizing phase, which was capped with action potentials (Fig. 4A2,A3, bottom trace, MN 1L5). As observed in other vertebrate species *(74,75)* the inhibitory synaptic volley was reversed at around –60 mV and was dependent on chloride ions. We found that the synaptic drive was, at least partly, monosynaptic from the L1–L2 locomotor network to the L3–L5 motoneurons, since it persisted in the presence of saline containing high concentrations of divalent ions or mephenesin, which largely reduced or blocked polysynaptic transmission *(31)*. We clearly never excluded the possibility that they may also receive in parallel a polysynaptic component that may originate in the caudal segments when they are appropriately activated. To identify the transmitters and receptors involved in this synaptic control of the motoneurons, we used various antagonists. Figure 4B shows that in the presence of blockers of the excitatory amino acid receptors (NMDA, AP-5) and non-NMDA receptors 6-cyano-7-nitroguinoxaline-2-3-dione (CNQX), the depolarizing phase was completely abolished. Only the inhibitory part of the drive remained (Fig. 4B2, bottom trace, circle), which was sensitive to the glycinergic blocker strychnine *(31)*. As expected, the rostral network functioning was not affected by the bath application of the various blockers to the caudal cord, although all synaptic activity completely disappeared in the caudal segments *(31)*.

It has been shown in various studies that when isolated, the caudal segments (L3–L6) can produce a rhythmic motor pattern *(47,61)*. Therefore the following

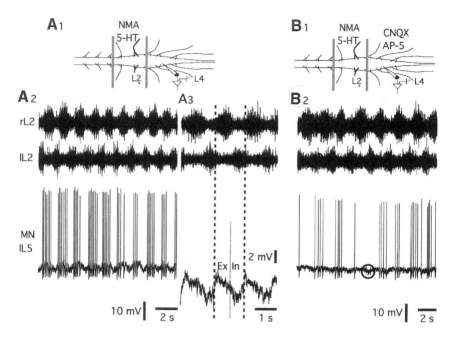

Fig. 4. Synaptic drive from the rostral locomotor network. In a partitioned spinal cord (**A1** and **B1**), locomotor-like activity was induced by bath application of NMA/5-HT. (**A**) Control condition. (**A2**) The intracellularly tested motoneurons (MN lL5, motoneuron from the left L5) emits a burst of action potentials that are out of phase with the burst recorded in the left lumbar ventral root 2 (lL2). Positive current (+1.2 nA) was permanently injected into the motoneuron. (**A3**), Same recording but at higher time scale to visualize the excitatory (Ex) and inhibitory (In) part of the drive. The amount of positive current injected (+0.7 nA) was reduced compared with A2, to avoid action potential firing at the crest of the depolarizing phase. (**B**). Addition to the caudal pool of the excitatory amino acid receptor blockers CNQX and AP-5 (see text) blocked the depolarizing phase of the drive. It only remains glycinergic inhibitory events (shown for one cycle by the circle in **B2**). In **A2** and **B2**, the same amount of positive current (+1.2 nA) was injected through the microelectrode to elicit spiking activity in the intracellularly recorded motoneuron.

question can be addressed: what is the relative contribution of the locomotor synaptic input arising in the L1–L2 segments versus the synaptic input originating in L3–L6 segments? One can assume that if the motor output in the caudal segment is only subserved by the endogenous rhythmogenic capabilities of these segments, disconnecting the upper and lower lumbar cord will not modify the motor pattern. In contrast, if the caudal motoneuron activity is synaptically driven, an interruption in the pathway should interrupt their activity. Figure 5 tested these alternatives. Locomotor-like activity was induced in a thoracolum-

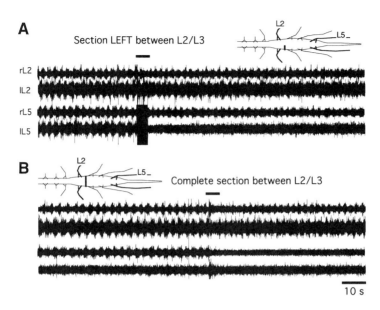

Fig. 5. The activity of caudal motoneurones is driven by descending influences. Bath application of NMA/5-HT elicited locomotor-like activity recorded in the L2 and L5 ventral roots. (**A**) The transverse section of the left hemispinal cord close to the L2 level (black bar) provoked the immediate cessation of rhythmic activity in the caudal L5 ventral roots, without major changes in activity on the contralateral side and the upper segments. (**B**) Subsequent transection of the remaining intact hemicord provoked the cessation of rhythmic activity in the L5 ventral roots.

bar cord (see diagram, Fig. 5A) by bath application of NMA/5-HT. A section of the left hemispinal cord (black bar) was performed between the L2 and L3 segments, to avoid lesioning of the L4 and L5 segments. This section provoked an immediate cessation of the rhythmic activity in the caudal part, while it remained unchanged in the contralateral side and above the section (Fig. 5A). The subsequent sectioning of the right spinal cord immediately interrupted the activity in the right caudal side (Fig. 5B). It can be concluded that under pharmacologic activation of the whole spinal cord, the activity of the caudal motoneurons depends on the rostral segments.

One drawback, however, when studying the locomotor drive by activating only the rostral segments, is its relative weakness compared with the drive elicited by pharmacologic activation of the whole lumbar cord. Therefore, the caudal motoneurons do not generally reach the threshold for spiking, and it is likely that the interneuronal processes at the caudal level may reinforce the rostral drive. Alternatively, the motoneurons may not discharge because of a lack of tonic input from superior centers, since it has been shown in the cat that during mesencephalic locomotor region (MLR)-induced locomotion the motoneuron received both excitatory and inhibitory postsynaptic potentials (EPSPs and IPSPs) *(76)*.

Recently, Kjaerulff and Kiehn *(77)* have used a comparable strategy to study the right and left coordinating synaptic inputs to motoneurons; for the first time, they directly identify the cross-inhibition in mammals. A barrier was built longitudinally along the midline instead of transversally. Then they pharmacologically activated the hemispinal cord while simultaneously recording motoneurons from the contralateral side, which was bathed in normal saline. These motoneurons received a rhythmic synaptic inhibitory input with the same characteristics we previously observed *(59)*, i.e., Cl⁻ dependent and sensitive to strychnine. In most, but not all, the motoneurons, the IPSP bursts disappeared in the presence of AP-5 and CNQX indicating that they were mediated at least partly through a polysynaptic pathway.

Together, these data show that the biphasic drive received by the motoneurons is complex *(31,77)*. One can schematically distinguished two main components: (1) the synaptic input involved in the flexor/extensor phase settings; and (2) the synaptic input regulating the right and left coordination. A motoneuronal group is simultaneously inhibited by its ipsilateral antagonist group (flexor versus extensor) and by its contralateral homologous (right flexor versus left flexor) during an alternating gait as the one we are recording in the isolated spinal cord. Therefore, the inhibitory drive observed (and probably also the excitatory drive; *see* ref. *77*) will result from the mixture of both ipsi- and contralateral synaptic influences.

5. THE SPINAL LOCOMOTOR NETWORK: LOCALIZED OR DISTRIBUTED

Following our initial work in neonatal rat *(31,59)*, an apparent controversy has emerged as a result of a misunderstanding or overinterpretation of our results. Part of the debate is semantic, since the answer to the "localized" or "distributed" problem will depend on how the question is formulated. One question is: "what is the minimal portion of the spinal cord capable of generating an activity that is unchanged when compared with intact nervous system?" All the studies have shown that the portion of the cord including part of the T13–L1–L2 segments is necessary to generate this activity. In fact, in our previous study, the term "localized" was used to point out that in contrast to what occurs with the motoneurons whose location spans six segments (from L1 to L6), the network that we were studying overlapped less than three segments.

On the basis of our findings, we concluded (1) that there is a network located in L1–L2 that was sufficient by itself to drive the entire caudal motoneuronal pool; and (2) that in an intact partitioned spinal cord only a descending drive from rostral to caudal could be observed. To date, these conclusions have not been challenged. What the remaining authors found complementary to our results is that: (1) cross-connections, although weaker *(46,47,61)*, also exist in thoracic and lumbar segments other than L1–L2 (Fig. 6); in our study we did not

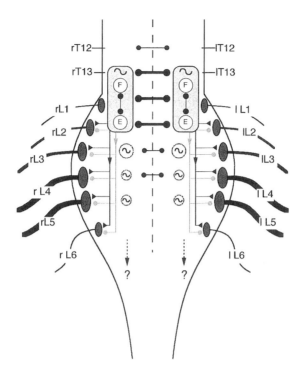

Fig. 6 Summary diagram of spinal locomotor circuitry. "The grey rectangle in the rostral part (T13-L2) of each hemi spinal cord represents the networks organizing flexor (F) and extensor (E) alternation." They reciprocally inhibit via glycinergic pathways, which run through the ventral commissure from the T12 to the L4 segments. The rostral network drives the activity of caudal motoneurons via glycinergic (dots) and acid aminergic (triangles) connections. Additional networks with rhythmogenic potentialities are located in the caudal segments L3–L6. The line thickness schematizes the relative ventral root size.

explore this possibility; and (2) slow and irregular bursting activities *(47,61)*, which were nonlocomotor according to Cowley and Schmidt *(46)*, were also observed in the caudal lumbar segments after sectioning. In this section, based on careful analysis of the results from these various studies, we believe that the discrepancies concern only minor points.

Figure 6 presents a synthesized view of the data collected to date. In comparison with our previous schema, it integrates the cross-pathways coordinating right and left alternation at various segmental levels. This appears at all segmental levels between T12 and L4 although the rostral coupling is more dominant than the caudal *(46,47,61)*. The second point (shown in Fig. 4) is the presence of rhythmogenic capabilities in the caudal segments, which decline in importance from the rostral to the caudal segments. All the authors agree that the rostral

T13–L2 part of the spinal cord is largely dominant for the generation of locomotor activity *(46,47,61)*. Concerning this latter feature, however, various
points should be discussed.

The role played by the caudal segments and the respective contribution of
each individual segment remains unclear. The main difference between our
study and those of other authors is that we are working on an intact spinal cord.
Even if some slow and irregular rhythmic motor patterns are expressed in caudal segments after sectioning *(46,47,61;* Cazalets, unpublished observations),
the problem is knowing whether this is relevant for "normal" functioning and
the extent of participation of these bursting activities during normal processes.
In fact, several lines of evidence led us to believe that T13–L2 is the crucial
area for organizing locomotion in neonatal rat, even though "latent" networks
(47) can be expressed in the caudal spinal cord. When the spinal cord is left
intact but partitioned (Figs. 3 and 4), we did not succeed under "normal" pharmacologic activation in activating the caudal part *(31,59)*. Under appropriate
conditions this may be possible, but to date it has not been achieved. We also
performed section experiments to try to resolve this discrepancy. We found
that in intact spinal cord even the sacral ventral roots displayed a rhythmic
activity in relationship with the one recorded in the lumbar roots (*see* also ref.
(78). More surprising is the observation that following section at the L3–L4
level, the sacral roots continued to exhibit rhythmic activity, which had virtually disappeared in L5–L6 (Cazalets, unpublished results). This activity was,
however, much more irregular and slower than before the section. At the same
time, it remained unperturbed in the L1–L2 segments. As the sacral ventral
roots do not convey axons to the hindlimbs, questions remain about the relevance of these rhythmic activities *(78)*. Whatever the case, these findings also
confirm that such activity is not expressed in the caudal segments during normal functioning.

Lesion experiments may also reveal redundant circuits that are not
expressed under normal locomotor processes. From the study of Ho and
O'Donovan *(58)*, who analyzed changes in rhythmogenic capabilities along
with development in the chick spinal lumbar cord, it appears that from d 8 to d
13 the caudal segments lose 80% of their bursting capabilities compared with
the rostral segments. "Although the authors did not go further in development,
it is not excluded that there is an additional reduction in bursting capabilities
with time, and that the rhythmogenesis is only ensured by rostral segments in
the adult."

It is likely that similar processes occur in the rat and that the circuits activated
in the caudal segments represent some residual vestige of development. This
possibility was also raised by Kjaerulff and Kiehn, who worked on 0–2-d-old
animals.

Another possible explanation can be proposed in the light of the results found
by Magnusson and Trinder *(64)*. To avoid nonspecific drug-induced activation
of neurons, these authors elicited locomotor-like activity in the neonatal rat

spinal cord by electrical stimulation of the ventrolateral funiculus at the cervical level, instead of bath-applying neurotransmitters. Under these conditions, they found that partial lesioning of the midline rostral lumbar cord (T13–L3) resulted in the loss of rhythmic capability for the whole lumbar cord, while lesioning of the caudal midline had no effects. In addition, they found that the amplitude of the field potential recorded bilaterally in response to the ventrolateral funiculus stimulation was maximal in the rostral (L1–L2) segments. This finding suggests that during physiologic initiation of locomotor activity, the rostral lumbar area is the preferential target, again reinforcing its preponderancy during locomotor processes.

Rhythmicity in caudal segments could also be interpreted on the basis of intrinsic neuronal properties. It has been shown that motoneurons may express oscillatory properties induced by NMDA under isolating conditions *(62,79)*. Therefore it is possible that part of the nonlocomotor pattern recorded by the same authors *(46)* may rely on these properties.

Another problem pertains to the relationships established between the rostral and the caudal segments. It has been shown in Figure 3, and also reported by Kjaerulff and Kiehn, that following a hemitransverse section at the junction between L2–L3, activity immediately disappeared in the ventral roots caudal to the section. This finding suggests that the activity in the caudal roots is dependent on a descending pathway and that ascending pathways have only minor influences. We also reported that alterations of neuronal functioning in the caudal bath (by bath-applying AP-5, CNQX, strychnine, etc.; Fig. 4B) *(59)* did not modify the activity in the rostral cord. Therefore the role of the rostral cord is beginning to be well understood, while the role played by the caudal segments needs to be clarified. It remains to be determined how rostral and caudal segments interact to generate the final motor output.

6. CONCLUSIONS

Altogether, these studies have produced complementary findings. They indicate that the mammalian lumbar spinal cord is an extremely heterogeneous structure. In lower vertebrates, undulatory swimming arises from the interaction of many segments, which have equipotent functional roles *(5,6)*, but this is unlikely to occur for walking in tetrapods. If this is now clear in neonatal rat, as shown here, it is also known in other vertebrate species, although it has not been studied in such detail. In the cat, it appears that the ability to generate fictive locomotion or scratching is more distributed than in the neonatal rat, although the rostral segments present much greater capacities for generating activity *(54,55)*. In the chick embryo, the capacity for bursting largely decreased in caudal lumbosacral segments along with development, and separate mechanisms for rhythm and pattern generation are involved *(58)*. The recently developed preparation of isolated in vitro mudpuppy spinal cord *(80)* is an attractive model for studying quadrupedal walking in vertebrates. A flexor and an extensor center that can oscillate independently has been identi-

fied at two different points in the spinal cord *(57)*. Many data have been collected from various preparations, but it is difficult to perform a synthesis from
all these different models.

From the results collected in the newborn rat, three important points should
be underlined. First, the fact one can activate the T13–L2 network independently from the caudal motoneurons, which receive a monosynaptic connection,
provides a useful tool for studying the functioning of the motor output without
interfering with the generation of locomotion *(81)*. Second, from a theoretical
point of view, because the main interneuronal elements are not uniformly distributed at the various segmental levels, previous models, like the one proposed
by Miller and Scott *(82)* in which the Ia interneurons and Renshaw cells play an
important role in the generation of locomotion, are no longer compatible with
the present findings.

Finally, many efforts directed toward promotion of functional recovery of
motor functions following spinal cord injury are based on the knowledge that
spinal networks are able to generate locomotor activity *(83)*. Pharmacologic
tools have been used to trigger or modulate a locomotor activity in a lumbar
cord that has been disconnected from superior centers. Another strategy aimed
at restoring locomotion involves transplantation of embryonic neurons in the
sublesional site *(see* Chap. 8 and refs. *84* and *85)*, or bridging lesions with multiple nerve grafts to direct specific rostral inputs from white to gray matter *(86)*.
Detailed knowledge of the network organization and location will facilitate the
operation at this level whatever the strategy used. Drug injection could be performed at a precise site, therefore avoiding indirect effects on other targets. In
the same way, we need to know the location of the target network, to position
bridging grafts accurately.

In conclusion, although strong evidence shows that the L1–L2 area is a key
region for spinal locomotor processes, more investigations are needed before
any definitive conclusion may be drawn. To date, the best model of the spinal
hindlimb locomotor systems appears to be that shown in Figure 6. There is no
doubt, however, that in the near future new details will be revealed, thus providing a more accurate view.

REFERENCES

1. Delcomyn, F. (1980) Neural basis of rhythmic behavior in animals. *Science* **210,**
   492–498.
2. Brown, T. G. (1911) The intrinsic factors in the act of progression in mammal. *Proc.
   Soc. B.* **84,** 308–319.
3. Brown, T. G. (1914) On the nature of the fundamental activity of the nervous centres,
   together with an analysis of the conditioning of rhythmic activity in progression, and
   a theory of evolution of function in the nervous system. *J. Physiol. (Lond.)* **48,** 18.
4. Grillner, S. (1981) Control of locomotion in bipeds, tetrapods and fish, in *Handbook
   of physiology. The nervous system II* (Brookhart, J. M. and Mountcastle, V. B, eds.),
   American Physiological Society, Bethesda, MD, pp. 1179–1236.

5. Stein, P. S. G., Grillner, S., Selverston, A. I., and Stuart, D. G., eds (1997) *Neurons, networks, and motor behavior. Bradford Book.* MIT Press, Cambridge.

6. Grillner, S. and Matsushima, T. (1991) The neural network underlying locomotion in lamprey-synaptic and cellular mechanisms. *Neuron* **7,** 1–15.

7. Roberts, A., Soffe, S. R., and Perrins, R. (1997) Spinal networks controlling swimming in hatching *Xenopus* tadpoles, in *Neurons, networks, and motor behavior. Bradford Book.* (Stein, P. S. G., Grillner, S., Selverston, A., and Stuart, D. G., eds.), MIT Press, Cambridge, pp. 83–89.

8. Monteau, R. and Hilaire, G. (1991) Spinal respiratory motoneurons. *Prog. Neurobiol.* **37,** 83–144.

9. Aoki, M., Mori, S., Kawahara, K., Watanabe, H., and Ebata, N. (1980) Generation of spontaneous respiratory rhythm in high spinal cats. *Brain Res.* **202,** 51–63.

10. Viala, D. and Freton, E. (1980) Demonstration of generators of the locomotor and respiration rhythms in the cervico-thoracic spinal cord of rabbits. *C.R. Acad. Sci.* **291,** 573–576.

11. Viala, D. and Vidal, D. (1978) Evidence for distinct spinal locomotion generators supplying respectively fore- and hindlimbs in the rabbit. *Brain Res.* **155,** 182–186.

12. Miller, S. and van der Meché, F. G. A. (1976) Coordinated stepping of all four limbs in the high spinal cat. *Brain Res.* **109,** 395–398.

13. Smith, J. C., Feldman, J. L., and Schmidt, B. J (1988) Neural mechanisms generating locomotion studied in mammalian brain stem-spinal cord *in vitro. FASEB J.* **2,** 2283–2288.

14. Devolvé, I., Bem, T., and Cabelguen, J-M. (1997) Epaxial and limb muscle activity during swimming and terrestrial stepping in the adult newt, *Pleurodeles waltl. J. Neurophysiol.* **78,** 638–650.

15. Ritter, D. (1995) Epaxial muscle function during locomotion in a lizard *(Varanus salvator)* and the proposal of a key innovation in the vertebrate axial musculoskeletal system. *J. Exp. Biol.* **198,** 2477–2490.

16. Thorstensson, A., Carlson, H., Zomlefer, M. R., and Nilsson, J. (1982) Lumbar back muscle activity in relation to trunk movements during locomotion in man. *Acta Physiol. Scand.* **116,** 13–20.

17. English, A. W. (1980) The functions of the lumbar spine during stepping in the cat. *J. Morphol.* **165,** 55–66.

18. Geisler, H. C., Westerga, J., and Gramsbergen, A. (1996) The function of the long back muscles during postural development in the rat. *Behav. Brain Res.* **80,** 211–215.

19. Koehler, W. J., Schomburg, E. D., and Steffens, H. (1984) Phasic modulation of trunk muscle efferents during fictive spinal locomotion in cats. *J. Physiol.* **353,** 187–197.

20. Viala, D. and Freton, E. (1983) Evidence for respiratory and locomotor pattern generators in the rabbit cervico-thoracic cord and for their interactions. *Exp. Brain Res.* **49,** 247–256.

21. Kawahara, K., Nakazono, Y., Yamauchi, Y., and Miyamoto Y. (1989) Coupling between respiratory and locomotor rhythms during fictive locomotion in decerebrate cats. *Neurosci. Lett.* **103,** 326–332.

22. Romanes, G. J. (1964) The motor pools of the spinal cord. *Prog. Brain Res.* **11,** 93–119.

23. Cazalets, J. -R., Menard, I., Cremieux, J., and Clarac, F. (1990) Variability as a characteristic of immature motor systems: an electromyographic study of swimming in the newborn rat. *Behav. Brain Res.* **40,** 215–225.

24. Geisler, H. C., Westerga, J., and Gramsbergen, A. (1993) Development of posture in the rat. *Acta Neurobiol. Exp.* **53,** 517–523.

25. Fady, J. -C., Jamon, M., and Clarac, F. (1998) Early olfactory-induced rhythmic limb activity in the newborn rat. *Dev. Brain Res.* **108,** 111–123.

26. Bekoff, A. and Trainer, W. (1979) The development of interlimb coordination during swimming in postnatal rats. *J. Exp. Biol.* **83,** 1–11.

27. Menard, I., Crémieux, J., and Cazalets, J. R. (1991) Evolution non-linéaire de la fréquence des mouvements de nage pendant l'ontogenèse chez le rat. *C. R. Acad. Sci. Paris* **312,** 233–240.

28. Van Hartesveldt, C., Sickles, A. E., Porter, J. D., and Stehouwer, D. J. (1991) L-DOPA-induced air-stepping in developing rats. *Dev. Brain Res.* **58,** 251–255.

29. Cazalets, J. -R., Sqalli-Houssaini Y., and Clarac, F. (1992) Activation of the central pattern generators for locomotion by serotonin and excitatory amino acids in neonatal rat. *J. Physiol. (Lond.)* **455,** 187–204.

30. Sqalli-Houssaini, Y., Cazalets, J.-R., Martini, F., and Clarac, F. (1993) Induction of fictive locomotion by sulphur-containing amino acids in an in vitro newborn rat preparation. *Eur. J. Neurosci.* **5,** 1226–1232.

31. Cazalets, J.-R., Borde, M., and Clarac, F. (1995) Localization and organization of the central pattern generator for hindlimb locomotion in newborn rat. *J. Neurosci.* **15,** 4943–4951.

32. Cohen, A. H. and Gans C. (1975) Muscle activity in rat locomotion: movement analysis and electromyography of the flexors and extensors of the elbow. *J. Morphol.* **146,** 177–196.

33. Gruner, J. A. and Altman, J. (1980) Swimming in the rat: analysis of locomotor performance in comparison to stepping. *Exp. Brain Res.* **40,** 374–382.

34. Nicolopoulos-Stournaras, S. and Iles, J. F. (1984) Hindlimb muscle activity during locomotion in the rat *(Rattus norvegicus)* (Rodentia:Muridae). *J. Zool. Lond.* **203,** 427–440.

35. Goudard, I., Orsal, D., and Cabelguen, J.-M. (1992) An electromyographic study of the hindlimb locomotor movements in the acute thalamic rat. *Eur. J. Neurosci.* **4,** 1130–1139.

36. Westerga, J. and Gramsbergen, A. (1993) Changes in the electromyogram of two major hindlimb muscles during locomotor development in the rat. *Exp. Brain Res.* **92,** 479–488

37. Kiehn, O. and Kjaerulff, O. (1996) Spatiotemporal characteristics of 5-HT and dopamine-induced rhythmic hindlimb activity in the *in vitro* neonatal rat. *J. Neurophysiol.* **75,** 1472–1482.

38. Nicolopoulos-Stournaras, S. and Iles, J. F. (1983) Motor neuron columns in the lumbar spinal cord of the rat. *J. Comp. Neurol.* **217,** 75–85.

39. Landmesser, L. (1977) The distribution of motoneurones supplying chick hind limb muscles. *J. Physiol.* **284,** 371–389.
40. Kudo, N. and Yamada, T. (1987) N-Methyl-D, L-aspartate-induced locomotor activity in a spinal cord-hindlimb muscles preparation of the newborn rat studied *in vitro*. *Neurosci. Lett.* **75,** 43–48.
41. Smith, J. C. and Feldman, J. L. (1987). *In vitro* brainstem-spinal cord preparations for study of motor systems for mammalian respiration and locomotion. *J. Neurosci. Methods* **21,** 321–333.
42. Sqalli-Houssaini, Y., Cazalets, J. R., and Clarac, F. (1991) A cooling/heating system for use with *in vitro* preparations: study of temperature effects on newborn rat rhythmic activities. *J. Neurosci. Methods* **39,** 131–139.
43. Sqalli-Houssaini, Y., Cazalets, J.-R., and Clarac, F. (1993a) Oscillatory properties of the central pattern generator for locomotion in neonatal rats. *J. Neurophysiol.* **70,** 803–813.
44. Kjaerulff, O., Barajon, I., and Kiehn, O. (1994) Sulphorhodamine-labelled cells in the neonatal rat spinal cord following chemically induced locomotor activity *in vitro*. *J. Physiol. (Lond.)* **478,** 265–273.
45. Magnuson, D. S. K., Schramm, M. J., and MacLean, J. N. (1995) Long-duration, frequency-dependent motor responses evoked by ventrolateral funiculus stimulation in the neonatal rat spinal cord. *Neurosci. Lett.* **192,** 97–100.
46. Cowley, K. C. and Schmidt, B. J. (1997) Regional distribution of the locomotor pattern-generating network in the neonatal rat spinal cord. *J. Neurophysiol.* **77,** 247–259.
47. Kremer, E. and Lev-Tov, A. (1997) Localization of the spinal network associated with generation of hindlimb locomotion in the neonatal rat and organization of its transverse coupling system. *J. Neurophysiol.* **77,** 1155–1170.
48. Cowley, K. C. and Schmidt, B. J. (1994) A comparison of motor patterns induced by N-methyl-D-aspartate, acetylcholine and serotonin in the *in vitro* neonatal rat spinal cord. *Neurosci. Lett.* **171,** 147–150.
49. Barthe, J. Y. and Clarac, F. (1997) Modulation of the spinal network for locomotion by substance P in the neonatal rat. *Exp. Brain Res.* **115,** 485–492.
50. Keifer, J., Vyas, D., and Houk, J. C. (1992) Sulforhodamine labeling of neural circuits engaged in motor pattern generation in the in vitro turtle brainstem-cerebellum. *J. Neurosci.* **12,** 3187–3199.
51. Barajon, I., Gossard, J. P., Hultborn, H. (1992) Induction of fos expression by activity in the spinal rhythm generator for scratching. *Brain Res.* **588,** 168–172.
52. Carr, P. A., Huang, A., Noga, B. R., and Jordan, L. M. (1995) Cytochemical characteristics of cat spinal neurons activated during fictive locomotion. *Brain Res. Bull.* **37,** 213–218.
53. Lev-Tov, A. and O'Donovan, M. J. (1995) Calcium imaging of motoneuron activity in the en-bloc spinal cord preparation of the neonatal rat. *J. Neurophysiol.* **74,** 1324–1334.
54. Grillner, S. and Zangger, P. (1979) On the central generation of locomotion in the low spinal cat. *Exp. Brain Res.* **34,** 241–261.
55. Deliagina, T. G, Orlovsky, G. N, and Pavlova, G. A. (1983) The capacity for generation of rhythmic oscillations is distributed in the lumbosacral spinal cord of the cat. *Exp. Brain Res.* **53,** 81–90.

56. Cowley, K. C., Schmidt, B. J. (1995) Effects of inhibitory amino acid antagonists on reciprocal inhibitory interactions during rhythmic motor activity in the in vitro neonatal rat spinal cord. *J. Neurophysiol.* **74,** 1109–1117.

57. Cheng, J., Stein, R. B., Jovanovic, K., Yoshida, K., Bennett, D. J., and Han, Y. (1998) Identification, localization, and modulation of neural networks for walking in the mudpuppy *(Necturus maculatus)* spinal cord. *J. Neurosci.* **18,** 4295–4304.

58. Ho, S. and O'Donovan, M. J (1993) Regionalization and intersegmental coordination of rhythm-generating networks in the spinal cord of the chick embryo. *J. Neurosci.* **13,** 1354–1371.

59. Cazalets, J.-R., Borde, M., and Clarac, F. (1996) The synaptic drive from the locomotor network to the motoneurons in newborn rat. *J. Neurosci.* **16,** 298–306.

60. Viala, D., Buisseret-Delmas, C., and Portal, J. J. (1988) An attempt to localize the lumbar locomotor generator in the rabbit using 2-deoxy-(^{14}C) glucose autoradiography. *Neurosci. Lett.* **86,** 139–143.

61. Kjaerulff, O., and Kiehn, O. (1996) Distribution of networks generating and coordinating locomotor activity in the neonatal rat spinal cord *in vitro:* a lesion study. *J. Neurosci.* **16,** 5777–5794.

62. Hochman, S., Jordan, L., and Schmidt, B. J. (1994) TTX-resistant NMDA receptor-mediated voltage oscillations in mammalian lumbar motoneurons. *J. Neurophysiol.* **72,** 2559–2562.

63. McLean, J. N., Hochman, S., and Magnuson, D. S. K. (1995) Lamina VII neurons are rhythmically active during locomotor-like activity in the neonatal rat spinal cord. *Neurosci. Lett.* **197,** 9–12.

64. Magnuson, D. S. K. and Trinder, T. C. (1997) Locomotor rhythm evoked by ventrolateral funiculus stimulation in the neonatal rat spinal cord *in vitro. J. Neurophysiol.* **77,** 200–206.

65. Kiehn, O., Johnson, B., and Raastad, M. (1996) Plateau properties in mammalian spinal interneurons during transmitter-induced locomotor activity. *Neuroscience* **75,** 263–273.

66. Dubayle, D., and Viala, D. (1996) Localization of the spinal respiratory rhythm generator by an *in vitro* electrophysiological approach. *Neuroreport* **7,** 1175–1180

67. Dubayle, D. and Viala, D. (1998) Effects of CO_2 and pH on the spinal respiratory rhythm generator *in vitro. Brain Res. Bull.* **45,** 83–87

68. Cazalets, J.-R., Squalli-Houssaini, Y., and Clarac, F. (1994). GABAergic inactivation of the central pattern generators for locomotion in isolated neonatal rat spinal cord. *J. Physiol.* **474,** 173–181.

69. Bracci, E. (1996) Localization of rhythmogenic networks responsible for spontaneous bursts induced by strychnine and bicuculline in the rat isolated spinal cord. *J. Neurosci.* **16,** 7063–7076.

70. Shapiro, BI., Wang, C. M., and Narahashi, T. (1974) Effects of strychnine on ionic conductances of squid axon membrane. *J. Pharmacol. Exp. Ther.* **188,** 66–76.

71. Dale, N. (1995) Experimentally derived model for the locomotor pattern generator in the *Xenopus* embryo. *J. Physiol.* **489,** 489–510.

72. Debarbieux, F., Brunton, J., and Charpak, S. (1998) Effect of bicuculline on thalamic activity: a direct blockade of IAHP in reticularis neurons. *J. Neurophysiol.* **79,** 2911–2918.

73. Cazalets, J. R., Bertrand, S., Sqalli-Houssaini, Y., and Clarac, F. (1998) GABA-ergic control of spinal locomotor networks in the neonatal rat. *Ann. Acad. Sci.* NY, **860,** 168–180.

74. Russel, D. F. and Wallen, P. (1983) On the control of myotonal motoneurons during fictive swimming in the lamprey spinal cord in vitro. *Acta Physiol. Scand.* **117,** 161–170.

75. Orsal, D, Perret, C., and Cabelguen, J. M. (1986) Evidence of rhythmic inhibitory synaptic influences in hindlimb motoneurons during fictive locomotion in the thalamic cat. *Exp. Brain Res.* **64,** 217–224.

76. Shefchik, S. J. and Jordan, L. M. (1985) Excitatory and inhibitory postsynaptic potentials in α-motoneurons produced during fictive locomotion by stimulation of the mesencephalic locomotor region. *J. Neurophysiol.* **53,** 1345–1355.

77. Kjaerulff, O. and Kiehn, O. (1997) Crossed rhythmic synaptic input to motoneurons during selective activation of the contralateral spinal locomotor network. *J. Neurosci.* **17,** 9433–9447.

78. Bonnot, A., Morin, D., and Viala, D. (1998) Genesis of spontaneous rhythmic motor patterns in the lumbosacral spinal cord of neonatal mouse. *Dev. Brain Res.,* **108,** 89–99.

79. MacLean, J., Schmidt, J., and Hochman, S. (1997) NMDA receptor activation triggers voltage oscillations, plateau potentials and bursting in neonatal rat lumbar motoneurons *in vitro. Eur. J. Neurosci.* **9,** 2702–2711.

80. Wheatley, M., Jovanovic K., Stein, R. B., and Lawson, V. (1994) The activity of interneurons during locomotion in the *in vitro* necturus spinal cord. *J. Neurophysiol.* **71,** 2025–2032.

81. Bertrand, S. and Cazalets, J.-R. (1998) Postinhibitory rebound during locomotor-like activity in neonatal rat motoneurons in vitro. *J. Neurophysiol.* **79,** 342–351.

82. Miller, S. and Scott, P. D. (1977) The spinal locomotor generator. *Exp. Brain Res.* **30,** 387–403.

83. Bussel B., Roby-Brami A., Yakovleff A., and Bennis N. (1989) Late flexion reflex in paraplegic patients. Evidence for a spinal stepping generator. *Brain Res. Bull.* **22,** 53–56.

84. Feraboli-Lohnherr, D., Orsal, D., Yakovleff, A., Gimenez y Ribotta, M., and Privat, A. (1997) Recovery of locomotor activity in the adult chronic spinal rat after sublesional transplantation of embryonic nervous cells: specific role of serotonergic neurons. *Exp. Brain Res.* **113,** 443–454.

85. Gimenez y Ribotta, M., Orsal, D., Feraboli-Lohnherr, D., and Privat, A. (1998) Recovery of locomotion following transplantation of monoaminergic neurons in the spinal cord of paraplegic rats. *Ann. NY Acad. Sci.,* **860,** 393–411.

86. Cheng H., Cao Y., and Olson L. (1996) Spinal cord repair in adult paraplegic rats: partial restoration of hind limb function. *Science* **273,** 510–513.

5

Strategies for Spinal Cord Repair
Clues from Neurodevelopment

John D. Steeves and Wolfram Tetzlaff

1. INTRODUCTION

We conceive of three essential goals that must be achieved for effective functional repair after traumatic injury of the adult spinal cord. These goals are as follows: (1) to protect neural tissue after injury and limit the secondary damage and death of those cells that "escaped" the initial neurotrauma; to (2) to provide extrinsically or stimulate the expression of factors or substrates that facilitate axonal regrowth; and (3) to block the expression of factors inherent to the adult central nervous system (CNS) that inhibit neural repair. Furthermore, it cannot be overemphasized that it is essential to provide appropriate sensorimotor activity through active rehabilitation both to enhance plasticity within surviving circuits and to consolidate any induced anatomic repair or regeneration.

In the following sections, we address these spinal cord injury (SCI) repair goals, with some emphasis on the contributions from developmental neurobiology. Because of space considerations, we must necessarily limit most of the illustrations to recent findings concerning descending brainstem-spinal neurons, only one of the candidates for any repair strategy. Of course, corticospinal pathways, ascending spinal projections, and propriospinal axons are undoubtedly as important to functional recovery.

2. NEURAL CELL DEATH AFTER INJURY

The SCI site contains large numbers of both apoptotic and necrotic neurons and glia. Typically, the center of an SCI is prominently characterized by necrotic death, with the surrounding tissue forming a penumbra, which may be analogous to the area surrounding a focal ischemic lesion, where apoptotic death predominates *(1)*. Several mechanisms may contribute to this pattern of destruction, including anoxia, excitotoxicity, free radical generation, inflammatory responses, macrophage invasion, and proteases released from neighboring necrotic cells,. The secondary damage induced by these processes contributes to the formation of cyst cavities *(2)*, which can further impair surviving axonal

From: *Neurobiology of Spinal Cord Injury*
Edited by: R. G. Kalb and S. M. Strittmatter © Humana Press Inc., Totowa, NJ

function and also limit any subsequent repair. For further discussion of the mechanisms of cell death, we refer the reader to other chapters in this volume.

3. FACTORS AND MECHANISMS PROMOTING AXONAL REGNERATION

A myriad of developmental factors and mechanisms could contribute to a successful outcome for functional regeneration after SCI. Recent evidence suggests that many descending brainstem-spinal neurons (e.g., reticulospinal and rubrospinal cells) do not die after SCI, but nevertheless undergo substantial atrophy that is not clearly detectable histologically *(3)*. Although rubrospinal neurons survive in an atrophied state, local infusion of specific growth factors, such as brain-derived neurotrophic factor (BDNF), prevents the atrophied appearance and also upregulates gene expression patterns associated with axon growth *(4,5)*. BDNF treatment effectively reverses the atrophy of axotomized rubrospinal neurons even several months after injury *(3)*. Thus, the long-term survival of CNS projection neurons after SCI suggests that future treatment of both acute and chronic spinal cord injuries is a distinct possibility.

3.1. Regeneration-Associated Gene Expression Within Axotomized Neurons

Since brainstem-spinal neurons survive after an SCI, it is important to characterize the known components of axonal development. Some proteins and their related genes, such as growth-associated protein 43 (GAP-43) *(6)* and Tα1-tubulin *(7)*, have already been implicated in axonal outgrowth during development, downregulating their expression after axonal growth is completed. These same genes and proteins have also been observed to increase their expression in response to axonal injury within the adult CNS *(4,8,9)*. Thus, increased expression of these and other genes may be required for axonal regeneration. Elucidating the genetic programs necessary for axonal development is thus an important step for designing future in vivo gene or biochemical therapies for driving regeneration.

These genes will be referred to collectively as regeneration-associated genes (RAGs). They include transcription factors such as c-jun *(10)*, cytoskeletal proteins, especially the developmentally regulated tubulin isotype *(11,12)*, microtubule-associated proteins *(13)*, and the fast axonally transported growth-associated proteins, e.g., GAP-43 *(14–16)*. In addition, changes in the expression of cell adhesion molecules (CAMs), e.g., N-CAM, L1, and TAG-1 *(17)*, neurotrophic factors *(18–20)*, and cytokines *(21)* are likely to play an important role in axonal regeneration as well.

The capacity of mammalian CNS neurons to regenerate their axons varies considerably with the type and age of the neuron *(22,23)* but invariably correlates with a cell body response that includes the expression of RAGs after axotomy *(4,5)*. The importance of "intrinsic neuronal properties" for axonal regeneration is highlighted by the substantial growth of axons from embryonic hippocampal or adult dorsal root ganglion cells (DRGs) transplanted into the CNS *(24)*. The importance of the intrinsic neuronal capacity to mount an injury response is further supported by the classic finding that the central axon of DRGs does not regenerate into peripheral nerve (PN) transplants when inserted into the dorsal column, unless the peripheral process has been axotomized at the same time *(25)*. Axotomy of the peripheral process, but not the central axon, induces a vigorous RAG expression response in the parent DRGs *(26)*. The central role of neuronal gene expression is also evident in the correlation of GAP-43 expression with the ability of CNS neurons to regenerate successfully into PN transplants *(5,27)*. For instance, some rubrospinal neurons are able to regenerate into PN transplants if axotomized relatively close to the cell body of origin (i.e., at the cervical, but not thoracic, level of the spinal cord) *(4,5,28)*. We have shown that this correlates with a transient expression of RAGs found after cervical, but not after thoracic axotomy *(4)*. Similarly, other CNS neurons increase their expression of RAGs, but only after an axonal injury very close to their cell bodies, which means within 6 mm for retinal ganglion cells *(29)* or as little as 200 µm in the case of corticospinal neurons *(4)*. Thus, despite the well-known permissive nature of PN transplants, regeneration is only observed if the PN implant is provided to a CNS neuron with an appropriate regenerative response.

3.2. Trophic Factor Stimulation of Regenerative Outgrowth

Other animal studies have suggested that some trophic factors may also have roles in stimulating the regeneration of axons *(30–34)*. For example, when combined with an antibody (IN-1) that blocks myelin inhibitory proteins, neurotrophin-3 (NT-3) facilitates increased axonal growth (or sprouting) in some axotomized corticospinal fibers *(30)*. Thus, it is important to identify which specific growth factors will influence the differentiation or regeneration (i.e., axonal outgrowth) of each CNS neuronal population having a spinal projection (i.e., each specific neuronal phenotype). Various trophic factors maintain the survival of developing neurons and may also be important for the survival and regeneration of injured adult neurons. They include nerve growth factor (NGF), BDNF, NT-3, NT-4, ciliary neurotrophic factor (CNTF), epidermal growth factor (EGF), fibroblast growth factors (FGFs), glial-derived neurotrophic factor (GDNF) and other members of the transforming growth factor family, platelet-derived growth factor (PDGF), and insulin-like growth factor (IGF).

For example, we have recently shown that injured cervical rubrospinal neurons, which express full-length trkB neurotrophin receptors *(5)*, can be treated with BDNF infusion in the vicinity of their cell bodies to prevent their atrophy, sustain RAG expression, and increase by three-fold the number of regenerating rubrospinal axons *(3,5)*. Interestingly, the cell body application of BDNF to thoracic axotomized rubrospinal neurons successfully stimulated an appropriate cell body response, which permitted axonal regeneration into thoracic PN transplants for the first time, *(3)*. By allowing these regenerating axons to pick up a retrograde marker within the PN graft, we subsequently demonstrated, on a cell-by-cell basis, that only those neurons that maintained high levels of RAG expression (c-jun, GAP-43, Tα1) also regenerated (Fan and Tetzlaff, unpublished observations). Thus, trophic factor application or transplantation of a cellular source of growth factors to the parent cell body can promote regeneration of the axonal process *(5,35)*. In our hands, local application of BDNF to the spinal cord was ineffective at concentrations that were able to stimulate robust RAG expression when applied in the vicinity of the rubrospinal neuronal cell bodies. This may be due to the abundance of truncated trkB receptors *(36)*, the differential subcellular distribution of neurotrophin receptors, or the limited retrograde signaling from the axon to the cell body *(37)*. These limitations may not apply when other growth factors are infused directly into the injured spinal cord *(30–34)*.

3.2.1. Growth Factor Effects on Brainstem-Spinal Neuron Survival and Neurite Outgrowth

For each brainstem-spinal phenotype, it is necessary to differentiate the components that facilitate brainstem-spinal neuron survival and/or axonal outgrowth. This is important because even though growth factors have preferential high-affinity receptors, they can also bind nonspecifically. Furthermore, the influence of a growth factor on adjacent glial cells (i.e., indirect or paracrine effects) may be equally as important as direct neuronal effects. To address these questions, we have devised in vitro assays, based on prior in vivo retrograde labeling from the developing spinal cord that permits rapid screening of specific trophic factors for their influence on particular brainstem-spinal populations *(38)*. We chose to start with the developing chick embryo because the organization of descending projections is similar to mammalian brainstem-spinal pathways *(39,40)*, as well as for ease of access during development to identify brainstem-spinal neurons with a retrograde tracer (e.g., the carbocyanine dye DiI) implanted within the cervical cord prior to culturing. Effects on neuronal survival or neurite outgrowth were respectively measured using either a dissociated brainstem cell culture or explantation of a brainstem-spinal nucleus (e.g., vestibulospinal nucleus).

In vitro assays, derived from whole brainstems, are useful for identifying factors with more widespread effects on brainstem-spinal neurons *(38)*. However, these mass dissociated cultures can mask specific influences on discrete neuronal populations. For example, our data suggests that NT-3 had significant influences on isolated reticulospinal neurons that were masked (i.e., not evident) in dissociated cell cultures derived from whole brainstems. Perhaps more importantly, it is difficult to differentiate or quantify growth factor effects on neurite outgrowth using a dissociated cell culture. We were able to examine both survival (using dissociated cultures from a specific brainstem-spinal nucleus) and neurite outgrowth (using explant cultures from an identified brainstem-spinal nucleus). We found that FGF-2 effectively promoted both vestibulospinal neuron survival and neurite outgrowth from vestibulospinal neurons. In contrast, FGF-1 was ineffective at promoting vestibulospinal neuron survival but was equally as effective as FGF-2 in stimulating neurite outgrowth. Conversely, FGF-9 enhanced vestibulospinal neuron survival but was less effective in promoting neurite outgrowth than FGF-1 or FGF-2. These data suggest that differential effects on the survival and outgrowth of distinct CNS populations can be distinguished using in vitro assays *(38)*.

More detailed knowledge of the trophic factor influences on a specific neuronal population may benefit the development of more effective in vivo therapies to promote repair of the damaged CNS. Using in vivo approaches, several investigators *(30–34)* have already begun to use specific trophic factors to stimulate improved repair after a spinal cord injury. For example, after a spinal cord hemisection, application of NT-3 at the spinal injury site enhanced reticulospinal and rubrospinal axon outgrowth but did not promote raphe-spinal, or vestibulospinal regeneration *(33)*. Most interesting in the context of the above in vitro findings was the recent report that FGF-1 application to the severed adult spinal cord facilitated increased brainstem-spinal and propriospinal regeneration, when combined with directed peripheral nerve bridges across the site of injury *(31)*. Our in vitro finding also suggests that FGF-1 has a distinct role in promoting brainstem-spinal axonal outgrowth. As more growth factors are identified and characterized, the application of appropriate factor(s) after an adult spinal cord injury may benefit repair of severed brainstem-spinal projections, especially when combined with treatments to neutralize axonal growth inhibitors *(34,41)*.

3.3. Cell Adhesion Molecules, Semaphorins, Netrins, and Ephrins

CAMs and axonal guidance molecules play an important role in the development of the peripheral and central nervous systems that has been subject to extensive reviews *(42)*. The expression of CAMs changes considerably in regenerating peripheral nerves *(43,44)*. L1, a member of the immunoglobulin superfamily, has been observed to increase in regenerating PNs *(41,45,46)*. More than one binding mechanism seems to account for the broad range of functions of L1, including homophilic binding, binding to integrin or axogenin 1, and FGF receptor dimerization and activation. Antibodies to FGF receptors or dominant negative FGF receptors inhibit the L1 stimulation of neurite outgrowth *(42)*. In vivo, L1 knock-out mice show hypoplasia of the corticospinal tract, which also fails to decussate in the pyramid *(47)*. While the development of peripheral nerves in these mice shows only discrete impairment of axon ensheathment by Schwann cells *(48)*, it remains to be shown whether regeneration after peripheral nerve injury is hampered or whether other CAMs compensate for the deficit. In spinal cord projection neurons of lower vertebrates, the expression of L1 corresponds well with their regenerative propensity *(49)*. For example, zebrafish brainstem-spinal neurons with high levels of L1, N-CAM, and GAP-43 regenerate successfully, whereas other descending cells (e.g., Mauthner cells), which lack one or more of these molecules, exhibit poor regeneration *(49)*. Thus, the combined expression of RAGs (see above) and growth-promoting cell adhesion molecules may correlate with better regenerative success. The functional distinction between the aforementioned CAMs and the families of axonal guidance molecules (semaphorins, netrins, and ephrins) may be somewhat artificial. Both play a crucial role in axonal guidance during neural development. However, much less is known about the latter group during axonal regeneration. Once again, we refer the reader to other chapters in this volume.

4. INHERENT FACTORS AND MECHANISMS OF THE ADULT CNS INHIBITING FUNCTIONAL REGENERATION

The poor regenerative capacity of the mature CNS of higher vertebrates may also be due to the presence of growth-inhibitory factors within the CNS, a situation reciprocal to that of the developing CNS *(50)*. Since many of the events underlying adult axonal regeneration are a recapitulation of axonal development, investigation of the regenerative capacity of the injured embryonic spinal cord has helped to identify some of the developmental events impinging on functional regeneration by adult CNS axons.

4.1. Stabilization of Axonal Projections

The stabilization of axon sprouting (i.e., prevention of subsequent collateral branching) after functional axonal connections have been established is a role that has been suggested for CNS myelin during development *(41)*. This is suggested by the temporal correlation between the developmental onset of myelination and the completion of primary axonal growth. For example, the onset of myelination in the developing chick spinal cord always begins on embryonic (E) 13 after the completion (on E11) of axonal projections from the brainstem to even the most caudal targets within the lumbosacral spinal cord *(51)*. In the mammalian visual system, oligodendrocytes appear in the hamster optic tract and rostral superior colliculus on postnatal d 5 and spread throughout most of the stratum opticum by postnatal d 9; once again, myelination begins after the retinotectal axons have innervated the superior colliculus *(52)*.

Further evidence for a role of myelin in the stabilization of axonal projections comes from studies indicating that neuronal plasticity involving axonal sprouting is inversely related to the degree of CNS myelination. Axonal sprouting within the adult CNS has been reported to occur more readily in lightly myelinated regions such as the olfactory system *(53)*, the hippocampus *(54)*, the septum *(55)*, the cerebellum *(56)*, and the substantia gelatinosa of the spinal cord *(57)*. In contrast, regions that are heavily myelinated in the adult CNS only show evidence for axonal sprouting prior to myelin formation. For example, in the highly myelinated trigeminal nuclear complex of the rat, lesion-induced sprouting can be observed at fetal stages when myelin is not present *(58)*, but rarely postnatally when myelin becomes associated with these axons *(59)*. Likewise, the corticospinal tract of the spinal cord becomes severely restricted in its ability to sprout after myelination *(60)*.

Although very little is known about morphologic changes of primary axonal projections within the adult mammalian CNS, an impressive degree of functional plasticity of the cerebral cortical gray matter has been demonstrated *(61)*. Changes in synaptic morphology and density have also been demonstrated in the hippocampus after long-term potentiation or kindling *(62,63)*, and in animals exposed to different environments *(64)*. It is an intriguing possibility that the low level of myelination within these regions is a facilitating factor in their ability to engage in such striking plasticity. It would follow, therefore, that either unmyelinated or lightly myelinated axons within in the adult CNS may be better candidates for enhanced plasticity and/or regeneration.

4.2. Influence of CNS Myelin on Axonal Regeneration Within Lower Vertebrates and the Developing CNS of Higher Vertebrates

Studies of developing chick embryos have indicated that their ability to recover functionally from SCI is lost at a critical point during the 21-d *in ovo* developmental period. Severing the thoracic spinal cord, prior to E13, results in relatively complete neuroanatomic axonal regeneration and functional recovery *(39,40,65)*. Conversely, transection after E13 results in sparse axonal regrowth and little or no functional recovery; upon hatching, an animal is as paralyzed as a bird or mammal transected as an adult *(39,40,65,66)*. Thus, the transition from a permissive to a restrictive period for CNS repair correlates with the developmental onset of myelination in the embryonic chick spinal cord *(67)*.

To assess a possible inhibitory role of myelin in the regeneration of brainstem-spinal projections, an immunologic method for selectively delaying the developmental onset of avian myelination has been examined. On E9–E11, a direct intraspinal injection of serum complement proteins along with galactocerebroside (GalC) or 04 (i.e., myelin surface-specific) antibodies targets newly differentiated premyelinating oligodendrocytes for destruction. This treatment delays the normal onset of spinal cord myelination [i.e., E17 *(67,68)*] until well into the restrictive period for spinal cord repair. The subsequent transection of such a myelin-suppressed cord resulted in substantial brainstem-spinal axonal regeneration and functional behavioral recovery *(67)*. This was in sharp contrast to normally myelinated control embryos, also transected on E15, which never exhibited axonal regeneration and were completely incapable of voluntary locomotion after hatching. These findings demonstrated that developmental myelin suppression extends the permissive period for functional CNS repair.

Similar data have been reported for the embryonic rat *(69)* and the neonatal Brazilian opossum *(70)*. Once again, the transition from a permissive to restrictive period for axonal regeneration corresponded to the developmental appearance of CNS myelin. Furthermore, if myelination is delayed with X-irradiation of the neonatal rat spinal cord, cortico spinal tract (CST) axons regenerated for several millimeters following spinal cord transection (2 weeks after birth), whereas normally myelinated animals of the same age exhibited little or no regeneration *(71)*. Likewise, treatment of the injured neonatal opossum spinal cord with the myelin-specific IN-1 monoclonal antibody facilitated improved functional axonal regeneration *(70)*. IN-1 was raised against the partially purified neurite growth-inhibitory proteins NI-35/250 found in myelin and selected for its ability to neutralize the inhibitory affects of myelin on neurite growth *(72)*.

Functional regeneration within the severed or injured spinal cord of the primitive lamprey is also well documented. One of the more distinguishing characteristics of the lamprey CNS is the lack of compact myelin *(73)*. Teleosts (bony fish) also show a remarkable degree of axonal regeneration after CNS injury *(74,75)*, as do anurans (frogs), most notably as tadpoles, prior to metamorphosis into adults *(76)*. These studies suggest an evolutionary transition in the composition of vertebrate myelin *(76)*. Nevertheless, the extensive repair of lower vertebrate CNS injuries may not only be due to differences in CNS myelin structure. Since fish and frogs exhibit continual (indeterminate) body growth throughout their life, many of the intrinsic and extrinsic CNS conditions important for maintaining the accompanying essential axonal growth may also endure and thus facilitate axonal regeneration. Thus, continued comparative studies could identify many of the important elements for developing effective repair therapies for CNS injuries in higher vertebrates, including humans *(73)*.

3.3. Myelin Suppression and Axonal Regeneration After Injury of the Mature CNS

The preceding experimental evidence on the inhibitory nature of myelin toward axonal regeneration in the developing CNS supports the idea that similar strategies might also facilitate functional regeneration in the injured adult CNS of birds and mammals. For example, the neuroanatomic regeneration and functional recovery observed in myelin-suppressed embryonic chicks prompted the development of a similar protocol for the mature avian spinal cord. A similar immunologic treatment (infused intraspinally with an osmotic pump over 14) also resulted in transient demyelination within the mature rat cord *(77)*. The myelin suppression extended rostrocaudally over several spinal segments away from the infusion site and also appears to involve activated macrophages or microglia *(78)*.

More significantly, the immunologic protocol in birds was initiated after (not prior to) transection of the thoracic spinal cord. Neuroanatomic regeneration of avian brainstem-spinal projections was assessed 2–4 weeks after injury. In the mature avian spinal cord, myelin suppression facilitated axonal regeneration by approximately 6–19% of the severed brainstem-spinal neurons, and this axonal regrowth was accompanied by functional synaptogenesis *(40)*. Although hatchling chicks that had undergone spinal transection and subsequent myelin disruption did not show any voluntary signs of locomotor recovery, focal electrical stimulation of a brainstem locomotor region (in a decerebrate animal preparation) evoked rhythmic motor activity in the right and left legs *(40).*

Other studies have observed axonal regeneration of the spinal cord following neutralization of myelin-associated neurite growth inhibitors. IN-1 antibody treatment of young adult rats with a transected CST resulted in massive sprouting at the lesion site, and fine axons and fascicles were observed 7–11 mm caudal to the lesion site *(79).* Following IN-1 treatment of young adult rats with an over-hemisection of the spinal cord, serotonergic brainstem-spinal and CST axons also regenerated, and behavioral analysis suggested that the improved recovery of some reflexes might be due to the regrowth of these descending projections *(80).*

More recent studies of adult SCIs indicated that transient focal demyelination of the rat spinal cord using serum complement proteins and myelin-specific, complement-fixing (IgG) antibodies (e.g., GalC) also facilitates axonal regeneration *(77,78).* Ultrastructural analysis also revealed regenerating growth cones that were restricted to demyelinated regions of the injured cord in numbers that correlated with the number of axons severed *(77).* Using retrograde tract tracing techniques, transient immunologic demyelination after an SCI facilitates some rubrospinal axon regeneration when compared with control treated animals *(78).* Preliminary evidence also suggests that delaying immunologic demyelination treatment for as much as 2 months after a spinal injury can still facilitate regeneration of some (but not all) brainstem-spinal projections (Dyer and Steeves, unpublished observations). Once again, this underscores the need to understand the specific regeneration requirements for each particular neuronal phenotype affected by a spinal injury. The potential importance of activated macrophages or microglia to the beneficial effects of the immunologic demyelination cannot be ignored. Recent results have suggested that transplantation of peripherally activated macrophages has beneficial effects on functional spinal cord regeneration *(81).*

4.4. Astrocytic Responses to CNS Injury and Their Influences on Axonal Regeneration

After SCI, other studies implicate a select population of astrocytes, known as reactive astrocytes, as contributing to an inhibitory environment within the injured CNS *(24,50,82,83).* After a penetrating CNS injury, there is rapid local inflammation and edema, with a plethora of cytokines and chemokines being released at the injury site, all of which have pleurotrophic effects on many of the surrounding glial cells *(84–86).* There is a subsequent proliferation and hypertrophy of astrocytes around the injury site, which is one of the characteristic gliotic injury responses in all mammals (2). Specifically, reactive astrogliosis is denoted by increased immunoreactivity to glial fibrillary acidic protein (GFAP), a distinct cellular marker of astrocytes *(85).* The increased number of astrocytes may be due to migration or cell division *(87).* Reactive astrocytes have been suggested to form an astroglial scar that acts as a physical and/or chemical barrier to axonal regeneration *(24,81–83)*. Alternatively, astrogliosis may be a CNS response to reduce the magnitude of SCI by walling off the CNS from further damage, but also raising a barrier to regenerating axons *(83)*.

Whether astrocytes are inhibitory or supportive for CNS repair has been difficult to resolve and may depend on the developmental state of the CNS. For example, it has been noted that immature (or embryonic) astrocytes can be a supportive substrate for axonal growth by adult retinal ganglion cells, whereas adult astrocytes or reactive astrocytes were inhibitory to axonal growth *(83)*. Throughout development, the expression patterns change for several astrocyte-related molecules [e.g., tenascin, phosphacan, heparin sulphate proteoglycan (HSPG), chondroitin sulphate proteoglycan (CSPG), dermatan/keratan sulphate proteoglycan (KSPG) *(83,87–89)*.]. Tenascin appears to be a permissive molecule for axonal outgrowth during early stages of CNS development, but it may have a nonpermissive or repulsive role later in ontogeny *(90)*. Likewise, we have noted an initially high ratio of HSPG to CSPG expression during early stages of spinal cord development (the permissive period for functional regeneration), which subsequently reverses during later stages of development (the restrictive period for regeneration) *(89)*. Recent findings suggest that in the adult CNS, tenascin, CSPG, and KSPG may be inhibitory, whereas HSPG is a supportive substrate for neuronal differentiation *(24,81–83,89)*. Chondroitinase and keratanase enzyme pretreatment cleaves specific glycosaminoglycan side chains from CSPG or KSPG and reduces the inhibition of CSPG and KSPG for axonal growth in vitro *(89,91)*. Nevertheless, the inhibitory nature of phosphacan is only slightly reduced by such cleavage, suggesting that the core glycoprotein retains substantial inhibitory influences on axonal outgrowth.

Astrocytes can potentially express molecules that are both growth permissive and growth inhibitory at the same time. Evidence suggests that not only is the expression of inhibitory and permissive molecules important but also that the precise ratio of these molecules within the extraneuronal environment may be important. For example, in vitro rat forebrain neurons, chicken retinal ganglions cells and DRGs are able to grow on high CSPG substrates if they grow along an increasing ratio of CSPG to laminin as opposed to a single stripe of CSPG and laminin *(92)*. As mentioned above, we have noted a shift in proteoglycan expression from the more permissive HSPG form to the more inhibitory CSPG during the transition from the permissive to restrictive periods for embryonic avian spinal cord regeneration *(89)*. In vivo, only when there is a lack of an astrogliotic response, with the absence of CSPG expression, do a subset of DRG neurons have long neurite growth after transplantation in the cortex *(24)*. Since astrocytes are responsible for the expression of many of these inhibitory molecules *(83)*, the creation of a more permissive environment for axonal regeneration might be achieved by the selective removal of astrocytes from the site of injury. Thus, the changes in the expression of proteoglycans as potential inhibitors of CNS regeneration are very compelling, yet it is important to replicate these findings with candidate neurons other than DRG cells, such as axotomized brainstem-spinal neurons.

5. CONCLUDING REMARKS

Most researchers agree it is unlikely that any single therapeutic intervention will facilitate complete functional repair whether it promotes neuron survival, stimulates axonal outgrowth, or transiently blocks inhibitory signals. Furthermore, delicately altering the balance between those factors promoting regeneration with those endogenous mechanisms inhibiting adult CNS repair may be sufficient to drive recovery in a functionally positive direction. Finally, determining the spatial and temporal organization of experimental therapies is critical to designing an integrated approach to neurotrauma recovery. The rapidly expanding information regarding CNS development will continue to provide valuable information for creating the necessary conditions within the injured adult CNS that are more conducive to functional repair, perhaps in part by approximating the favorable growth state of initial development.

REFERENCES

1. Choi, D. W. and Gage, F. H. (1996) Disease, transplantation and regeneration—editorial overview. *Curr. Opin. Neurobiol.* **6**, 635–637.
2. Dusart, I. and Schwab, M. E. (1994) Secondary cell death and the inflammatory reaction after dorsal hemisection of the rat spinal cord. *Eur. J. Neurosci.* **6**, 712–724.
3. Kobayashi, N. (1998) Neurotrophins and the neuronal response to axotomy. *Ph.D. Thesis, University of British Columbia.*
4. Tetzlaff, W., Kobayashi, N. R., Giehl, K. M., Tsui, B. J., Cassar, S. L., and Bedard, A. M. (1994) Response of rubrospinal and corticospinal neurons to injury and neurotrophins. *Prog. Brain Res.* **103**, 271–286.
5. Kobayashi, N. R., Fan, D. P., Giehl, K. M., Bedard, A. M., Wiegand, S. J., and Tetzlaff, W. (1997) BDNF and NT-$4/5$ prevent atrophy of rat rubrospinal neurons after cervical axotomy, stimulate GAP-43 and Talphal-tubulin mRNA expression, and promote axonal regeneration. *J. Neurosci.* **17**, 9583–9595.
6. Skene, J. H. (1989) Axonal growth-associated proteins. *Annu. Rev. Neurosci.* **12**, 127–156.
7. Miller, F. D., Tetzlaff, W., Bisby, M. A., Fawcett, J. W., and Milner, R. J. (1989) Rapid induction of the major embryonic alpha-tubulin mRNA, T alpha 1, during nerve regeneration in adult rats. *J. Neurosci.* **9**, 1452–1463.
8. Tetzlaff, W., Alexander, S. W., Miller, F. D., and Bisby, M. A. (1991) Response of facial and rubrospinal neurons to axotomy: changes in mRNA expression for cytoskeletal proteins and GAP-43. *J. Neurosci.* **11**, 2528–2544.
9. Fournier, A. E. and McKerracher, L. (1997) Expression of specific tubulin isotypes increases during regeneration of injured CNS neurons, but not after the application of brain-derived neurotrophic factor (BDNF). *J. Neurosci.* **17**, 4623–4632.

10. Herdegen, T., Skene, P., and Bahr, M. (1997) The c-Jun transcription factor—bipotential mediator of neuronal death, survival and regeneration. *Trends Neurosci.* **20,** 227–231.

11. Gloster, A., Wu, W., Speelman, A., et al. (1994) The T alpha 1 alpha-tubulin promoter specifies gene expression as a function of neuronal growth and regeneration in transgenic mice. *J. Neurosci.* **14,** 7319–7330.

12. Bisby, M. A. and Tetzlaff, W. (1992) Changes in cytoskeletal protein synthesis following axon injury and during axon regeneration. *Mol. Neurobiolo.* **6,** 107–123.

13. Ernfors, P., Rosario, C. M., Merlio, J. P., Grant, G., Aldskogius, H., and Persson, H. (1993) Expression of mRNAs for neurotrophin receptors in the dorsal root ganglion and spinal cord during development and following peripheral or central axotomy. *Brain Res. Mol. Brain Res.* **17,** 217–226.

14. Aigner, L. and Caroni, P. (1993) Depletion of 43-kD growth-associated protein in primary sensory neurons leads to diminished formation and spreading of growth cones. *J. Cell Biol.* **123,** 417–429.

15. Aigner, L., Arber, S., Kapfhammer, J. P., et al. (1995) Overexpression of the neural growth-associated protein GAP-43 induces nerve sprouting in the adult nervous system of transgenic mice. *Cell* **83,** 269–278.

16. He, Q., Dent, E. W., and Meiri, K. F. (1997) Modulation of actin filament behavior by GAP-43 (neuromodulin) is dependent on the phosphorylation status of serine 41, the protein kinase C site. *Jo. Neurosci.* **17,** 3515–3524.

17. Jung, M., Petrausch, B., and Stuermer, C. A. (1997) Axon-regenerating retinal ganglion cells in adult rats synthesize the cell adhesion molecule L1 but not TAG-1 or SC-1. *Mol. Cell. Neurosci.* **9,** 116–131.

18. Koliatsos, V. E., Price, D. L., Gouras, G. K., Cayouette, M. H., Burton, L. E., and Winslow, J. W. (1994) Highly selective effects of nerve growth factor, brain-derived neurotrophic factor, and neurotrophin-3 on intact and injured basal forebrain magnocellular neurons. *J. Comp. Neurol.* **343,** 247–262.

19. Kobayashi, N. R., Bedard, A. M., Hincke, M. T., and Tetzlaff, W. (1996) Increased expression of BDNF and trkB mRNA in rat facial motoneurons after axotomy. *Eur. J. Neurosci.* **8,** 1018–1029.

20. Grothe, C., Meisinger, C., Hertenstein, A., Kurz, H., and Wewetzer, K. (1997) Expression of fibroblast growth factor-2 and fibroblast growth factor receptor 1 messenger RNAs in spinal ganglia and sciatic nerve: regulation after peripheral nerve lesion. *Neuroscience* **76,** 123–135.

21. Murphy, P. G., Grondin, J., Altares, M., and Richardson, P. M. (1995) Induction of interleukin-6 in axotomized sensory neurons. *J. Neurosci.* **15,** 5130–5138.

22. Dusart, I., Airaksinen, M. S., and Sotelo, C. (1997) Purkinje cell survival and axonal regeneration are age dependent: an in vitro study. *J. Neurosci.* **17,** 3710–3726.

23. Fawcett, J. W. (1997) Astrocytic and neuronal factors affecting axon regeneration in the damaged central nervous system. *Cell Tissue Res.* **290,** 371–377.

24. Davies, S. J., Fitch, M. T., Memberg, S. P., Hall, A. K., Raisman, G., and Silver, J. (1997) Regeneration of adult axons in white matter tracts of the central nervous system. *Nature* **390,** 680–683.

25. Richardson, P. M. and Issa, V. M. (1984) Peripheral injury enhances central regeneration of primary sensory neurones. *Nature* **309,** 791–793.
26. Schreyer, D. J. and Skene, J. H. (1993) Injury-associated induction of GAP-43 expression displays axon branch specificity in rat dorsal root ganglion neurons. *J. Neurobiol.* **24,** 959–970.
27. Vaudano, E., Campbell, G., Anderson, P. N., et al. (1995) The effects of a lesion or a peripheral nerve graft on GAP-43 upregulation in the adult rat brain: an in situ hybridization and immunocytochemical study. *J. Neurosci.* **15,** 3594–3611.
28. Richardson, P. M., McGuinness, U. M., and Aguayo, A. J. (1980) Axons from CNS neurons regenerate into PNS grafts. *Nature* **284,** 264–265.
29. Doster, S. K., Lozano, A. M., Aguayo, A. J., and Willard, M. B. (1991) Expression of the growth-associated protein GAP-43 in adult rat retinal ganglion cells following axon injury. *Neuron* **6,** 635–647.
30. Schnell, L., Schneider, R., Kolbeck, R., Barde, Y. A., and Schwab, M. E. (1994) Neurotrophin-3 enhances sprouting of corticospinal tract during development and after adult spinal cord lesion [see comments]. *Nature* **367,** 170–173.
31. Cheng, H., Cao, Y., and Olson, L. (1996) Spinal cord repair in adult paraplegic rats: partial restoration of hind limb function [see comments]. *Science* **273,** 510–513.
32. Grill, R. J., Blesch, A., and Tuszynski, M. H. (1997) Robust growth of chronically injured spinal cord axons induced by grafts of genetically modified Ngf-secreting cells. *Exp. Neurol.* **148,** 444–452.
33. Ye, J. H. and Houle, J. D. (1997) Treatment of the chronically injured spinal cord with neurotrophic factors can promote axonal regeneration from supraspinal neurons. *Exp. Neurol.* **143,** 70–81.
34. Guest, J. D., Hesse, D., Schnell, L., Schwab, M. E., Bunge, M. B., and Bunge, R. P. (1997) Influence of IN-1 antibody and acidic FGF-fibrin glue on the response of injured corticospinal tract axons to human Schwann cell grafts. *J. Neurosci. Res.* **50,** 888–905.
35. Berry, M., Carlile, J., and Hunter, A. (1996) Peripheral nerve explants grafted into the vitreous body of the eye promote the regeneration of retinal ganglion cell axons severed in the optic nerve. *J. Neurocytol.* **25,** 147–70.
36. Fryer, R. H., Kaplan, D. R., and Kromer, L. F. (1997) Truncated Trkb receptors on nonneuronal cells inhibit Bdnf-induced neurite outgrowth in vitro. *Exp. Neurol.* **148,** 616–627.
37. Toma, J. G., Rogers, D., Senger, D. L., Campenot, R. B., and Miller, F. D. (1997) Spatial regulation of neuronal gene expression in response to nerve growth factor. *Dev. Biol.* **184,** 1–9.
38. Pataky, D. M., Borisoff, J., Fernandes, K. L., Tetzlaff, W., and Steeves, J. D. (1999) Differential effects of fibroblast growth factors on survival and neurite outgrowth from identified brainstem-spinal neurons in vitro, submitted.
39. Hasan, S. J., Keirstead, H. S., Muir, G. D., and Steeves, J. D. (1993) Axonal regeneration contributes to repair of injured brainstem-spinal neurons in embryonic chick. *J. Neurosci.* **13,** 492–507.

40. Keirstead, H. S., Dyer, J. K., Sholomenko, G. N., McGraw, J., Delaney, K. R., and Steeves, J. D. (1995) Axonal regeneration and physiological activity following transection and immunological disruption of myelin within the hatchling chick spinal cord. *J. Neurosci.* **15,** 6963–6974.

41. Keirstead, H. S. and Steeves, J. D. (1998) CNS myelin: does a stabilizing role in neurodevelopment result in inhibition of neuronal repair after injury. *Neuroscientist* **4,** 273–284.

42. Walsh, F. S. and Doherty, P. (1997) Neural cell adhesion molecules of the immunoglobulin superfamily: role in axon growth and guidance. *Annu. Rev. Cell Dev. Biol.* **13,** 425–456.

43. Schachner, M., Martini, R., Hall, H., and Orberger, G. (1995) Functions of the L2/HNK-1 carbohydrate in the nervous system. *Prog. Brain Res.* **105,** 183–188.

44. Martini, R., Xin, Y., and Schachner, M. (1994) Restricted localization of L1 and N-CAM at sites of contact between Schwann cells and neurites in culture. *Glia* **10,** 70–74.

45. Kamiguchi, H. and Lemmon, V. (1997) Neural cell adhesion molecule L1—signaling pathways and growth cone motility [Review]. *J. Neurosci. Res.* **49,** 1–8.

46. Bernhardt, R. R., Tongiorgi, E., Anzini, P., and Schachner, M. (1996) Increased expression of specific recognition molecules by retinal ganglion cells and by optic pathway glia accompanies the successful regeneration of retinal axons in adult zebrafish. *J. Comp. Neurol.* **376,** 253–264.

47. Cohen, N. R., Taylor, J. S. H., Scott, L. B., Guillery, R. W., Soriano, P., and Furley, A. J. W. (1998) Errors in corticospinal axon guidance in mice lacking the neural cell adhesion molecule L1. *Curr. Biol.* **8,** 26–33.

48. Dahme, M., Bartsch, U., Martini, R., Anliker, B., Schachner, M., and Mantei, N. (1997) Disruption of the mouse L1 gene leads to malformations of the nervous system. *Nature Genet.* **17,** 346–349.

49. Becker, T., Bernhardt, R. R., Reinhard, E., Wullimann, M. F., Tongiorgi, E., and Schachner, M. (1998) Readiness of zebrafish brain neurons to regenerate a spinal axon correlates with differential expression of specific cell recognition molecules. *J. Neurosci.* **18,** 5789–5803.

50. McKeon, R. J., Schreiber, R. C., Rudge, J. S., and Silver, J. (1991) Reduction of neurite outgrowth in a model of glial scarring following CNS injury is correlated with the expression of inhibitory molecules on reactive astrocytes. *J. Neurosci.* **11,** 3398–3411.

51. Okado, N. and Oppenheim, R. W. (1985) The onset and development of descending pathways to the spinal cord in the chick embryo. *J. Comp. Neurol.* **232,** 143–161.

52. Jhaveri, S., Erzurumlu, R. S., Friedman, B., and Schneider, G. E. (1992) Oligodendrocytes and myelin formation along the optic tract of the developing hamster: an immunohistochemical study using the Rip antibody. *Glia* **6,** 138–148.

53. Devor, M. (1976) Neuroplasticity in the rearrangment of olfactory tract fibers after neonatal transections in hamsters. *J. Comp. Neurol.* **166,** 49–72.

54. Cotman, C. W. (1981) Synapse replacement in the nervous system of adult vertebrates. *Physiol. Rev.* **61,** 684–784.

55. Raisman, G. and Field, P. M. (1973) A quantitative investigation of the development of collateral reinnervation after partial deafferentation of the septal nuclei. *Brain Res.* **50,** 241–264.

56. Rossi, F., Bravin, M., Buffo, A., Fronte, M., Savio, T., and Strata, P. (1997) Intrinsic properties and environmental factors in the regeneration of adult cerebellar axons. *Prog. Brain Res.* **114,** 283–296.

57. McMahon, S. B. and Kett-White, R. (1991) Sprouting of peripherally regenerating primary sensory neurones in the adult central nervous system. *J. Comp. Neurol.* **304,** 307–315.

58. Rhoades, R. W., Chiaia, N. L., Macdonald, G. J., and Jacquin, M. F. (1989) Effect of fetal infraorbital nerve transection upon trigeminal primary afferent projections in the rat. *J. Comp. Neurol.* **287,** 82–97.

59. Renehan, W. E., Crissman, R. S., and Jacquin, M. F. (1994) Primary afferent plasticity following partial denervation of the trigeminal brainstem nuclear complex in the postnatal rat. *J. Neurosci.* **14,** 721–739.

60. Kuang, R. Z. and Kalil, K. (1990) Specificity of corticospinal axon arbors sprouting into denervated contralateral spinal cord. *J. Comp. Neurol.* **302,** 461–472.

61. Recanzone, G. H., Merzenich, M. M., and Dinse, H. R. (1992) Expansion of the cortical representation of a specific skin field in primary somatosensory cortex by intracortical microstimulation. *Cereb. Cortex* **2,** 181–196.

62. Cavazos, J. E., Golarai, G., and Sutula, T. P. (1991) Mossy fiber synaptic reorganization induced by kindling: time course of development, progression, and permanence. *J. Neurosci.* **11,** 2795–2803.

63. Geinisman, Y., Morrell, F., and deToledo-Morrell, L. (1992) Increase in the number of axospinous synapses with segmented postsynaptic densities following hippocampal kindling. *Brain Res.* **569,** 341–347.

64. Patel, S. N., Rose, S. P., and Stewart, M. G. (1988) Training induced dendritic spine density changes are specifically related to memory formation processes in the chick, *Gallus domesticus. Brain Res.* **463,** 168–173.

65. Shimizu, I., Oppenheim, R. W., O'Brian, M., and Schneiderman, A. (1990) Anatomical and functional recovery following spinal cord transection in the chick embryo. *J. Neurobiol.* **21,** 918–937.

66. Steeves, J. D., Keirstead, H. S., Ethell, D. W., et al. (1994) Permissive and restrictive periods for brainstem-spinal regeneration in the chick. *Prog. Brain Res.* **103,** 243–262.

67. Keirstead, H. S., Hasan, S. J., Muir, G. D., and Steeves, J. D. (1992) Suppression of the onset of myelination extends the permissive period for the functional repair of embryonic spinal cord. *Proc. Natl. Acad. Sci. USA* **89,** 11664–11668.

68. Keirstead, H. S., Pataky, D. M., McGraw, J., and Steeves, J. D. (1997) In vivo immunological suppression of spinal cord myelin development. *Brain Res. Bull.* **44,** 727–734.

69. Saunders, N. R., Balkwill, P., Knott, G., et al. (1992) Growth of axons through a lesion in the intact CNS of fetal rat maintained in long-term culture. *Proc. R. Soc. Lond. Biol.* **250,** 171–180.

70. Varga, Z. M., Bandtlow, C. E., Erulkar, S. D., Schwab, M. E., and Nicholls, J. G. (1995) The critical period for repair of CNS of neonatal opossum *(Monodelphis domestica)* in culture: correlation with development of glial cells, myelin and growth-inhibitory molecules. *Eur. J. Neurosci.* **7,** 2119–2129.

71. Savio, T. and Schwab, M. E. (1990) Lesioned corticospinal tract axons regenerate in myelin-free rat spinal cord. *Proc. Natl. Acad. Sci. USA* **87,** 4130–4133.

72. Caroni, P. and Schwab, M. E. (1988) Antibody against myelin-associated inhibitor of neurite growth neutralizes nonpermissive substrate properties of CNS white matter. *Neuron* **1,** 85–96.

73. McClellan, A. D. (1998) Spinal cord injury: lessons from locomotor recovery and axonal regeneration in lower vertebrates. *Neuroscientist* **4,** 250–263.

74. Sharma, S. C., Jadhao, A. G., and Rao, P. D. (1993) Regeneration of supraspinal projection neurons in the adult goldfish. *Brain Res.* **620,** 221–228.

75. Becker, T., Wullimann, M. F., Becker, C. G., Bernhardt, R. R., and Schachner, M. (1997) Axonal regrowth after spinal cord transection in adult zebrafish. *J. Comp. Neurol.* **377,** 577–595.

76. Lang, D. M., Rubin, B. P., Schwab, M. E., and Stuermer, C. A. (1995) CNS myelin and oligodendrocytes of the *Xenopus* spinal cord—but not optic nerve—are nonpermissive for axon growth. *J. Neurosci.* **15,** 99–109.

77. Keirstead, H. S., Hughes, H. C., and Blakemore, W. F. (1998) A quantifiable model of axonal regeneration in the demyelinated adult rat spinal cord. *Exp. Neurol.* **151,** 303–313.

78. Dyer, J. K., Bourque, J. A., and Steeves, J. D. (1998) Regeneration of brainstem-spinal axons after lesion and immunological disruption in adult rat. *Exp. Neurol.* **154,** 12–22.

79. Schnell, L. and Schwab, M. E. (1990) Axonal regeneration in the rat spinal cord produced by an antibody against myelin-associated neurite growth inhibitors. *Nature* **343,** 269–272.

80. Bregman, B. S., Kunkelbagden, E., Schnell, L., Dai, H. N., Gao, D., and Schwab, M. E. (1995) Recovery from spinal cord injury mediated by antibodies to neurite growth inhibitors. *Nature* **378,** 498–501.

81. Rapalino, O., Lazarov-Spiegler, O., Agranov, E., et al. (1998) Implantation of stimulated homologous macrophages results in partial recovery of paraplegic rats. *Nature Med.* **4,** 814–821.

82. Fawcett, J. (1994) Astrocytes and axon regeneration in the central nervous system. *J. Neurol.* **242,** S25–S28.

83. Fitch, M. T. and Silver, J. (1997) Glial cell extracellular matrix: boundaries for axon growth in development and regeneration. *Cell Tissue Res.* **290,** 379–384.

84. Logan, A., Berry, M., Gonzalez, A. M., Frautschy, S. A., Sporn, M. B., and Baird, A. (1994) Effects of transforming growth factor beta 1 on scar production in the injured central nervous system of the rat. *Eur. J. Neurosci.* **6,** 355–363.

85. Fitch, M. T. and Silver, J. (1997) Activated macrophages and the blood-brain barrier: inflammation after CNS injury leads to increases in putative inhibitory molecules. *Exp. Neurol.* **148,** 587–603.

86. Ridet, J. L., Malhotra, S. K., Privat, A., and Gage, F. H. (1997) Reactive astrocytes: cellular and molecular cues to biological function [published erratum appears in *Trends Neurosci.* 1998;21:80]. *Trends Neurosci.* **20,** 570–577.

87. Norenberg, M. D. (1994) Astrocyte responses to CNS injury. *J. Neuropathol. Exp. Neurol.* **53,** 213–220.

88. Snow, D. M., Lemmon, V., Carrino, D. A., Caplan, A. I., and Silver, J. (1990) Sulfated proteoglycans in astroglial barriers inhibit neurite outgrowth in vitro. *Exp. Neurol.* **109,** 111–130.

89. Dow, K. E., Ethell, D. W., Steeves, J. D., and Riopelle, R. J. (1994) Molecular correlates of spinal cord repair in the embryonic chick: heparan sulfate and chondroitin sulfate proteoglycans. *Exp. Neurol.* **128,** 233–238.

90. Bartsch, S., Husmann, K., Schachner, M., and Bartsch, U. (1995) The extracellular matrix molecule tenascin: expression in the developing chick retinotectal system and substrate properties for retinal ganglion cell neurites in vitro. *Eur. J. Neurosci.* **7,** 907–916.

91. Meiners, S., Powell, E. M., and Geller, H. M. (1995) A distinct subset of tenascin/CS-6-PG-rich astrocytes restricts neuronal growth in vitro. *J. Neurosci.* **15,** 8096–8108.

92. Snow, D. M. and Letourneau, P. C. (1992) Neurite outgrowth on a step gradient of chondroitin sulfate proteoglycan (CS-PG). *J. Neurobiol.* **23,** 322–336.

6

Transduction of Inhibitory Signals by the Axonal Growth Cone

Li-Hsien Wang, Alyson Fournier, Fumio Nakamura, Takuya Takahashi, Robert G. Kalb, and Stephen M. Strittmatter

1. INTRODUCTION

The distal tip of the growing axons is a specialized structure termed the growth cone, consisting of a lamelipodium with numerous filopodial extensions *(1,2)* (Fig. 1). The growth cone is largely responsible for determining the direction as well as the extent of axon outgrowth. Obviously, axonal growth cone function is critical for neuronal development and hence the proper functioning of the adult nervous system. The same mechanisms are thought to determine whether adult axons regenerate (as in peripheral nerve injury) or fail to regenerate (as in spinal cord injury). When regeneration does occur, adult axons are thought to utilize the same guidance cues that developing axons use to identify appropriate synaptic partners among a myriad of possibilities. In this review, the molecular cues known to inhibit axonal outgrowth are briefly reviewed and the mechanisms of their action on growth cones considered in detail.

1.1. Factors Stimulating Axonal Growth

The guidance of growing axons has been a primary issue in neuronal development throughout this century. Much work has focused on molecules capable of stimulating axonal growth. Numerous trophic factors, e.g., nerve growth factor (NGF), brain-derived neurotrophic factor (BDNF), neurotrophin (NT)-3, NT-4/5, and ciliary neurotrophic factor (CNTF), play roles in the differentiation and survival of particular neurons *(3)*. However, their effects are largely chronic, transcription dependent, and not mediated locally within growth cones. In contrast, a number of molecules are known to function acutely and locally to stimulate motility at the neuronal growth cone. The latter group includes soluble, cell surface, and matrix-bound extracellular proteins. The cell adhesion molecules (CAMs) function largely via a homophilic mecha-

From: *Neurobiology of Spinal Cord Injury*
Edited by: R. G. Kalb and S. M. Strittmatter © Humana Press Inc., Totowa, NJ

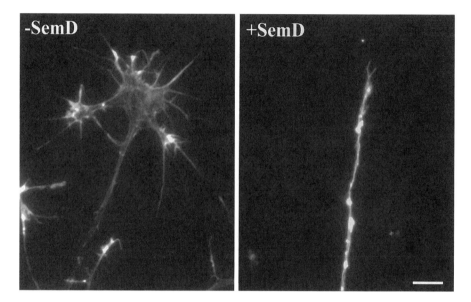

Fig. 1. SemD-induced growth cone collapse. The basal spread state of DRG growth cone morphology is shown at left (-semD). Within 10 min of 1 n*M* semD addition to the culture medium, growth cones are collapsed (+semD). Bar = 100 μm.

nism, linking two adjacent cells *(4)*. Structurally, they can be segregated into two general groups: the calcium-dependent cadherins and the calcium-independent immunoglobulin superfamily members such as N-CAM. CAMs may have primary roles in neuronal cell migration, but recent evidence from humans with the mental retardation, aphasia, shuffling gait, and adducted thumbs (MASA) syndrome *(5)* and from mice with targeted gene deletions *(6)* indicates that the L1 protein functions in axonal guidance as well. Laminin is an example of a matrix-bound protein with strong neurite-outgrowth-promoting effects in tissue culture *(7)*. These effects are largely mediated via cell surface integrin receptors. The role of such molecules in vertebrate central nervous system (CNS) axonal guidance has not been clearly documented. The netrins are diffusible axonal outgrowth stimulators *(8)* that act via the growth cone receptor protein termed deleted in colorectal carcinoma (DCC) *(9,10)*. The slit/robo ligand/receptor pair has a documented role in axonal midline crossing decisions in the fly *(11)*, but its mechanism of action is complicated, involving receptor internalization and altered cellular phenotype *(12)*. The expression of these growth-promoting molecules has gone largely unstudied in spinal cord injury (SCI). The focus of therapeutic intervention in SCI has been on providing NT support to injured spinal cord, with little emphasis on axonal growth stimulators to date.

Table 1
Inhibitors of Axonal Growth Cone Extension

Ligand	Type	Receptor	Transduction
Semaphorin	Secreted and membrane associated	NP for class III semaphorins	CRMP, rac1, rho activity
Ephrins	Transmembrane and GPI-linked	Eph tyrosine kinases	Tyrosine kinase activity
Netrin	Secreted	DCC plus UNC-5	cAMP modulated
Neurotransmitters	Small soluble molecules	G protein-coupled receptors	Heterotrimeric G proteins
Thrombin, LPA	Soluble	G protein-coupled receptors	Heterotrimeric G proteins, rho activity
NI-35	Membrane-associated component of myelin	???	Ca_I^{2+} concentration, rho activity
MAG	Membrane-associated component of myelin	???	???

1.2. Factors Inhibiting Axonal Growth

Over the last 5–10 years, the critical importance of axonal outgrowth inhibitors for both neuronal development and regeneration has been appreciated (Table 1). This revolution in the axonal guidance field has been supported to a large degree by two tissue culture assays. The first is the growth cone collapse assay of Raper and Kapfhammer *(13)*, which is based on the ability of a number of outgrowth inhibitors to collapse the lamelipodium of the growth cone acutely (Fig. 1). The second assay is that of Bonhoeffer and colleagues *(14)*, in which a growing axon is presented with a choice between alternating 90-μm stripes of two different membrane carpets. Through the use of these two assays, numerous investigators have demonstrated that inhibitory signals play a major, if not primary, role in determining the direction and extent of axonal growth during development. Some examples are discussed below.

1.2.1. Semaphorins

The growth cone collapse assay was utilized by Raper and colleagues *(15)* to purify a protein from brain extracts that potently induces growth cone collapse of primary sensory neurons (Fig. 1). The primary sequence of this protein, collapsin-1, was found to be similar to a group of insect semaphorins shown to influence axonal guidance by genetic and immunologic methods *(16,17)*. It is now clear that the semaphorin family of proteins consists of close to 20 members in vertebrate organisms *(18,19)*. All members share a conserved "sema" domain of about 500 amino acid residues. The association of other structural features such as immunoglobulin motifs, basic residue-rich segments, thrombospondin repeats, transmembrane segments, or glycophosphatidyl inositol linkage sites with the sema domain defines several semaphorin subclasses. It is presumed that many or all of these proteins play a role in axonal guidance, although this is clearly documented only for the class III semaphorin collapsin-1/semD/semaIII *(20–23)*, and for insect semaphorins *(16,24–26)*. The semD molecule is expressed in areas of the developing embryo that are avoided by the peripheral extensions of dorsal root ganglion (DRG) neurons *(27,28)*, and deletion of the semD gene results in overexuberant sprouting of sensory axons in the periphery *(20,23)*. In the cerebral cortex, semD expression in superficial layers of the cortex contributes to the initial trajectory of pyramidal cell axons away from the cortical surface and toward the underlying white matter *(29)*. The roles of other semaphorins are now under intensive scrutiny and may extend beyond axonal guidance to include vasculogenesis *(23)*, immune function *(30,31)*, viral invasion *(32)*, and neoplastic transformation *(33)*.

1.2.2. Ephrins

The role of the ephrins in axonal guidance has been explored most extensively in the development of the retinotectal system *(34)*. Bonhoeffer and colleagues demonstrated that the optic tectum possesses a gradient of repulsive molecules situated in a fashion that could direct the formation of a retinotopic map *(14,35,36)*. The repulsive molecule from the optic tectum was identified as retinal axon guidance signal (RAGS) *(37)*. RAGS (now termed ephrin-A5) is a member of a family of structurally related, membrane-associated ephrin proteins that bind to tyrosine kinase receptors of the ephrin receptor (EphR) family. RAGS possesses growth cone collapsing activity for retinal axons and can repel their outgrowth. Flannagan and co-workers demonstrated that whereas an ephrin ligand is distributed in a gradient across the tectum, its EphR exhibits a complimentary gradient in the retina *(38,39)*. The retinotectal map formation is complicated by the presence of more than one ephrin/EphR pair in the system. Overexpression *(39)* and gene knock-out studies *(40)* have supported the hypothesis that these systems contribute to retinotectal map generation in vivo. The in vivo studies have also revealed the importance of such ligand/receptor systems in the formation of axonal pathways such as the anterior commissure *(41)*.

1.2.3. Netrins

The netrins were initially identified by Tessier-Lavigne and colleagues as diffusible molecules released by the floor plate that attract commissural axons to the midline of the developing spinal cord *(42)*. However, for certain neurons the netrins possess axon-repulsive activity. In the brainstem, the axons of oculumo-tor and abducens motoneurons extend ventrally and exit the brain ipsilaterally. In contrast, trochlear axons project dorsally and exit contralaterally. In part, this appears to be caused by inhibitory action of floor plate netrin on trochlear axons *(43)*. Thus, this guidance molecule can function either as an attractive or as a repulsive cue. A netrin homologue in *Caenorhabditis elegans,* UNC-6, also exhibits both attractive and repulsive activity *(44,45)*. It appears that the DCC protein (UNC-40 in *C. elegans;* frazzled in *Drosophila*) is a netrin receptor *(9,10,46)*. The repulsive action of netrin requires both DCC/UNC-40 and a sec-ond transmembrane protein, UNC-5, or its vertebrate homologs *(47)*.

1.2.4. Components of CNS Myelin

In the CNS of adult vertebrates, axons regenerate poorly, if at all, whereas in the peripheral nervous system (PNS) axonal regeneration is generally success-ful. The peripheral nerve grafting studies of Aguayo and colleagues *(48,49)* have demonstrated that CNS axons can regenerate if given an appopriate envi-ronment. Comparison of CNS and PNS myelin by Schwab and colleagues has revealed that the CNS white matter is inhibitory for axonal growth *(50)*. Two components of the CNS white matter with inhibitory activity for axon growth have been characterized *(51)*. Fractionation of myelin under denaturing condi-tions allowed Schwab and co-workers to identify the protein NI-35 and an anti-genically related NI-250 as the primary inhibitory components of myelin. Partially purified NI-35 can induce growth cone collapse and halt axon exten-sion. The IN-1 antibody directed against these proteins partially blocks the axon-repulsive properties of CNS myelin in vitro *(52)* and in vivo *(53)*. A 220-kDa neurite growth inhibiting an IN-1-immunoreactive protein has recently been purified to homogeneity from bovine spinal cords *(54)*. The derived pep-tides do not display similarity with any published protein sequences, indicating that it is a previously unidentified protein.

The myelin-associated glycoprotein (MAG) has been shown by several groups to possess axon outgrowth-inhibiting activity *(55,56)*. This activity can be demonstrated after fractionation of CNS myelin under nondenaturing condi-tions. Studies of axon regeneration in MAG -/- mice have suggested that MAG contributes to CNS myelin inhibition of axon growth in vivo *(57)*. The relative contribution of NI-35/250 and MAG to myelin's inhibitory properties remains unclear. Evidence suggests that there are likely to be other inhibitory compo-nents of CNS myelin.

1.2.5. Relevance to Spinal Cord Injury

SCI severs axons in the spinal cord but leaves distal cell bodies intact. These axons exhibit little if any regeneration. There is strong reason to believe that the inhibitory components of CNS myelin contribute to this failure of regeneration *(58)*. The role of axon-repulsive molecules with major roles in neuronal devel-opment, such as the semaphorins, ephrins, and netrins, has not been studied to any significant degree in SCI. Such proteins may or may not contribute to the

Table 2
Intracellular Axonal Growth Cone Signal Transduction Proteins

Protein	Identification	Pathway
GAP-43	Induced by nerve regeneration	G protein, protein kinase C, calmodulin
G_o protein	Major growth cone membrane protein	GPCR action
CRMP	Expression cloning of semD signaling cascade	SemD signaling
Rho	Monomeric GTP binding protein	Required for growth cone sensitivity to inhibitors
Rac1	Rho homology	SemD
IP$_3$ receptor	IP$_3$ binding protein; release of Ca stores	Required for basal outgrowth

failure of regeneration after SCI. However, they are very likely to guide the direction and specificity of any regeneration that does occur. If CNS myelin inhibition could be overcome by some therapeutic intervention, then semaphorins, ephrins, and netrins might be critical for the establishment of functionally useful synaptic circuitry in the spinal cord.

2. GROWTH CONE SIGNALING MOLECULES

After growth cones encounter various extracellular guidance cues, a signal transduction process occurs at the growth cone membrane and within the growth cone to produce alterations in motility. Knowledge of signal transduction mechanisms (Table 2) has lagged behind the identification of axon guidance signals.

2.1. GAP-43

One method of identifying proteins contributing to growth cone function is to characterize those proteins whose expression is induced during periods of active axonal extension. Growth-associated protein (GAP-43, 43 kDa; neuromodulin; B50) was identified in studies of rapidly transported axonal proteins after sciatic nerve injury *(59,60)*. Its expression is induced 10–100-fold in adult sensory neurons during nerve regeneration *(61)*. The protein is also highly expressed during brain development and is localized to the axonal growth cone membrane *(62,63)*. In vivo mouse studies demonstrate that lack of GAP-43 results in axonal guidance errors *(64)*. These errors are consistent with intra-growth-cone GAP-43 being required for the transduction of specific signals into alterations in axonal extension. Overexpression of GAP-43 alters the normal balance between axonal sprouting and synaptic stabilization at the neuromuscular junction *(65)*.

What molecular mechanisms might underlie GAP-43 participation in growth cone signal transduction? We have shown that GAP-43 can interact with heterotrimeric G proteins, increasing the proportion of G protein in the activated state *(66)*. The effect of GAP-43 is synergistic with receptor activation *(67)*, so that GAP-43 provides a means for increasing the "gain" for growth cone detection of extracellular guidance cues. GAP-43 also interacts with calmodulin in a fashion independent of calcium, but sensitive to protein kinase C *(68,69)*. Thus, GAP-43 may provide a coordinating site for kinase and guanosine triphosphate (GTP) binding protein regulation in growth cone signal transduction.

2.2. Heterotrimeric G Proteins

A second method for identifying proteins contributing to growth cone signal transduction is to analyze the major constituents of the growth cone membrane. Pfenninger and colleagues developed a method for the subcellular fractionation of growth cones from developing brain *(70)*. We demonstrated that the heterotrimeric G protein, G_o, is a major constituent of the growth cone membrane, comprising close to 10% of the growth cone membrane, with only actin and tubulin in higher concentrations *(66)*. This implies that G_o activation state can regulate growth cone motility. We showed that activated mutants of G_o α-subunit can enhance neurite outgrowth from PC12 pheochromocytoma cells *(71)*. The involvement of G_o in growth cone regulation is also supported by the interaction of GAP-43 with G_o *(66)*. Mice with a targeted mutation in the G_o α-gene have been created, and most such mice die in the first few weeks after birth *(72,73)*. Behavior is abnormal is several respects, but whether neurodevelopmental axonal guidance defects underlie these abnormalities is not known.

Heterotrimeric G proteins such as G_o primarily function to couple seven membrane-spanning receptors, G protein-coupled receptors (GPCR) or serpentine receptors, to intracellular second messenger systems. This predicts that ligands for such receptors should control growth cone motility. Indeed, several such ligands (including bradykinin, serotonin, acetylcholine, thrombin, and lysophosphatidic acid) do collapse growth cones *(74–76)*. GAP-43 in turn enhances the coupling of these receptors to heterotrimeric G proteins *(67)*.

2.3. Intracellular Calcium Levels

A third approach to exploring signal transduction pathways in the growth cone is to measure alterations in the levels of potential second messengers directly as growth cones receive extracellular signals. Intracellular calcium ion concentrations have been implicated in many signaling pathways and can be conveniently measured in real time via ratiometric fluorescence measurements *(77)*. Kater and co-workers have recognized that a number of axonal guidance molecules do modulate calcium ion concentrations as they alter growth cone motility *(78,79)*. Such changes occur locally in the growth cone and temporally precede alterations in growth cone structure.

Intracellular calcium concentrations can be regulated by alterations in plasma membrane ion channels or by release from intracellular stores. There is evidence that both of these mechanisms contribute to growth cone regulation. Discrete "hot spots" of voltage-gated calcium channel activity have been found in the growth cone *(80)*. Voltage-gated calcium channel activity appears to be required for the regulation of growth cone motility by certain cell adhesion molecules *(81)*. Recent studies have demonstrated that release of calcium from inositol (1,4,5)-trisphosphate (IP$_3$)-sensitive intracellular stores is an integral part of basal growth cone advance *(82)*. The role of the IP$_3$ receptor required local inactivation of the protein in growth cones by the chromophore-assisted laser inactivation (CALI) method.

2.4. Cyclic Nucleotides as Modulators

In many signal transduction systems, cyclic nucleotides function as the primary second messengers. In the growth cone, these molecules appear to have secondary and modulatory roles in signal transduction. The work of Moo-Ming Poo and colleagues has demonstrated that cyclic adenosine monophosphate (cAMP) or cyclic guanosine monophosphate (cGMP) can switch a growth cone from attraction to repulsion or vice versa *(83,84)*. For example, *Xenopus* retinal ganglion cell axons are attracted to areas of increased netrin concentration under basal conditions, but are repulsed from netrin-rich areas after treatments that elevate cAMP levels *(83)*. Similarly, *Xenopus* motoneurons are repulsed by semD under basal conditions, but are attracted to semD after elevation of cGMP *(84)*.

3. GROWTH CONE STRUCTURE AND MOTILITY

After the growth cone encounters extracellular signals, appropriate receptors and signal cascades are activated. These molecular changes in turn lead to alterations in the growth cone structure and activity. The structure of the axon shaft is largely dependent on tubulin, but the growth cone itself appears to derive its structure from actin filaments and from membranous organelles.

3.1. Actin Filaments

Both the growth cone filopodia and lamelipodia contain very high concentrations of actin filaments. The dynamics of these filaments have been studied at high resolution by Forscher et al. in *Aplysia* bag cell neurons *(85)*. Filaments assemble distally in the growth cone and move toward the central domain of the growth cone *(86)*. Distal assembly appears to be a constitutive process, while contact with target cells regulates the rate of retrograde flow. Thus, the rate of growth cone advance across a substratum is inversely proportional to the rate of retrograde actin filament flow in the growth cone *(87)*. The retrograde flow rate depends on the action of myosin-type motors *(88)*. Thus, proteins regulating actin assembly, disassembly, crosslinking, and interaction with myosin are all predicted to modulate growth cone behavior *(89)*.

The rho subfamily of monomeric GTP binding proteins plays a significant role in determining actin filament structure in nonneuronal cells *(90)*. It is now clear that rho, rac, and cdc42 also participate in axonal guidance. Flies overexpressing constitutive active or dominant negative forms of these proteins exhibit impaired axonal outgrowth *(91)*. Dendritic branching is altered in mouse cerebellar Purkinjie neurons expressing mutant rac1 *(92)*. More acute changes in monomeric G protein activity have been shown to modulate growth behavior directly in tissue culture *(93,94)*.

Cofilin, or actin depolymerizing factor, is another major determinant of actin filament dynamics in nonneuronal cells *(95,96)*. This protein may function in concert with monomeric G proteins to determine growth cone motility. LIM-kinase I (lin-II, ISL1, mec-3, domain-containing kinase I) *(97,98)* and WAVE (Wiskoff-Aldrich syndrome family, verprolin homologous protein) *(99)* are two identified proteins linking activated rac1 to cofilin and hence actin filaments. Dysregulation of cofilin, via loss of LIM-kinase I, appears to underlie the visuospatial recognition deficits of patients with the Williams syndrome *(100)*.

3.2. Membrane Turnover

During axonal growth there is a dramatic increase in the total cell surface of the neuron. The major site of new membrane addition is the distal tip of the growth cone *(101)*. Once delivered from the cell soma to the growth cone through vesicular transport and exocytosis, plasma membrane flows in a retrograde fashion toward the cell soma. The growth cone is also a site of extensive endocytosis *(102,103)*. Thus, the balance of exo- and endocytosis at the growth cone may regulate axonal extension and growth cone motility. The growth cone itself contains extensive stacks of intraneuronal membranes and large vacuolar structures active in endocytosis *(104)*. Regulation of vesicle dynamics by axonal guidance cues has not been studied.

Several molecules known to participate in synaptic vesicle dynamics are localized to the growth cone prior to mature synapse formation. Such protiens include vesicle-associated membrane protein (VAMP), synaptosome-associated protein of $M_r25,000$ (SNAP-25), syntaxin, Munc-18, rab3A, N-ethylmaleimide-sensitive factor (NSF), and soluble NSF attachment protein (β-SNAP) *(105)*. Cleavage of syntaxin by neurotoxin C1 of *Clostridium botulinum* induces growth cone collapse and the cessation of axonal extension *(106)*. Suppression of SNAP-25 with antisense oligonucleotides reduces axonal extension *(107)*. Two proteins that play a role in synaptic vesicle endocytic recycling also appear to be required for neurite extension. Suppression of either dynamin *(108)* or amphiphysin *(109)* expression by antisense oligonucleotides inhibits neurite extension in cultured neurons.

3.3. Axon Repulsion and Growth Cone Collapse

How are the actin filament and membrane vesicle determinants of growth cone structure altered in response to extracellular guidance cues? For molecularly defined axonal guidance signals, there are almost no data in this regard. One experiment with partially purified c-semD (collapsin-1) has documented a moderate decrease in the concentration of actin filaments in the growth cone

(110). The control of membrane expansion and contraction in the growth cone by extracellular cues is unstudied.

4. SEMAPHORIN SIGNALING PATHWAYS

The semaphorins are the prototypic examples of axon-repulsive molecules. Chick semD (c-semD, collapsin-1) was the first repulsive molecule purified and characterized on the basis of a functional assay for axonal growth inhibition. Although a number of components of the semD signaling cascade have been identified (Fig. 2), key components remain uncharacterized.

4.1. Neuropilin

The first step in semD action is high-affinity (10–200 pM) binding to the growth cone membrane *(15).* To characterize these binding sites, semD-alkaline phosphatase (AP) fusion proteins have been utilized in our laboratory and others *(111–114).* Binding sites for this fusion protein can be detected by a simple histologic reaction; they exhibit a tissue and cell type specificity consistent with their being physiologically relevant sites *(111,114).* Expression cloning methods were utilized by Kolodkin and co-workers *(113),* and by He and Tessier-Lavigne *(112),* to identify neuropilin-1 as the binding site for semD-AP. Expression of neuropilin-1 in nonneuronal cells is sufficient to create high-affinity binding sites for semD-AP.

It is now clear that neuropilin-1 is necessary for semD-mediated axon repulsion. Antibodies to NP-1 block semD-induced inhibition of axonal outgrowth *(112).* Mice lacking NP-1 *(115)* have abnormalities in peripheral sensory axon projections that are nearly identical to those found in mice lacking semD *(20,23).* Cultured neurons from NP-1 -/- mice do not respond to semD. We have recently shown that virus-mediated expression of NP-1 in retinal neurons converts these neurons from a semD-insensitive to a semD-sensitive state *(116,117).*

Neuropilin has four protein domains *(118).* The amino terminal region (CUB) bears sequence similarity to Tolloid, BMP-1, and the complement components C1r/s. This is followed by a region with homology to coagulation factors V and VIII, and then a MAM domain with homology to a protease merpin and several receptor phosphatases, RPTPµ and RPTPκ. The transmembrane and short (40-amino-acid residue) cytoplasmic tail are unique to neuropilins. We have recently shown that the CUB domain is the primary sema binding domain, and the basic tail of semD interacts with both the CUB and factor V/VIII regions *(117).* The MAM domain appears to be a major site mediating neuropilin homooligomerization. It is not known whether if oligomerization is necessary for activity.

How are signals transmitted from NP-1 to the growth cone interior? By expressing deletion mutants of NP-1 in retinal neurons, we have demonstrated that only the CUB and MAM portions of the ectodomain of NP are required at

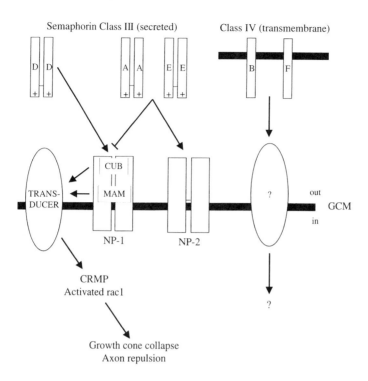

Fig. 2. A model for semaphorin action at the neuronal growth cone membrane (GCM). See text for further description.

the cell surface *(117)*. The transmembrane sequence and the cytoplasmic tail are not required. This implies that the NP-1 must interact with a second transmembrane protein to transduce signals to the cell interior. The identity of this transducing protein is not known.

Two NPs and more than twenty semaphorins have been identified *(119)*. It now appears that only the class III semaphorins, such as semA, semD/III, semE, and semIV, interact with neuropilins. Within this class of semaphorins, semD binds selectively to NP-1, whereas the other molecules bind to both NP-1 and NP-2 with high affinity *(120)*. NP-1 binding of semD, or NP-2 binding of semA, semE, or semIV, induces growth cone collapse and axon repulsion *(116,121,122)*. However, NP-1 binding of semA or semE does not induce axon repulsion. This binding appears to possess the properties of classic pharmacologic antagonism, creating no direct effect but competitively blocking the action of semD *(116)*. DRG neurons largely express NP-1 receptors, and sympathetic neurons express NP-1 and NP-2 receptors *(120)*. A detailed review of the anatomic distribution of semaphorin and NP expression is beyond the scope of this chapter. Suffice it to say that, whereas adult DRG neurons are known to express NP-1 *(61)*, the role of NPs and semaphorins in SCI has not been investigated.

4.2. CRMPs

Studies of NP itself have not yet provided evidence as to downstream signal transduction events. We have employed an expression cloning strategy to identify potential components of semD signal transduction cascades *(123)*. Sensory neurons responsive to semD were utilized to prepare mRNA that might express such components. This mRNA was injected into *Xenopus* oocytes to allow protein expression, and then semD was applied to the oocyte surface. Membrane conductance was monitored during semD application as a measure of semD responsiveness. Using this assay, a single mRNA species was identified that converted oocytes from a semD-unresponsiveness to a state in which semD induces a transient inward current flux. The protein encoded by this clone, collapsin response mediator protein (CRMP-62, 62 kDa), is expressed exclusively in developing neurons and is localized to growth cones. The protein is cytoplasmic, indicating that it is not part of the semaphorin receptor complex, but rather a component of an intracellular transduction cascade. Antibodies to CRMP-62 block semD-induced growth cone collapse in DRG neurons, demonstrating the physiologic relevance of the oocyte findings. Consistent with a role in growth cone regulation, CRMP-62 shares sequence with unc-33, a *C. elegans* protein required for accurate axonal extension in worms. CRMP-62 also shares homology with liver dihydropyrmidinase, but possesses no dihydropyrmidinase activity and lacks some highly conserved dihydropyrmidinase amino acid residues *(124)*.

In mammals, there are four highly related CRMP genes *(125)*. All are expressed primarily in the nervous system, but different neurons express different complements of CRMPs. It is conceivable that different CRMPs are involved in the response to different semaphorins. At a structural level, the CRMPs exist as mixed tetramers, with heteromeric CRMP tetramers being preferred over homotetramers *(124)*. It remains unknown how CRMPs relate to upstream NP receptor complexes and downstream changes in actin and membrane dynamics during the growth cone response to semD.

4.3. Rac1

Since semD reduces the level of growth cone actin filaments *(126)*, and rho family proteins are regulators of the actin cytoskeleton *(90)*, we considered the role of these GTP binding proteins in semD action. Introduction of a dominant negative form of rac1 (DN-rac) has a highly specific effect on semD action *(93)*. Basal axon extension is not altered by DN-rac, but semD-induced changes in growth cone function are completely blocked. Of the rho family proteins, only rac1 has such an effect. When introduced into sensory neurons, constitutively active rac1 has a weak growth cone collapsing- and outgrowth-inhibiting activity. These results suggest that semD binding to NP activates a cascade involving both CRMPs and rac1 activation.

5. EPHRIN SIGNALING PATHWAYS

The ephrins are a second class of axon-repulsive molecule with clearly documented roles in neuronal development *(34)*. The first step in ephrin action is quite clear since EphR is known to be a tyrosine kinase. As for some other tyrosine kinase receptors, clustering plays an important role in activation *(127)*. This can be achieved naturally by receptor interaction with membrane-bound ephrin ligands. It can also be achieved experimentally by antibody-induced clustering of soluble ligand. Receptor clustering and the activation of tyrosine kinase activity are likely to initiate the creation of a receptor complex containing transduction proteins such as Grb2 (128) and P13-kinase *(129)*. However, this has not been studied in detail for the EphRs, as it has for other tyrosine kinase receptors.

Since the transmembrane ephrin ligands possess conserved cytoplasmic tails, they might participate in bidirectional signaling, serving as receptors as well as ligands. In vitro studies showing Eph-induced phosphorylation in ephrin-expressing cells *(130,131)* and in vivo gene knock-out studies both support the bidirectional signaling hypothesis for ephrin/Eph interaction in axonal development *(41)*.

6. CNS MYELIN AND AXONAL GROWTH CONES

Although the semaphorins and the ephrins are well-characterized axonal guidance signals in neuronal development, they may play no role in SCI. In contrast, neurite outgrowth-inhibiting components of CNS myelin are thought to be a central cause of failed axonal regeneration in SCI. By what molecular mechanisms do NI-35 and MAG block recovery after SCI?

6.1. NI-35 Signaling

After partial purification and reconstitution into liposomes, the NI-35 preparation induces axonal growth cone collapse and the cessation of axonal extension *(132)*. These effects are morphologically indistinguishable from semD effects on sensory neurons *(133)*. However, several molecular features distinguish NI-35-induced collapse from semD-induced alteration in growth cone motility. Real-time calcium imaging has revealed that the intracellular concentration of calcium increases dramatically immediately after NI-35 application, before growth cone collapse occurs *(132)*. No such change in calcium ion concentration occurs during semD-induced growth cone collapse. Whereas DN-rac blocks semD-induced growth cone collapse, it has no effect on CNS myelin-mediated inhibition of axonal extension *(93)*.

Pharmacologic manipulation has revealed the source of the calcium mobilized by NI-35. Most of the calcium appears to derive from a caffeine-sensitive intracellular storage pool *(132)*. Removal of extracellular calcium has no effect on the response, ruling out the opening of plasma membrane calcium channels as the mechanism of NI-35 action. It remains unknown how NI-35 interaction with the growth cone plasma membrane initiates this calcium release from intracellular storage pools.

Due to the high concentrations of G_o protein in the growth cone membrane, we considered whether this GTP binding protein might be involved in NI-35-induced growth cone collapse. Pertussis toxin inactivates G_o and G_i proteins by ADP-ribosylating the α-subunits of these proteins. Treatment of DRG neurons with pertussis toxin reduces the potency of NI-35 by a factor of 10 *(133)*. These data suggest that G_o or G_i participates in NI-35 signal transduction. However, it remains possible that these effects are mediated by the cell surface binding subunit of pertussis toxin, rather than the catalytic domain *(134)*. Further studies remain difficult because the molecular identity of NI-35 has not yet been determined.

6.2. MAG Signaling

MAG inhibits axonal extension from a variety of neuronal types, although inhibition of DRG axonal growth is seen only in neurons cultured from rats older than 4 d *(56)*. Essentially nothing is known about signal transduction events initiated by MAG in growth cones. Filbin and co-workers have shown that MAG binds at least in part to neuronal sialic acid *(135)*. Desialated neurons are less responsiveness to MAG-mediated inhibition of axonal growth. However, binding to neuronal sialic acid is insufficient for outgrowth inhibition, since a MAG fragment that binds to neuronal sialic acid does not reduce axonal extension. Neuronal receptors for MAG other than sialic acid have not yet been identified.

6.3. Monomeric G Proteins

Actin and actin filament-regulating monomeric rho family G proteins play a major role in determining growth cone function. Inhibition of rho activity with the C3 exoenzyme of *C. botulinum* has demonstrated that active rho is required for CNS myelin inhibition of outgrowth *(93)*. When the C3 protein is introduced into neurons, rho is ADP-ribosylated and inactivated *(136)*. Axons then exhibit small growth cones, which extend more rapidly than in control cultures. Such C3-treated, rapidly extending growth cones are insensitive to myelin. They are also insensitive to other growth cone inhibitors such as semD, thrombin, and lysophosphatidic acid *(93)*. Whereas the rho protein is unlikely to participate directly in NI-35 signal transduction, a basal level of activated rho protein is required for growth cone responsiveness to myelin.

7. PROSPECTS FOR MODULATING GROWTH CONE SIGNALING IN SCI

Growth cone signal transduction mechanisms are becoming much better understood. As such knowledge accumulates, the probability of therapeutic intervention to augment axonal regeneration after SCI increases. If receptors for inhibitory components of CNS myelin are identified, they will be obvious targets for pharmaceutical intervention. If roles for semaphorins, ephrins, or netrins are identified in SCI, then these molecules may also have therapeutic potential in SCI. Intracellular signal transduction pathways may also provide opportunities for enhancing recovery after SCI. For example, the ability of rho protein inhibition by C3 exoenzyme to overcome CNS myelin inhibition provides an obvious potential intervention. Here the issue is identifying methods for intraneuronal delivery of the C3 protein to neurons in vivo. Current research makes clear the fact that axonal growth cones have the capacity to alter their growth properties in response to changes in the extracellular environment.

ACKNOWLEDGMENTS

This work was supported by grants to S.M.S. and R.G.K. from the National Institutes of Health, and to S.M.S. from the American Paralysis Association. S.M.S. is a John Merck Scholar in the Biology of Developmental Disorders in Children, and an Investigator of the Patrick and Catherine Weldon Donaghue Medical Research Foundation. L.H.W. is a Coxe-Brown Fellow of Yale University. F.N. is supported by a research fellowship from the Spinal Cord Research Fund of the Paralyzed Veterans of America. A.F. is supported by an F.C.A.R. research fellowship.

7. REFERENCES

1. Ramón y Cajal, S. (1890) A quelle époque apparaissent les expansions des cellules nerveuses de la moelle epinère du poulet. *Anat. Anzerger* **5,** 609–613.
2. Strittmatter, S. M. (1995) Neuronal guidance molecules: inhibitory and soluble factors. *Neuroscientist* **1,** 255–258.
3. Snider, W. D. (1994) Functions of the neurotrophins during nervous system development: what the knockouts are teaching us. *Cell* **77,** 627–638.
4. Dodd, J. and Jessell, T. M. (1988) Axon guidance and the patterning of neuronal projections in vertebrates. *Science* **242,** 692–699.
5. Jouet, M., Rosenthal, A., Armstrong, G., et al. (1994) X-linked spastic paraplegia (SPG1), MASA syndrome and X-linked hydrocephalus result from mutations in the L1 gene. *Nature Genet.* **7,** 402–407.
6. Cohen, N. R., Taylor, J. S. H., Scott, L. B., et al. (1997) Errors in corticospinal axon guidance in mice lacking the neural cell adhesion molecule L1. *Curr. Biol.* **8,** 26–33.
7. Reichardt, L. F. and Tomaselli, K. J. (1991) Extracellular matrix molecules and their receptors: functions in neural development. *Annu. Rev. Neurosci.* **14,** 531–570.
8. Serafini, T., Kennedy, T. E., Falko, N. J., et al. (1994) The netrins define a family of axon outgrowth-promoting proteins homologous to *C. elegans* UNC-6. *Cell* **78,** 409–424.
9. Chan, S. S.-Y., Zheng, H., Su, M.-W., et al. (1996) UNC-40 a *C. elegans* homolog of DCC (deleted in colorectal cancer) is required in motile cells responding to UNC-6 netrin cues. *Cell* **87,** 187–195.

10. Keino-Masu, K., Masu, H., Hinck, L., et al. (1996) Deleted in colorectal cancer (DCC) encodes a netrin receptor. *Cell* **87,** 175–185.
11. Kidd, T., Brose, K., Mitchell, K. J., et al. (1998) Roundabout controls axon crossing of the CNS midline and defines a novel subfamily of evolutionarily conserved guidance receptors. *Cell* **92,** 205–215.
12. Kidd, T., Russell, C., Goodman, C. S., and Tear, G. (1998) Dosage-sensitive and complementary functions of roundabout and commissureless control axon crossing of the CNS midline. *Neuron* **20,** 25–33.
13. Raper, J. A. and Kapfhammer, J. P. (1990) The enrichment of a neuronal growth cone collapsing activity from embryonic chick brain. *Neuron* **2,** 21–29.
14. Walter, J., Kern-Veits, B., Huf, J., Stolze, B., and Bonhoeffer, F. (1987) Recognition of position-specific properties of tectal cell membranes by retinal axons in vitro. *Development* **101,** 685–696.
15. Luo, Y., Raible, D., and Raper, J. A. (1993) Collapsin: a protein in brain that induces the collapse and paralysis of neuronal growth cones. *Cell* **75,** 217–227.
16. Kolodkin, A. L., Matthews, D. J., O'Connor, T. P., et al. (1992) Fasciclin IV: sequence, expression, and function during growth cone guidance in the grasshopper embryo. *Neuron* **9,** 831–845.
17. Kolodkin, A. L., Matthews, D. J., and Goodman, C. S. (1993) The *semaphorin* genes encodes a family of transmembrane and secreted growth cone guidance molecules. *Cell* **75,** 1389–1399.
18. Kolodkin, A. (1996) Semaphorins: mediators of repulsive growth cone guidance. *Trends Cell Biol.* **6,** 15–22.
19. Püschel, A. W. (1996) The semaphorins: a family of axonal guidance molecules? *Eur. J. Neurosci.* **8,** 1317–1321.
20. Taniguchi, M., Yuasa, S., Fujisawa, H., et al. (1997) Disruption of semaphorin III/D gene causes severe abnormality in peripheral nerve projection. *Neuron* **19,** 519–530.
21. Messersmith, E. K., Leonardo, E. D., Shatz, C. J., et al. (1995) Semaphorin III can function as a selective chemorepellent to pattern sensory projections in the spinal cord. *Neuron* **14,** 949–959.
22. Püschel, A. W., Adams, R. H., and Betz, H. (1995) Murine semaphorin D/collapsin is a member of a diverse gene family and creates domains inhibitory for axonal extension. *Neuron* **14,** 941–948.
23. Behar, O., Golden, J. A., Mashimo, H., Schoen, F. J., and Fishman, M. C. (1996) Semaphorin III is needed for normal patterning and growth of nerves, bones, and heart. *Nature* **383,** 525–528.
24. Matthes, D. J., Sink, H., Kolodkin, A. L., and Goodman, C. S. (1995) Semaphorin II can function as a selective inhibitor of specific synaptic arborization. *Cell* **81,** 631–639.
25. Winberg, M. L., Mitchel, K. J., and Goodman, C. S. (1998) Genetic analysis of the mechanisms controlling target selection:complementary and combinatorial functions of netrins, semaphorins, and IgCAMs. *Cell* **93,** 581–591.

26. Yu, H.-H., Araj, H. H., Ralls, S. A., and Kolodkin, A. L. (1998) The transmembrane semaphorin sema I is required in *Drosophila* for embryonic motor and CNS axon guidance. *Neuron* **20**, 207–220.

27. Shepherd, I., Luo, Y., Raper, J. A., and Chang, S. (1996) The distribution of collapsin-1 mRNA in the developing chick nervous system. *Dev. Biol.* **173**, 185–199

28. Giger, R. J., Wolfer, D. P., De Wit, G. M. J., et al. (1996) Anatomy of rat semaphorinIII/collapsin-1 mRNA expression and relationship to developing nerve tracts during neuroembryogenesis. *J. Comp. Neurol.* **375**, 378–392.

29. Polleux, F., Giger, R. J., Ginty, D. D., Kolodkin, A. L., and Ghosh, A. (1998) Patterning of cortical efferent projections by semaphorin-neuropilin interactions. *Science* **282**, 1904–1906.

30. Furuyama, T., Inagaki, S., Kosugi, A., et al. (1996) Identification of a novel transmembrane semaphorin expressed on lymphocytes. *J. Biol. Chem.* **271**, 33376–33381.

31. Hall, K. T., Boumsell, L., Schultze, J., et al. (1996) Human CD100, a novel leukocyte semaphorin that promotes B-cell aggregation and differentiation. *Proc. Natl. Acad. Sci. USA* **93**, 11780–11785.

32. Comeau, M. R., Johnson, R., DuBose, R. F., et al. (1998) A poxvirus-encoded semaphorin induces cytokine production from monocytes and binds to a novel cellular semaphorin receptor, VESPR. *Immunity* **8**, 473–482.

33. Sekido, Y., Bader, S., Latif, F., et al. (1996) Human semaphorin A(V) and IV reside in the 3p21.3 small cell lung cancer deletion region and demonstrate distinct expression patterns. *Proc. Natl. Acad. Sci. USA* **93**, 4120–4125.

34. Flanagan, J. G. and Vanderhaeghen, P. (1998) The ephrins and Eph receptors in neural development. *Annu. Rev. Neurosci.* **21**, 309–345.

35. Bonhoeffer, F. and Huf, J. (1982) In vitro experiments on axon guidance demonstrationg an anterior-posterior gradient on the tectum. *EMBO J.* **4**, 427–431.

36. Walter, J., Henke-Fahle, S., and Bonhoeffer, F. (1987) Avoidance of posterior tectal membranes by temporal retinal axons. *Development* **101**, 909–913.

37. Drescher, U., Kremoser, C., Handwrecker, C., et al. (1995) In vitro guidance of retinal ganglion cell axons by RAGS, a 25 kDa tectal protein related to ligands for Eph receptor tyrosine kinases. *Cell* **82**, 359–370.

38. Cheng, H.-J., Nakamoto, M., Bergemann, A. D., and Flanagan, J. G. (1995) Complementary gradients in expression and binding of ELF-1 and Mek4 in development of the topographic retintectal projection map. *Cell* **82**, 371–381.

39. Nakamoto, M., Cheng, H.-J., Friedman, G. C., et al. (1996) Topographically specific effects of ELF-1 on retinal axon guidance in vitro and retinal axon mapping in vivo. *Cell* **86**, 755–766.

40. Frisen, J., Yates, P. A., McLaughlin, T., et al. (1998) Ephrin-A5 (AL-1/RAGS) is essential for proper retinal axon guidance and topographic mapping in the mammalian visual system. *Neuron* **20**, 235–243.

41. Henkemeyer, M., Orioli, D., Henderson, J. T., et al. (1996) Nuk controls pathfinding of commissural axons in the mammalian central nervous system. *Cell* **86**, 35–46.

42. Kennedy, T. E., Serafini, T., de la Torre, J. R., and Tessier-Lavigne, M. (1994) Netrins are diffusible chemotropic factors for commissural axons in the embryonic spinal cord. *Cell* **78,** 425–435.

43. Colamarino, S. A. and Tessier-Lavigne, M. (1995) The axonal chemoattractant netrin-1 is also a chemorepellent for trochlear motor axons. *Cell* **81,** 621–629.

44. Hedgecock, E. M., Culotti, J. G., and Hall, D. H. (1990) The unc-5, unc-6 and unc-40 genes guide circumferential migrations of pioneer axons and mesodermal cells on the epidermis in *C. elegans. Neuron* **4,** 61–85.

45. Ishii, N., Wadsworth, W. G., Stern, B. D., Culotti, J. G., and Hedgecock, E. M. (1992) UNC-6, a laminin-related protein, guides cells and pioneer axon migrations in *C. elegans. Neuron* **9,** 873–881.

46. Kolodziej, P. A., et al. (1996) Frazzled encodes a *Drosophila* member of the DCC immunoglobulin subfamily and is required for CNS and motor axon guidance. *Cell* **87,** 197–204.

47. Hamelin, M., Zhou, Y., Su, M.-W., Scott, I. M., and Culotti, J. G. (1993) Expression of the unc-5 guidance gene in the touch neurons of *C. elegans* steers their axons dorsally. *Nature* **364,** 327–330.

48. David, S. and Aguayo, A. J. (1981) Axonal elongation in peripheral nervous system bridges after central nervous system injury in adult rats. *Science* **214,** 391–393.

49. Vidal-Sanz, M., Bray, G. M., Villegas-Perez, M. P., Thanos, S., and Aguayo, A. J. (1987) Axonal regeneration and synapse formation in the superior colliculus by retinal ganglion cells in the adult retina. *J. Neurosci.* **7,** 2894–2909.

50. Schwab, M. E. and Thoenen, H. (1985) Dissociated neurons regenerate into sciatic but not optic nerve explants in culture irrespective of neurotrophic factors. *J. Neurosci.* **5,** 2415–2423.

51. Caroni, P. and Schwab, M. E. (1988) Two membrane protein fractions from rat central myelin with inhibitory properties for neurite growth and fibroblast spreading. *J. Cell Biol.* **106,** 1281–1288.

52. Caroni, P. and Schwab, M. E. (1988) Antibody against myelin-associated inhibitor of neurite growth neutralizes nonpermissive substrate properties of CNS white matter. *Neuron* **1,** 85–96.

53. Schnell, L. and Schwab, M. E. (1990) Axonal regeneration in the rat spinal cord produced by an antibody against myelin-associated neurite growth inhibitors. *Nature* **343,** 269–272.

54. Spillmann, A. A., Bandtlow, C. E., Lottspeich, F., Keller, F., and Schwab, M. E. (1998) Identification and characterization of a bovine neurite growth inhibitor (bNI220). *J. Biol. Chem.* **273,** 19283–19293.

55. McKerracher, L., David, S., Jackson, D. L., et al. (1994) Identification of myelin-associated glycoprotein as a major myelin-derived inhibitor of neurite growth. *Neuron* **13,** 805–811.

56. Mukhopadhyay, G., Doherty, P., Walsh, F. S., Crocker, R., and Filbin, M. T. (1994) A novel role for myelin-associated glycoprotein as an inhibitor of axonal regeneration. *Neuron* **13,** 805–811.

57. Bartsch, S., Montag, D., Schachner, M., and Bartsch, U. (1997) Increased number of unmyelinated axons in optic nerves of adult mice deficient in the myelin-associated glycoprotein (MAG). *Brain Res.* **11,** 231–234.

58. Schwab, M. E., Kapfhammer, J. P., and Bandtlow, C. E. (1993) Inhibitors of neurite outgrowth. *Annu. Rev. Neurosci.* **16,** 565–595.

59. Benowitz, L. I. and Rottenberg, A. (1987) A membrane phosphoprotein associated with neural development, axonal regeneration, phospholipid metabolism, and synaptic plasticity. *Trends Neurosci.* **10,** 527–532.

60. Skene, J. H. P. (1989) Axonal growth-associated proteins. *Annu. Rev. Neurosci.* **12,** 127–156.

61. Pasterkamp, R. J., Giger, R. J., and Verhaagen, J. (1998) Regulation of semaphorin III/collapsin-1 gene expression during peripheral nerve regeneration. *Exp. Neurol.* **153,** 313–327.

62. Skene, J. H. P., Jacobson, R. D., Snipes, G. J., et al. (1986) A protein induced during nerve growth (GAP-43) is a major component of growth-cone membranes. *Science* **233,** 783–785.

63. Zuber, M., Strittmatter, S. M., and Fishman, M. C. (1989) A membrane-targeting signal in the amino terminus of the neuronal protein GAP-43. *Nature* **341,** 345–348.

64. Strittmatter, S. M., Fankhauser, C., Huang, P. L., Mashimo, H., and Fishman, M. C. (1995) Neuronal pathfinding is abnormal in mice lacking the neuronal growth cone protein GAP-43. *Cell* **80,** 445–452.

65. Aigner, L., Arber, S., Kapfhammer, J. P., et al. (1995) Overexpression of the neural growth-associated protein GAP-43 induces nerve sprouting in the adult nervous system of transgenic mice. *Cell* **83,** 269–278.

66. Strittmatter, S. M., Valenzuela, D., Kennedy, T. E., and Fishman, M. C. (1990) G_o is a major growth cone protein subject to regulation by GAP-43. *Nature* **344,** 836–841.

67. Strittmatter, S. M., Cannon, S. C., Ross, E. M., Higashijima, T., and Fishman, M. C. (1993) GAP-43 augments G protein-coupled receptor transduction in *Xenopus laevis* oocytes. *Proc. Natl. Acad. Sci. USA* **90,** 5327–5331.

68. Alexander, K. A., Cimler, B. M., Meier, K. E., and Storm, D. R. (1987) Regulation of calmodulin binding to P-57. A neurospecific calmodulin binding protein. *J. Biol. Chem.* **262,** 6108–6113.

69. Nakamura, F., Strittmatter, P., and Strittmatter, S. M. (1998) GAP-43 augmentation of G protein-mediated signal transduction is required by both phosphorylation and palmitoylation. *J. Neurochem.* **70,** 983–992.

70. Pfenninger, K. H., Ellis, L., Johnson, M. P., Friedman, L. B., and Somlo, S. (1983) Nerve growth cones isolated from fetal rat brain. Subcellular fractionation and characterization. *Cell* **35,** 573–584.

71. Strittmatter, S. M., Fishman, M. C., and Zhu, X. P. (1994) Activated mutants of the alpha subunit of G(o) promote an increased number of neurites per cell. *J. Neurosci.* **14,** 2327–2338.

72. Jiang, M., Gold, M. S., Boulay, G., et al. (1998) Multiple neurological abnormalities in mice deficient in the G protein Go. *Proc. Natl. Acad. Sci. USA* **95,** 3269–3274.

73. Valenzuela, D., Han, X., Mende, U., et al. (1997) G alpha(o) is necessary for muscarinic regulation of Ca^{2+} channel in mouse heart. *Proc. Natl. Acad. Sci. USA* **94,** 1727–1732.

74. Haydon, P. G., McCobb, D. P., and Kater, S. B. (1984) Serotonin selectively inhibits growth cone motility and synaptogenesis of specific identified neuron. *Science* **226,** 561–564.

75. Rodrigues, P. S. and Dowling, J. E. (1990) Dopamine induces neurite retraction in retinal horizontal cells via diacylglycerol and protein kinase C. *Proc. Natl. Acad. Sci. USA* **87,** 9693–9697.

76. Jalink, K., Corven, E. J., Hengeveld, T., et al. (1994) Inhibition of lysophosphatidate- and thrombin-induced neurite retraction and neuronal cell rounding by ADP ribosylation of the Small GTP-binding protein rho. *J. Cell Biol.* **126,** 801–810.

77. Clapham, D. E. (1995) Calcium signaling. *Cell* **80,** 259–269.

78. Kuhn, T. B., Schmidt, M. F., and Kater, S. B. (1995) Laminin and fibronectin guideposts signal sustained by opposite effects to passing growth cones. *Neuron* **14,** 275–285.

79. Kuhn, T. B., Williams, C. V., Dou, P., and Kater, S. B. (1998) Laminin directs growth cone navigation via two temporally and functionally distinct calcium signals. *J. Neurosci.* **18,** 184–194.

80. Silver, R. A., Lamb, A. G., and Bolsover, S. R. (1990) Calcium hotspots caused by L-channel clustering promote morphological changes in neuronal growth cones. *Nature* **343,** 751–754.

81. Doherty, P. and Walsh, F. S. (1994) Signal transduction events underlying neurite outgrowth stimulated by cell adhesion molecules. *Curr. Opin. Neurobiol.* **4,** 49–55.

82. Takei, K., Shin, R.-M., Inoue, T., Kato, K., and Mikoshiba, K. (1998) Regulation of nerve growth mediated by inositol 1,4,5-trisphosphate receptors in growth cones. *Science* **282,** 1705–1708.

83. Ming, G.-L., Song, H. J., Berringer, B., et al. (1997) cAMP-dependent growth cone guidance by netrin-1. *Neuron* **19,** 1225–1235.

84. Song, H.-J., Ming, G. L., He, Z., et al. (1998) Conversion of neuronal growth cone responses from repulsion to attraction by cyclic nucleotides. *Science* **281,** 515–518.

85. Forscher, P. and Smith, S. J. (1988) Actions of cytochalasins on the organization of actin filaments and microtubules in a neuronal growth cone. *J. Cell Biol.* **107,** 1505–1516.

86. Forscher, P., Lin, C. H., and Thompson, C. (1992) Novel form of growth cone motility involving site-directed actin filament assembly. *Nature* **357,** 515–518.

87. Lin, C. H. and Forscher, P. (1995) Growth cone advance is inversely proportional to retrograde F-actin flow. *Neuron* **14,** 763–771.

88. Lin, C. H., Espreafico, E. M., Mooseker, M. S., and Forscher, P. (1996) Myosin drives retrograde F-actin flow in neuronal growth cones. *Neuron* **16,** 769–782.

89. Tanaka, E. and Sabry, J. (1995) Making the connection: cytoskeletal rearrangements during growth cone guidance. *Cell* **83,** 171–176.

90. Hall, A. (1998) Rho GTPase and the actin cytoskeleton. *Science* **279,** 509–514.

91. Luo, L., Liao, Y. J., Jan, L. Y., and Jan, Y. N. (1994) Distinct morphogenetic functions of similar small GTPases: *Drosophila* Drac1 is involved in axonal outgrowth and myoblast fusion. *Genes Dev.* **8,** 1787–1802.

92. Luo, L., Heusch, T. K., Ackerman, L., et al. (1996) Differential effects of the rac1 GTPase on Purkinje cell axons and dendritic trunks and spines. *Nature* **379,** 837–840.

93. Jin, Z. and Strittmatter, S. M. (1997) Rac1 mediates collapsin-1-induced growth cone collapse. *J. Neurosci.* **17,** 6256–6263.

94. Kozma, R., Sarner, S., Ahmed, S., and Lim, L. (1997) Rho family GTPase and neuronal growth cone remodelling: relationship between increased complexity induced by Cdc42Hs, Rac1 and acetylcholine and collapse induced by RhoA and lysophosphatidic Acid. *Mol. Cell. Biol.* **17,** 1201–1211.

95. Nishida, E., Maekawa, S., and Sakai, H. (1984) Cofilin, a protein in porcine brain that binds to actin filaments and inhibits their interactions with myosin and tropomyosin. *Biochemistry* **23,** 5307–5313.

96. Bamburg, J. R. and Bray, D. (1987) Distribution and cellular localization of actin depolymerizing factor. *J. Cell Biol.* **105,** 2817–2825.

97. Yang, N., Higuchi, O., Ohashi, K., et al. (1998) Cofilin phosphorylation by LIM-kinase 1 and its role in rac-mediated actin reorganization. *Nature* **393,** 809–812.

98. Arber, S., Barbayannis, F. A., Hauser, H., et al. (1998) Regulation of actin dynamics through phosphorylation of cofilin by LIM-kinase. *Nature* **393,** 805–809.

99. Miki, H., Suetsugu, S., and Takenawa, T. (1998) WAVE, a novel WASP family protein involved in actin reorganization induced by rac. *EMBO J.* **17,** 6932–6941.

100. Frangiskakis, J. M., Ewart, A. K., Morris, C. A., et al. (1996) LIM-kinase 1 Hemizygosity implicated in impaired visuospatial constructive cognition. *Cell* **86,** 59–69.

101 Dai, J. and Sheetz, M. P. (1995) Axon membrane flows from the growth cone to the cell body. *Cell* **83,** 693–701.

102. Tsui, H. C., Ris, H., and Klein, W. L. (1983) Ultrastructural networks in growth cones and neurites of cultured central nervous system neurons. *Proc. Natl. Acad. Sci. USA* **80,** 5779–5783.

103. Lockerbie, R. O., Miller, V. E., and Pfenninger, K. H. (1991) Regulated plasmalemmal expansion in nerve growth cones. *J. Cell Biol.* **112,** 1215–1227.

104. Cheng, T. P. and Reese, T. S. (1987) Recycling of plasmalemma in chick tectal growth cones. *J. Neurosci.* **7,** 1752–1759.

105. Igarashi, M., Tagaya, M., and Komiya, Y. (1997) The soluble N-ethylmaleimide-sensitive factor attached protein receptor complex in growth cones: molecular aspects of the axon terminal development. *J. Neurosci.* **17,** 1460–1470.

106. Igarashi, M., Kozaki, S., Terakawa, S., et al. (1996) Growth cone collapse and inhibition of neurite growth by botulinum neurotoxin C1: a t-SNARE is involved in axonal growth. *J. Cell Biol.* **34,** 205–215.

107. Osen-Sand, A., Catsicas, M., Staple, J. K., et al. (1993) Inhibition of axonal growth by SNAP-25 antisense oligonucleotides in vitro and in vivo. *Nature* **364,** 445–448.

108. Torre, E., McNiven, M. A., and Urrutia, R. (1994) Dynamin I antisense oligonucleotide treatment prevents neurite formation in cultured hippocampal neurons. *J. Biol. Chem.* **269,** 32411–32417.

109. Mundigl, O., Ochoa, G. C., David, C., et al. (1998) Amphiphysin I antisense oligonucleotides inhibit neurite outgrowth in cultured hippocampal neurons. *J. Neurosci.* **18,** 93–103.

110. Fan, J. and Raper, J. A. (1995) Localized collapsing cues can steer growth cones without inducing their full collapse. *Neuron* **14,** 263–274.

111. Feiner, L., Koppel, A., Kobayashi, H., and Raper, J., A. (1997) Secreted chick semaphorins bind recombinant neuropilin with similar affinities but bind different subsets of neurons in situ. *Neuron* **19,** 539–545.

112. He, Z. and Tessier-Lavigne, M. (1997) Neuropilin is a receptor for the axonal chemorepellent semaphorin III. *Cell* **90,** 739–751.

113. Kolodkin, A. L., Levengood, D. V., Rowe, E. G., et al. (1997) Neuropilin is a semaphorin III receptor. *Cell* **90,** 753–762.

114. Takahashi, T., Nakamura, F., and Strittmatter, S. M. (1997) Neuronal and non-neuronal collapsin-1 binding sites in developing chick are distinct from other semaphorin binding sites. *J. Neurosci.* **17,** 9183–9193.

115. Kitsukawa, T., Shimizu, M., Saubo, M., et al. (1997) Neuropilin-semaphorin III/D-mediated chemorepulsive signals play a crucial role in peripheral nerve projection in mice. *Neuron* **19,** 995–1005.

116. Takahashi, T., Nakamura, F., Jin, Z., Kalb, R. G., and Strittmatter, S. M. (1998) Semaphorins A and E act as antagonists of neuropilin-1 and agonists of neuropilin-2 receptors. *Nature Neurosci.* **1,** 487–493.

117. Nakamura, F., Tanaka, M., Takahashi, T., Kalb, R. G., and Strittmatter, S. M. (1998) Neuropilin-1 extracellular domains mediate semaphorin D/III-induced growth cone collapse. *Neuron* **21,** 1093–1100.

118. Kawakami, A., Kitsukawa, T., Takagi, S., and Fujisawa, H. (1996) Developmentally regulated expression of a cell surface protein, neuropilin, in the mouse nervous system. *J. Neurobiol.* **29,** 1–17.

119. Kolodkin, A. L. and Ginty, D. D. (1997) Steering clear of semaphorins: neuropilins sound the retreat. *Neuron* **19,** 1159–1162.

120. Chen, H., Chedotal, A., He, Z., Goodman, C. S., and Marc, T.-L. (1997) Neuropilin-2, a novel member of the neuropilin family, is a high affinity receptor for the semaphorins sema E and sema IV but not sema III. *Neuron* **19,** 547–559.

121. Giger, R. J., Urquhart, E. R., Gillespie, S. K. H., et al. (1998) Neuropilin-2 is a receptor for semaphorin IV: insight into the structural basis of receptor function and specificity. *Neuron* **21,** 1079–1092.

122. Chen, H., He, Z., Bagri, A., and Tessier-Lavigne, M. (1998) Semaphorin-neuropilin interactions underlying sympathetic axon responses to class III semaphorins. *Neuron* **21,** 1283–1290.

123. Goshima, Y., Nakamura, F., Strittmatter, P., and Strittmatter, S. M. (1995) Collapsin-induced growth cone collapse mediated by an intracellular protein related to UNC-33. *Nature* **376,** 509–514.

124. Wang, L.-H. and Strittmatter, S. M. (1997) Brain CRMP forms heterotetramers similar to liver dihidropyrimidinase. *J. Neurochem.* **69,** 2261–2269.

125. Wang, L.-H. and Strittmatter, S. M. (1996) A family of rat CRMP genes is differentially expressed in the nervous system. *J. Neurosci.* **16,** 6197–6207.

126. Fan, J., Mansfield, S. G., Redmond, T., Gordon-Weeks, P. R., and Raper, J. A. (1993) The organization of F-actin and microtubules in growth cones exposed to a brain-derived collapsing factor. *J. Cell Biol.* **121,** 867–878.

127. Davis, S., Gale, N. W., Aldrich, T. H., et al. (1994) Ligands for EPH-related receptor tyrosine kinases that require membrane attachment or clustering for activity. *Science* **266,** 816–819.

128. Stein, E., Cerretti, D. P., and Daniel, T. O. (1996) Ligand activation of Elk receptor tyrosine kinase promotes its association with Grb10 and Grb2 in vascular endothelial cells. *J. Biol. Chem.* **271,** 23588–23593.

129. Pandey, A., Lazar, D. F., Saltiel, A. R., and Dixit, V. M. (1994) Activation of the Eck receptor protein kinase stimulates phosphatidylinositol 3 kinase activity. *J. Biol. Chem.* **269,** 30154–30157.

130. Holland, S. J., Gale, N. W., Mbamalu, G., et al. (1996) Bidirectional signalling through the EPH-family receptor Nuk and its transmembrane ligands. *Nature* **383,** 722–725.

131. Bruckner, K., Pasquale, E. B., and Klein, R. (1997) Tyrosine phosphorylation of transmembrane ligands for Eph receptors. *Science* **275,** 1640–1643.

132. Bandtlow, C. E., Schmidt, M. F., Hassinger, T. D., Schwab, M. E., and Kater, S. B. (1993) Role of intracellular calcium in NI-35-evoked collapse of neuronal growth cones. *Science* **259,** 80–83.

133. Igarashi, M., Strittmatter, S. M., Vartanian, T., and Fishman, M. C. (1993) Mediation by G proteins of signals that cause collapse of growth cones. *Science* **259,** 77–79.

134. Kindt, R. M. and Lander, A. D. (1995) Pertussis toxin specifically inhibits growth cone guidance by a mechanism independent of direct G protein inactivation. *Neuron* **15,** 79–88.

135. Collins, B. E., Yang, L. J.-S., Mukhopadhyay, G., et al. (1997) Sialic acid specificity of myelin-associated glycoprotein binding. *J. Biol. Chem.* **272,** 1248–1255.

136. Aktories, K. and Just, I. (1995) In vitro ADP-ribosylation of Rho by bacterial ADP-ribosyltransferases. *Methods Enzymol.* **256,** 184–195.

7

Elaboration of the Axonal Microtubule Array During Development and Regeneration

Peter W. Baas

1. INTRODUCTION

Axons grow over potentially enormous distances to reach their target tissues during the development of the nervous system and also during regeneration after injury. The growth of the axon is dependent on the elaboration of highly organized arrays of cytoskeletal elements (for reviews, *see* Baas and Yu, 1996; Baas, 1997). Chief among these cytoskeletal elements are the hollow filaments known as microtubules. Microtubules are polymers of alternating α- and β-tubulin subunits. They provide architectural support necessary for the growth and maintenance of the axon, and they also act as railways for the anterograde and retrograde transport of vesicular elements and associated proteins. The importance of the axonal microtubule array is reflected in the fact that it is tightly regulated in terms of its structure and organization. Each microtubule within the array has a consistent 13-protofilament lattice (13 tubulin subunits composing the perimeter of the tube and thereby constraining its diameter), and each microtubule is oriented with its assembly-favored end, termed its plus-end, distal to the cell body of the neuron. The spacing between the microtubules and the stability properties of the microtubules are also tightly regulated, and both of these are thought to result from the binding of various microtubule-associated proteins (MAPs) along the surface of the microtubule. An important issue in neuronal differentiation concerns the mechanisms by which the microtubule array of the axon is elaborated. Intimately related to this issue is how the axonal microtubule array is rebuilt during regeneration after injury to the axon.

The mechanisms underlying the elaboration of the axonal microtubule array have been a matter of great interest to cell biologists and neuroscientists for over two decades. Pioneering studies were performed principally by the Lasek laboratory, using what has now become a classic technique for studying axonal transport (for reviews, *see* Lasek, 1986, 1988; Baas and Brown, 1997). In this technique, radioisotopically labeled amino acids are injected into the vicinity of a neuronal cell body, which results in a "pulse" of newly synthesized

From: *Neurobiology of Spinal Cord Injury*
Edited by: R. G. Kalb and S. M. Strittmatter © Humana Press Inc., Totowa, NJ

radioactive proteins that are then conveyed down the axon by active transport mechanisms. Lasek (1986,1988) identified various "components" of axonal transport, groups of proteins that move down the axon at similar rates. Vesicular elements move in what was termed the fast component of axonal transport. Tubulin and the proteins that comprise another cytoskeletal element termed neurofilaments move much more slowly, in what was termed slow component A. The protein that comprises the third cytoskeletal element, actin filaments, moves somewhat more rapidly than tubulin or neurofilaments, in what was termed slow component B. The slow components (especially slow component B) also consist of additional proteins that are thought to associate with the cytoskeletal elements as they are actively transported. In some axons, a portion of the tubulin is transported in slow component B. Because of several specific features of the manner by which the proteins are transported, Lasek (1986,1988) proposed that the cytoskeletal proteins are transported in the form of assembled polymers, that is, in the form of microtubules, neurofilaments, and actin filaments, rather than as their subunit components.

At the time of its proposal and shortly thereafter, this idea was strongly supported by the kinetic data, several lines of morphologic evidence, and the physics underlying active transport events (for review, *see* Lasek, 1988). However, direct observations of polymer transport events in the neuron were lacking, as was any strong evidence for the identity of the "molecular transport machinery" that might convey cytoskeletal polymers down the axon. As a result, a great deal of controversy ensued, which continues to this day. Other researchers have proposed that microtubules (and the other cytoskeletal polymers) are entirely stationary in the neuron, and that it is their subunit components that are actively transported (*see,* for example, Hirokawa et al., 1997). In this view, the cytoskeletal polymers are assembled locally, using the actively transported subunits. This latter model has gained popularity in some scientific circles in recent years, but in my view, it is not supported by the same kind of rigorous experimental data or the logic of the original model put forth by Lasek's group. The debate concerning the two perspectives has recently been aired in several different formats and will not be the focus of the present article (*see,* for example, Baas and Brown, 1997; Hirokawa et al., 1997). More exciting is the fact that over the past few years, several studies have reported new evidence for the transport of microtubule polymers, and substantial progress has been made toward elucidating the molecules responsible for their transport.

In the present article, I summarize this progress. I focus on microtubule transport because, in my view, it is the key to understanding this complicated issue. After having presented this material, I shift the focus of the article to the topic of axonal regeneration following injury. Specifically, I try to address how the neuron modifies the mechanisms that establish the axonal microtubule array during the development of the axon, to permit the injured axon to regenerate.

2. ELABORATING MICROTUBULE ARRAYS WITHIN LIVING CELLS

In my opinion, much of the confusion and controversy surrounding the axonal microtubule array relates to the fact that the fields of neuroscience and cell biology have been too isolated from one another over the years. I believe that great insights can be obtained on the axonal microtubule array by putting the issue into the context of the broader issue of how all cell types organize their microtubules. This broader issue has not been without its own controversies, but in general, has progressed more rapidly than the narrower one of how microtubules organize in axons. Classically, microtubule organization within cells has been attributed to the attachment of individual microtubules to their sites of nucleation (for review, *see* Brinkley, 1985). In interphase cells, cytoplasmic microtubules are nucleated from the centrosome, which is located at the cell center, and an array of microtubules emanates in all directions. Each microtubule is nucleated with its plus-end elongating away from the centrosome, and its minus-end attached to the centrosome, resulting in a radial array of microtubules of uniform polarity orientation. During cell division, duplicated centrosomes migrate to opposite poles of the cell, where each centrosome nucleates an array of uniformly oriented microtubules. In the region of the cell between the two centrosomes, microtubules of opposite orientation overlap. This results in a microtubule array that is commonly referred to as a bipolar spindle. It is now clear that the organization of microtubules into these typical interphase and mitotic arrays cannot be explained solely on the basis of microtubule nucleation and assembly from centrosomes. In fact, a variety of important regulatory mechanisms and proteins are essential for imposing order on these microtubule arrays. In my view, contemplating how these mechanisms and proteins organize microtubules in other cell types can provide useful clues as to how the neuron organizes microtubules within the axon.

In the test tube, tubulin subunits can form polymers *de novo* if the concentration of subunits is sufficiently high. In living cell, however, *de novo* initiation of microtubules is suppressed in favor of nucleation from defined structures such as the centrosome. This ensures that microtubules are assembled in an organized rather than haphazard fashion. Another purpose for nucleation from defined structures is that these structures constrain the lattice structure of the microtubule to a consistent 13 protofilaments (that is, 13 subunits make up the perimeter of the microtubule). *De novo* initiation, by contrast, results in microtubules with a variety of protofilament numbers. Recent studies suggest that the capacity of the centrosome to nucleate microtubules may be its principal function, and that its role in organizing microtubules may have been exaggerated for many years. For example, it was recently shown that microtubules can form bipolar spindles in the absence of centrosomes, as long as the microtubules are exposed to mitotic extracts that contain molecular motor proteins (Heald et al., 1996). These and related studies indicate that motor proteins are capable of generating forces on microtubules that can transport and organize them into a variety of configurations, including the classic radial array of uniform orientation previously attributed to the centrosome. The need for such noncentrosomal mechanisms is further indicated by the fact that many microtubule configurations, such as those found within the mitotic spindle and the axon, cannot be explained in terms of attachment of the microtubules to centralized sites of nucleation.

It is now clear that the formation and functioning of the mitotic spindle, the most fundamental of all microtubule arrays, involve a variety of motor proteins that generate complementary and antagonistic forces that are essential for the migration of the duplicated centrosomes during prophase, the establishment of the bipolar spindle during metaphase, the separation of the chromosomes during early anaphase, and the separation of the half-spindles during late anaphase and telophase (for reviews, *see* Barton and Goldstein, 1996; Walzak and Mitchison, 1996). These motors generate forces on microtubules that spatially orient them and move them into specific configurations. What is truly amazing is that the growing body of knowledge concerning motor-driven microtubule organization in the mitotic spindle very closely parallels the theories on axonal microtubules proposed by Lasek over two decades ago. It is now clear from these studies on the mitotic spindle that cells do indeed contain the kinds of molecular transport machinery that Lasek envisioned. In addition, it would appear that models invoking motor-driven microtubule transport as a principal element in the establishment of the axonal microtubule array are not suggesting something at all unique, but in fact are based on themes that are consistent with the types of mechanisms used across cell types.

3. ELABORATING THE MICROTUBULE ARRAYS OF THE NEURON

Over the past 10 years, most efforts to understand the axonal microtubule array have shifted away from the classic radiolabeling paradigm and have instead used contemporary cell biologic approaches on cultured neurons. Among other things, these approaches have been useful for elucidating the sites of origin of the axonal microtubules. Do the microtubules within the axonal array originate within the axon itself or from elsewhere within the neuron? As noted above, it is a basic principle of cytoplasm that *de novo* initiation of microtubules is suppressed in favor of nucleation from defined structures such as the centrosome. Microtubules in the axon are known to have a consistent 13-protofilament lattice and are also uniformly oriented with their plus-ends away from cell center. However, the microtubules are not attached to the centrosome but rather stop and start along the length of the axon. Do axons contain some other kind of microtubule nucleating structures? Studies on cultured neurons have addressed this issue by documenting the sites at which microtubules reassemble during recovery from pharmacologic treatments that depolymerize microtubules. These studies indicate that microtubule assembly in the axon is limited to the elongation of existing polymers

(Baas and Heidemann, 1986; Baas and Ahmad, 1992). In other words, no bona fide nucleation of new microtubules occurs within the axon, and this is presumably how the polarity orientation and consistent lattice structure of the microtubule array is preserved. Therefore, by the process of elimination, one would have to conclude that all new microtubules must originate within the cell body of the neuron, perhaps at the centrosome.

At about the same time as these analyses were performed, a great deal of new information was emerging about how the centrosome nucleates microtubules. Studies from many laboratories established that the centrosome contains a novel member of the tubulin superfamily called γ-tubulin and that γ-tubulin is required for normal microtubule nucleation within the cytoplasm of living cells. Subsequent studies established that γ-tubulin is one component of a complex of proteins that form a ring structure within the pericentriolar region of the centrosome (Moritz et al., 1995; Zheng et al., 1995). This ring structure is ideally suited to nucleate a microtubule while simultaneously constraining its lattice to 13 protofilaments. In studies on cultured neurons, my laboratory found that there is no detectable γ-tubulin within the axon, and that, as is the case in other cell types, γ-tubulin is concentrated at the centrosome within the cell body of the neuron (Baas and Joshi, 1992). Moreover, we found that experimental inhibition of microtubule nucleation at the centrosome with a function-blocking antibody to γ-tubulin severely inhibits microtubule formation and axon outgrowth in these neurons (Ahmad et al., 1994). On the basis of these observations, we proposed that the neuronal centrosome acts as a kind of "generator" for the nucleation and release of microtubules destined for transport into the axon (for review, *see* Baas, 1996).

Another possibility, of course, is that a functional centrosome is required for axon outgrowth but not because the microtubules nucleated at the centrosome are transported into the axon. To investigate further, we developed a novel pharmacologic regime whereby we could document the redistribution over time of a "burst" of microtubules nucleated at the centrosome (Ahmad and Baas, 1995). We found that within about 30 min, virtually all these microtubules had been released and transported to the periphery of the neuronal cell body and into developing axons. In a related study, we used a similar pharmacologic regime to show that microtubules translocate from the neuronal cell body into the growing axon with the plus-end-distal polarity orientation that is characteristic of axonal microtubules (Baas and Ahmad, 1993). This result suggested that it is the manner by which the microtubules are transported that establishes their pattern of polarity orientation within the axon. More recently, the Borisy laboratory has directly observed (without the use of drugs) the transport of microtubules from the centrosome of living epithelial cells, close cousins to the neuron (Keating et al., 1997). As with neurons, the transport is consistently outward from cell center, and the microtubules are transported with their plus-ends leading. This result supports the idea that neurons are capitalizing on cellular mechanisms that are active within other cell types.

The next important issue is the molecular motor protein responsible for this transport. Studies from the Pfister laboratory have suggested that cytoplasmic dynein may be a likely candidate for this motor protein (Dillman et al., 1996a, b). Using the traditional pulse-label technique, these workers found that most of the cytoplasmic dynein that is anterogradely transported down the axon moves in slow component B. It was initially surprising that cytoplasmic dynein would be conveyed anterogradely in association with cytoskeletal elements, in light of the fact that cytoplasmic dynein had traditionally been viewed as the motor that transports vesicular organelles in the retrograde direction within the axon. On the basis of their observations, Pfister's group proposed that cytoplasmic dynein associates via its cargo domain with the actin filaments of slow component B, thus permitting its motor domain to associate intermittently with microtubules and thereby transport them somewhat more slowly, in slow component A. The association of the cytoplasmic dynein with the actin filament matrix is mediated through the dynactin complex of proteins and probably also requires an additional linker, given that dynactin and actin probably do not interact directly. This scenario is attractive because cytoplasmic dynein has the appropriate directionality to transport microtubules anterogradely with their plus-ends leading and to transport organelles retrogradely toward the minus-ends of the same microtubules. In this way, the axon economizes by assigning one important function to cytoplasmic dynein during its anterograde transport and another important function during its retrograde transport.

To test whether cytoplasmic dynein is the motor protein that transports microtubules from the neuronal centrosome into developing axons, my laboratory used a novel experimental technique for inhibiting the functions of cytoplasmic dynein. This approach is based on the disruption of dynactin, a complex of proteins that is essential for all known functions of cytoplasmic dynein (Echeverri et al., 1996). We found that inhibiting dynein function within cultured neurons suppresses axon outgrowth and inhibits the transport of microtubules from the centrosome (Ahmad et al., 1998). These observations strongly support the idea that cytoplasmic dynein is the motor protein that transports microtubules from the centrosome into the axon and down its length.

Polymer transport is only one microtubule behavior that is important for the elaboration of the axonal microtubule array. Microtubule length changes are also important, as evidenced by the fact that microtubules in the axon can obtain enormous lengths, many times the diameter of the neuronal cell body. Serial reconstruction analyses on microtubule lengths in developing axons have shown that the range of microtubule lengths increases as the axon grows (Yu and Baas, 1994). In other words, axon growth is accompanied by increases in the numbers of longer microtubules and concomitant decreases in the numbers of shorter microtubules. Such shifts in microtubule lengths within a population are consistent with the predictions of dynamic instability, a model that accurately predicts microtubule behaviors both within cytoplasm and the test tube. On the basis of these observations, it would appear that many short microtubules are transported into the axon from the centrosome. As these short microtubules continue to move down the axon, many of them become shorter or even completely depolymerize, to provide subunits for the elongation of their neighbors. In this fashion, microtubule transport and assembly/disassembly events work together to establish the axonal microtubule array.

As I suggested earlier, this scenario is analogous in many ways to the means by which dividing cells organize microtubules into a functional mitotic spindle. In both cases, the microtubules are nucleated from the centrosome. In the case of the neuron, the microtubules move very long distances from the centrosome after their release. Microtubules are probably also released from the centrosomes of the mitotic spindle so that they can exchange subunits at both ends, but in the case of the spindle, the minus-ends of the microtubules remain associated in some way with the centrosome. Perhaps there is another motor protein that tethers released microtubules to the centrosome during mitosis, and this motor protein is either not expressed or utilized in a different fashion within the neuron. Cytoplasmic dynein, the motor that we showed is crucial for transporting the released microtubules, also plays an essential role in organizing microtubules into a bipolar spindle (Heald et al., 1996). In addition, cytoplasmic dynein appears to generate forces between the astral microtubules that emanate toward cell periphery and the cell cortex. These forces help to elongate the spindle during anaphase. Given that the cortex is actin rich, I suspect that the interactions in this region of the mitotic cell are very similar to the interactions hypothesized to drive microtubules down the axon of the postmitotic neuron. Specifically, microtubules are transported with plus-ends leading by cytoplasmic dynein, which generates its forces against actin filaments. One could certainly imagine that the same forces that elongate the spindle by pulling on the astral microtubules would translocate the microtubules outward if they were not in some way tethered to the centrosome.

Another interesting similarity between the neuronal and mitotic microtubule arrays relates to dendrites, the other type of process extended by the neuron. Dendrites contain nonuniformly oriented microtubules, and it appears that this polarity pattern is achieved via the transport of a subpopulation of microtubules by the kinesin-related protein CHO1/MKLP1 (Sharp et al., 1997). CHO1/MKLP1 is the same motor protein that is enriched in the midzone region of the mitotic spindle and is thought to generate forces between oppositely oriented microtubules during late anaphase. In the neuron, CHO1/MKLP1 is thought to transport microtubules with their minus-ends leading specifically into developing dendrites. These microtubules join other microtubules that have an opposite orientation, thus resulting in a nonuniform microtubule polarity pattern. In addition, the motor protein known as Eg5 and the non-motor protein known as NuMA are both essential components of the mitotic spindle, and studies from my laboratory indicate that they are components of the neuronal microtubule arrays as well (Ferhat et al., 1998a, 1998b).

Of course, the neuronal microtubule arrays are profoundly different from the mitotic spindle in many fundamental respects. How does the postmitotic neuron build axonal and dendritic microtubule arrays from the same basic components used to make bipolar spindles in mitotic cells? I suspect that the answer to this question is that neurons express a small number of novel proteins that modify the manner by which the "molecular transport machinery" interacts with microtubules and other components of the cytoplasm. Good candidates for such proteins are the neuron-specific MAPs such as tau and MAP2. These proteins bind along the length of the microtubules and may stiffen them so that they are better suited for force generation, and stabilize them so that they do not depolymerize after release from the centrosome. In addition, these proteins interact with other nonmicrotubule cytoskeletal elements and hence may promote interactions between microtubules and actin filaments, the putative substrate against which many of the motor-driven forces are generated.

4. CHALLENGES DURING AXON DEVELOPMENT

As the axon develops, it is presented with many challenges. For example, the growth cone at the leading tip of the axon must forage through its environment and make appropriate navigational decisions. In addition, axons can form complex branching patterns, with many branches arising as interstitial or "collateral" branches from the sides of the axon. Thus an important question is how the microtubule array of the axon changes to accommodate the needs of axons undergoing growth cone turning or collateral branch formation. With regard to the former, studies from several groups indicate that microtubules undergo complex behaviors within the growth cone. Microtubules in certain regions of the growth cone are somehow captured or selectively stabilized, and as a result, the growth cone turns in the corresponding direction. The mechanism by which this occurs is unknown, but several lines of evidence suggest that it probably relates to the manner by which the microtubules interact with the cortical actin filaments (for example, *see* Challacombe et al., 1996). If microtubules are in fact transported by generating forces against actin filaments, as suggested by the Pfister group, perhaps the turning of the growth cone relates to the efficiency of the interactions among actin filaments, microtubules, and cytoplasmic dynein in different regions of the growth cone.

In terms of collateral branch formation, a critical challenge for the axon is to increase microtubule numbers rapidly and focally so that there are sufficient numbers of microtubules for the parent axon as well as the newly forming branch. It is difficult to imagine that these new microtubules could be generated at the centrosome within the cell body of the neuron and then be rapidly transported to the specific site of branch formation. A more efficient means to increase microtubule numbers focally would be to fragment the microtubules that are already present within the parent axon. Some of these fragments could then move into the new branch, while others remain within the parent axon.

Electron microscopic serial reconstruction studies from my laboratory support this idea, showing a marked increase in microtubule number and a marked decrease in microtubule length in the region of the parent axon from which a new branch starts to form (Yu et al., 1994). Current efforts are aimed at determining whether a specific microtubule severing protein is activated in this region, and if so, how this activation is regulated.

5. REESTABLISHING THE AXONAL MICROTUBULE ARRAY DURING REGENERATION

The tip of the growing axon eventually reaches its target tissue and forms a mature axon terminal. After this, the axon continues to elongate interstitially as the embryo grows. As the axon reaches its mature length, the transport of cytoskeletal elements slows, but does not cease. When the moving cytoskeletal elements reach the axon terminal, they are systematically degraded. This permits a continuous replenishment of cytoskeletal proteins essential for the maintenance of the axon. The fact that the transport machinery is never shut down is an enormous advantage to the axon when it is damaged and must mount a regenerative response. When the axon is severed, there is no longer a terminal at which polymers can be systematically degraded. Instead, a motile growth cone forms at the severed end of the axon, and microtubules are able to invade new regions of the axon as it regrows toward its target tissue. Nevertheless, regeneration does not simply utilize microtubule transport as it would otherwise occur within the adult axon. Axonal regeneration involves significant alterations to the cytoskeleton, including a marked increase in the synthesis of tubulin within the cell body and a marked increase in the rate at which microtubules are transported down the axon. In addition, the plasticity of the axonal cytoskeleton appears to revert back to a more juvenile status.

How are these changes orchestrated? For the past decade, attention has focused principally on MAPs as the key regulators of the neuronal microtubule arrays. MAPs bind to the lattice of the microtubule polymer and can affect its stability, spacing, and stiffness, as well as its interactions with other components of the cytoplasm. It is known that juvenile and adult neurons differ markedly in their expression profiles of the various isoforms of tau, MAP2, MAP1a, and MAP1b. As such, it is reasonable to hypothesize that there may be a shift during regeneration from the adult profile back to the juvenile profile. Although some studies suggest that this might be the case (for example, *see* Yin et al., 1995), most of the relevant work suggests otherwise (for example, *see* Fawcett et al., 1994; Nothias et al., 1995). Indeed, numerous studies have documented essentially similar MAP profiles in adult and regenerating axons (for review, *see* Nunez and Fischer, 1997). In considering these results, it should be noted that neurons of the peripheral and central nervous systems differ in their complements of certain MAPs, and hence may differ with regard to how MAP expression is altered during regenerative events. Nevertheless, the available data strongly suggest that a major alteration in MAP expression does not underlie the changes in the microtubule array that occur during axonal regeneration.

It is known that other key cytoskeletal genes to revert back to more juvenile expression patterns during axonal regeneration. In fact, this is true of the tubulin genes themselves. Vertebrate neurons express multiple isotypes (primary gene products) of both α- and β-tubulin, and the expression levels of some of these isotypes vary significantly between developing and adult neurons. For example, the isotypes known as β-III and T-alpha-1 are more highly expressed in developing neurons, and the expression levels of these proteins markedly increase (to juvenile levels) during regeneration after injury to the adult axon (for example, *see* Miller et al., 1989; Moskowitz et al., 1993; Moskowitz and Oblinger., 1995). Also of interest is the fact that expression of neurofilament proteins is higher in the adult neuron compared with the juvenile, and expression levels of these proteins markedly decrease during axonal regeneration (for example, *see* Oblinger and Lasek, 1988; Hoffman and Cleveland, 1988). It is unknown whether the alterations in tubulin isotypes result in functional differences in the microtubules within the neuron, or whether regulated gene expression is simply a means to synthesize higher levels of tubulin during development and regeneration. In addition, it is unknown whether the alterations in neurofilaments are directly relevant to the changes in the microtubule array.

The fundamental issue regarding regeneration is the cellular mechanism by which the neuron mounts its regenerative response. How does the neuron increase the rate and efficiency of microtubule transport to accommodate the needs of a regenerating axon? This is clearly a complex issue for which there is probably not a simple answer. Unfortunately, it would appear that many of the lines of investigation (such as those on MAPs) have not produced great insights into the issue. It occurs to me that the most important answers may lie within a better understanding of the microtubule transport machinery itself. As I explained earlier in this article, significant progress in recent years has increased our understanding of this machinery. On the basis of this recent work, it seems reasonable to propose the following scenario. During regeneration, there is a marked increase in the rate and efficiency of microtubule nucleation and release at the centrosome, and of the transport of these microtubules after their release (Moskowitz and Oblinger, 1995). The rapid clearing of new microtubules from the cell body triggers the synthesis of new tubulin, as tubulin synthesis is known to be "autoregulated" by the levels of free tubulin in the cytoplasm (Fournier and McKerracher, 1995). The alteration in the profile of tubulin isotypes may result in polymers that are more efficiently transported. In addition, cytoplasmic dynein, the motor protein that transports microtubules down the axon, undergoes a marked increase in synthesis during regeneration (Takemura et al., 1996), which can explain the increase in the rate of microtubule transport that occurs during regeneration. Finally, there may be changes either directly or indirectly to the actin filament cytomatrix, the putative substrate against which cytoplasmic dynein generates its forces to transport microtubules down the axon. Figure 1 schematically illustrates these various

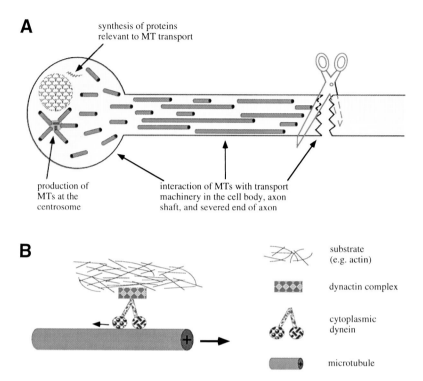

Fig. 1. Schematic illustration of a neuron that has been severed. The severing process is shown by the scissors on the right side of the diagram. Microtubules are shown as green cylinders, and the plus-end of each microtubule is indicated with a "+" sign. **(A)** The nucleus is shown as a red circle. The centrosome is shown as a gray blob with red "centrioles" within it. Polysomes are shown as a chain of small red beads. Potential sites where the generation of the axonal microtubule array might be altered during regeneration are labeled with arrows. **(B)** "Transport machinery" for an individual microtubule. The microtubule is transported with plus-end leading via cytoplasmic dynein, which interacts via dynactin with a substrate with a greater resistance to backward movement than the microtubule has to forward movement. A potential substrate would be actin filaments.

components of the microtubule transport system that may be affected during regeneration.

Unfortunately, it is difficult to speculate beyond this scenario because the molecules and mechanisms that regulate each element of the transport machinery are poorly understood. New information about microtubule nucleation and release from the centrosome, microtubule transport by cytoplasmic dynein, and the interactions among microtubules, actin filaments, and cytoplasmic dynein will be required to elucidate the factors that might increase or decrease the efficiency with which microtubules are transported down the axon. In light of the recent data suggesting that analogous mechanisms are used during mitosis and the establishment of the neuronal microtubule arrays, I would like to close this article with the following thought. Perhaps the microtubule array of a mature axon can be likened to that of a relatively quiescent interphase cell, whereas the microtubule array of a developing axon can be likened to one of the "half-spindles" of a cell undergoing mitotic division. If this is correct, it may be that the mechanisms responsible for shifting an adult axon into a regenerating axon are analogous to those that shift an interphase cell into mitosis. Thus the answers to the fundamental questions surrounding axonal regeneration may lie within the kinds of molecules and mechanisms that have traditionally been viewed as regulators of the cell cycle. Whether or not this is true remains to be seen, but if correct, this perspective could provide a better understanding of the events that underlie axonal regeneration as well as insights into potential clinical strategies for enhancing its efficiency.

ACKNOWLEDGMENTS

The work in my laboratory is supported by grants from the National Institutes of Health and the National Science Foundation. I thank Wenqian Yu for assistance in preparation of the figure and Monica Oblinger, Itzhak Fischer, and Mark Black for their critical comments on the manuscript.

REFERENCES

Ahmad, F. J. and Baas, P. W. (1995) Microtubules released from the neuronal centrosome are transported into the axon. *J. Cell Sci.* **108,** 2761–2769.

Ahmad, F. J., Joshi H. C., Centonze V. E., and Baas, P. W. (1994). Inhibition of microtubule nucleation at the neuronal centrosome compromises axon growth. *Neuron* **12,** 271–280.

Ahmad, F. J., Echeverri, C. J., Vallee R. B., and Baas, P. W. (1998) Cytoplasmic dynein and dynactin are required for the transport of microtubules into the axon. *J. Cell Biol.* **140,** 391–401.

Baas, P. W. (1996) The neuronal centrosome as a generator of microtubules for the axon. *Curr. Top. Dev. Biol.* **33,** 281–298.

Baas, P. W. (1997) Microtubules and axonal growth. *Curr. Opin. Cell Biol.* **9,** 29–36.

Baas, P. W. and Heidemann, S. R. (1986) Microtubule reassembly from nucleating fragments during the regrowth of amputated neurites. *J. Cell Biol.* **103,** 917–927.

Baas, P. W. and Ahmad, F. J. (1992) The plus ends of stable microtubules are the exclusive nucleating structures for microtubules in the axon. *J. Cell Biol.* **116,** 1231–1241.

Baas, P. W. and Joshi, H. C. (1992) Gamma-tubulin distribution in the neuron: implications for the origins of neuritic microtubules. *J. Cell Biol.* **119,** 171–178.

Baas, P. W. and Ahmad, F. J. (1993) The transport properties of axonal microtubules establish their polarity orientation. *J. Cell Biol.* **120,** 1427–1437.

Baas, P. W. and Yu, W. (1996) A composite model for establishing the microtubule arrays of the neuron. *Mol. Neurobiol.* **12,** 145–161.

Baas, P. W. and Brown, A. (1997) Slow axonal transport: the polymer transport model. *Trends Cell Biol.* **7,** 380–384.

Barton, N. R. and L. S. B. Goldstein, (1996) Going mobile: microtubule motors and chromosome segregation. *Proc. Natl. Acad. Sci. USA* **93,** 1735–1742.

Brinkley, B. R. (1985). Microtubule organizing centers. *Annu. Rev. Cell Biol.* **1,** 145–172.

Challacombe, J. F., Snow, D. M., and Letourneau, P. C. (1996) Actin filament bundles are required for microtubule reorientation during growth cone turning to avoid an inhibitory guidance cue. *J. Cell Sci.* **109,** 2031–2041.

Dillman J. F. III, Dabney, L. P., and Pfister, K. K. (1996a) Cytoplasmic dynein is associated with slow axonal transport. *Proc. Natl. Acad. Sci. USA* **93,** 141–144.

Dillman J. F. III, Dabney, L. P. Karki, S., Paschal, B. M. Holzbaur, E. L. F., and Pfister, K. K (1996b) Functional analysis of dynactin and cytoplasmic dynein in slow axonal transport. *J. Neurosci.* **16,** 6742–6752.

Echeverri, C. J., Paschal, B. M. Vaughan, K. T., and Vallee, R. B. (1996) Molecular characterization of the 50-kD subunit of dynactin reveals function for the complex in chromosome alignment and spindle organization during mitosis. *J. Cell Biol.* **132,** 617–633.

Fawcett, J. W., Matthews, G. Housden. E. Goedert, M., and Matus, A. (1994) Regenerating sciatic nerve axons contain the adult rather than the embryonic pattern of microtubule associated proteins. *Neuroscience* **61,** 789–804.

Ferhat, L., Cook, C., Chauviere, M., Harper, M., Kress M., Lyons, G. E., and Baas, P. W. (1998a) Expression of the mitotic motor protein Eg5 in postmitotic neurons: implications for neuronal development. *J. Neurosci.* **18,** 7822–7835.

Ferhat, L., Cook, C., Kuriyama, R., and Baas, P. W. (1998b) NuMA is a component of the somatodendritic microtubule arrays of the neuron. *J. Neurocytology* **27,** 887–899.

Fournier, A. E. and McKerracher, L. (1995) Tubulin expression and axonal transport in injured and regenerating neurons in the adult mammalian central nervous system. *Biochem. Cell Biol.* **73,** 659–664.

Heald, R., Tournebize, R., Blank, T., Sandaltzopoulos R., Becker P., Hyman A., and Karsenti, E. (1996) Self-organization of microtubules into bipolar spindles around artificial chromosomes in *Xenopus* egg extracts. *Nature (Lond.)* **382,** 420–425.

Hirokawa, N., Terada, S. Funakoshi, T., and Takeda, S. (1997) Slow axonal transport: the subunit transport model. *Trends Cell Biol.* **7,** 384–388.

Hoffman, P. N. and Cleveland, D. W. (1988) Neurofilament and tubulin expression recapitulates the developmental program during axonal regeneration: induction of a specific beta-tubulin isotype. *Proc. Natl. Acad. Sci. USA* **85,** 4530–4533.

Keating, T. J., Momcilovic, D., Rodionov, V. I., and Borisy, G. G. (1997) Microtubule release from the centrosome. *Proc. Natl. Acad. Sci. USA* **94,** 5078–5083.

Lasek, R. J. (1986) Polymer sliding in axons. *J. Cell Sci.* **5 (suppl),** 161–179.

Lasek, R. J. (1988) Studying the intrinsic determinants of neuronal form and function, in *Instrinsic Determinants of Neuronal Form and Function* (Lasek, R. J. and Black, M. M., eds.), Alan R. Liss, New York, pp. 1–60.

Miller, F. D., Tetzlaff, W., Bisby, M. A., Fawcett, J. W., and Milner, R. J. (1989) Rapid induction of the major embryonic alpha-tubulin mRNA, T alpha 1, during nerve regeneration in adult rats. *J. Neurosci.* **9,** 1452–1463.

Moritz, M., Braunfeld, M. B., Sedat, J. W., Alberts, B., and Agard, D. A. (1995) Microtubule nucleation by γ-tubulin-containing rings in the centrosome. *Nature (Lond.)* **378,** 638–640.

Moskowitz, P. F. and Oblinger, M. M. (1995) Sensory neurons selectively upregulate synthesis and transport of the beta III-tubulin protein during axonal regeneration. *J. Neurosci.* **15,** 1545–1555.

Moskowitz, P. F., Smith, R., Pickett, J., Frankfurter, A., and Oblinger, M. M. (1993) Expression of the class III beta-tubulin gene during axonal regeneration of dorsal root ganglion neurons. *J. Neurosci. Res.* **34,** 129–134.

Nothias, F., Boyne, L., Murray, M., Tessler, A., and Fischer, I. (1995) The expression and distribution of tau proteins and messenger RNA in rat dorsal root ganglion neurons during development and regeneration. *Neuroscience* **66,** 707–719.

Nunez, J. and Fischer, I. (1997). Microtubule-associated proteins (MAPs) in the peripheral nervous system during development and regeneration. *J. Mol. Neurosci.* **8,** 207–222.

Oblinger, M. M. and Lasek, R. J. (1988) Axotomy-induced alterations in the synthesis and transport of neurofilaments and microtubules in dorsal root ganglion cells. *J. Neurosci.* **8,** 1747–1758.

Sharp, D. J., Yu, W., Ferhat, L., Kuriyama, R., Rueger, D. C., and Baas, P. W. (1997) Identification of a motor protein essential for dendritic differentiation. *J. Cell Biol.* **138,** 833–843.

Takemura, R., Nakata, T., Okada, Y., Yamazaki, H., Zhang, Z., and Hirokawa, N. (1996) mRNA expression of KIF1A, KIF1B, KIF2, KIF3A, KIF3B, KIF4B, KIF5, and cytoplasmic dynein during axonal regeneration. *J. Neurosci.* **16,** 31–35.

Walezak, C. E. and Mitchison, T. J. (1996) Kinesin-related proteins at mitotic spindle poles: function and regulation. *Cell* **85,** 943–946.

Yin, H. S., Chou, H. C., and Chiu, M. M. (1995) Changes in the microtubule proteins in the developing and transected spinal cords of the bullfrog tadpole: induction of microtubule-associated protein 2c and enhanced levels of tau and tubulin in regenerating axons. *Neuroscience* **67,** 763–775.

Yu, W. and Baas, P. W. (1994) Changes in microtubule number and length during axon differentation. *J. Neurosci.* **14,** 2818–2829.

Yu, W., Ahmad, F. J., and Baas, P. W. (1994) Microtubule fragmentation and partitioning in the axon during collateral branch formation. *J. Neurosci.* **14,** 5872–5884.

Zheng Y., Wong, M. L., Alberts, B., and Mitchison, T. (1995) Nucleation of microtubule assembly by a γ-tubulin-containing ring complex. *Nature (Lond.)* **378,** 578–583.

8

Transplants and Neurotrophins Modify the Response of Developing and Mature CNS Neurons to Spinal Cord Injury

Axonal Regeneration and Recovery of Function

Barbara S. Bregman

1. INTRODUCTION

The consequences of spinal cord injury are complex. The circuitry underlying rhythmic alternating stepping movements such as locomotion are intrinsic within the spinal cord itself (*see* Chap. 4). Part of this segmental circuitry is known as the spinal pattern generator for locomotion (SPGL) *(1,2)*, which is capable of generating rhythmic alternating stepping movements. This segmental circuitry is normally under the influence of both descending and segmental afferent input (Fig. 1A). Spinal cord injury (SCI) during development or in the adult disrupts the supraspinal input to the segmental circuitry, leading to loss of function. The segmental circuitry itself, however, is still intact, but lacking descending modulation. This suggests that complete, point-to-point restoration of all damaged pathways after SCI may not be necessary, rather, even modest restoration of supraspinal input may be sufficient to yield recovery of patterned motor activity such as locomotion.

A number of factors appear to restrict growth in the central nervous system (CNS) after injury (Fig. 1B). It is likely that both intrinsic neuronal and extrinsic environmental influences contribute to the lack of regeneration that characterizes spinal cord and other CNS injury. Central and peripheral neurons differ in their intrinsic capacity for growth. During development both CNS and peripheral nervous system (PNS) neurons express a variety of growth-associated proteins. After connections have been established, however, the cellular expression of these proteins is downregulated. After axotomy of peripheral neurons, there is an alteration in the cell body of the injured neurons such that specific cellular and molecular programs of the injured neurons (transcription factors, immediate early genes, and genes for structural proteins associated with axonal regrowth) are upregulated, leading to successful regeneration. In contrast, after axotomy of

From: *Neurobiology of Spinal Cord Injury*
Edited by: R. G. Kalb and S. M. Strittmatter © Humana Press Inc., Totowa, NJ

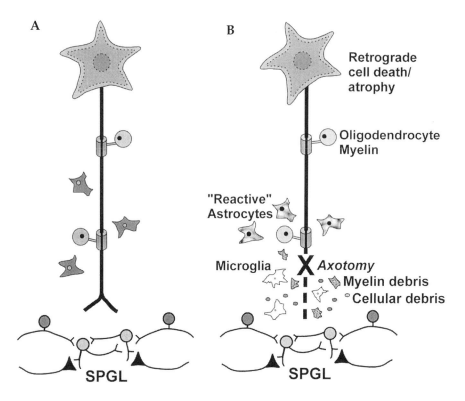

Fig. 1. Schematic diagram of normal input to the spinal pattern generator for locomotion (SPGL) **(A)** and the consequences of spinal cord injury **(B)**. See text for description.

adult CNS neurons, similar cellular programs are either not upregulated, or are upregulated only transiently. The axotomized neurons either atrophy (often observed after injury in the adult) or undergo retrograde cell death (usually observed after neonatal injury).

In addition to intrinsic neuronal factors, differences in the mature CNS environment also restrict axonal regeneration. During development, a number of molecules are expressed in the environment of the growing axons that support their growth and guide the axons toward their targets. As the CNS matures, the expression of molecules that support growth decreases and the expression of molecules that inhibit axonal growth increases. After peripheral nerve injury, Schwann cells contribute to a cellular and molecular environment, at and distal to the injury, that supports axonal regrowth. In contrast, after SCI, the environment at the injury site is altered dramatically, but is not conducive to growth. The blood-brain barrier is disrupted, there is activation of microglia and invasion of macrophages, and astrocytes become reactive and wall off the injured area. Recent studies suggest that the astrocytes not only present a physical barrier that restricts growth, but they also contribute to the hostile molecular environment at the injury site that interferes with axonal growth *(3,4).* In addition to the physical disruption at the injury site, a secondary injury cascade leads to increased local cell loss at the lesion site, cyst formation, and cavitation (*see* Chap. 1). Even if axons at the injury site are spared direct damage, conduction across the injury site is often abolished (*see* Chap. 10). Caudal to the lesion, the spinal cord circuitry is still present, but the supraspinal input has been decreased or abolished. Furthermore, the astrocyte and myelin-associated inhibitors of axonal elongation are still present below the injury site, presenting further limits on axonal regrowth (*see* Chap. 6).

In recent years, work in our laboratory and others has shown that transplants of fetal spinal cord tissue placed at the site of injury modify the response of both developing and mature CNS neurons to SCI. This review focuses on the influence of transplants of fetal spinal cord tissue and exogenous neurotrophic support after SCI during development and in the adult on (1) the cell body response to injury; (2) axonal growth at and below the injury site; and (3) the influence of such growth on the recovery of motor function after injury.

2. EXOGENOUS NEUROTROPHIC SUPPORT RESCUES IMMATURE AXOTOMIZED CNS NEURONS FROM RETROGRADE CELL DEATH

Neurons injured early in development are often more vulnerable to injury than if the same injury occurs in the mature neuron. Axotomy early in development leads to the rapid retrograde cell death of both central and peripheral neurons. In contrast, mature neurons are less likely to die after axotomy, especially if the injury occurs at some distance from the cell body. Instead, these mature neurons often persist, but in an atrophied state. We have shown that after spinal cord lesions in newborn rats, transplants of fetal spinal cord tissue modify the effect of neonatal injury by permanently rescuing the immature axotomized neurons from the retrograde cell death *(5–7).* The transplants may rescue the neurons by providing a substitute for the lost target-derived trophic support.

Much of our current understanding of the role that neurotrophins play during normal development comes from studies of the PNS, particularly the sensory dorsal root ganglia (DRG) *(8–16).* In vitro and in vivo studies, and more recently, transgenic studies of mice with targeted null mutations of the neurotrophin or neurotrophin receptor genes have established critical roles for members of the neurotrophin family [nerve growth factor (NGF), brain-derived neurotrophic factor (BDNF), and neurotrophin-3 (NT-3)] as classical target-derived survival factors for particular subclasses of developing sensory neurons *(15,17).* Not only are particular neurotrophins required for survival, but they have also been shown to contribute to axonal elongation during normal development. Specific neurotrophic factors appear to contribute to the development of particular pathways. For example, NT-3 and its specific receptor trk C are required for the development of large diameter (Ia) sensory muscle afferents projecting to motorneurons, whereas NGF contributes to the development of small-diameter sensory nociceptive afferents terminating in the superficial dorsal horn *(9,15,16,18–20)*.

Immature neurons projecting to the periphery are dependent on target-derived neurotrophic factors for their survival. Axotomy leads to massive retrograde cell death of the injured neurons; the exogenous administration of BDNF, ciliary neurotrophic factor (CNTF), or glial-derived neurotrophic factor (GDNF) can prevent this axotomy-induced death of immature motor neurons *(11,21–27)*. Neurons with axons contained entirely within the CNS are also dependent on target-derived factors for survival in vivo. Axotomy early in development results in the massive and rapid retrograde cell death of immature axotomized central neurons (for review, *see* ref. *8,* and refs. *5* and *28–30*). We have shown that the death of central neurons after axotomy is also due to loss of trophic support from target regions.

To test whether exogenous neurotrophic support *per se* could substitute for the loss of normal target-derived influences, spinal cord hemisection lesions were made in newborn (2–3 postnatal d) rats. Exogenous neurotrophic support was supplied directly to the injured axons at the spinal lesion site. These studies compared the influence of members of the neurotrophin family (NGF, BDNF, NT-3, NT-4) with that of the injury-associated growth factor CNTF *(31)*. The effect of exogenous neurotrophic support on cell survival was assayed qualitatively and quantitatively at 7 and at 30 d after injury in axotomized brainstem and spinal cord nuclei (red nucleus, locus coeruleus, raphe nuclei, and nucleus dorsalis) After spinal cord hemisection in newborn rats, the exogenous application of BDNF, NT-3, NT-4, or CNTF prevents the retrograde cell death of axotomized red nucleus neurons (and other brainstem spinal neurons) in vivo *(31,32)*. The rescue of red nucleus neurons was *permanent* in the presence of BDNF, but was only transient in the presence of NT-3, NT-4, NGF, or CNTF. Ascending sensory neurons within the nucleus dorsalis (Clarke's nucleus) of the spinal cord are also axotomized by this same lesion. The application of exogenous NT-3, but not NGF or BDNF, rescued Clarke's nucleus neurons. These observations indicate that neurotrophic factors play a crucial role in the survival of CNS neurons in vivo after injury in the developing animal. Furthermore, particular populations of neurons are dependent on *specific* neurotrophic support after injury. Quantitative analysis of the red nucleus neurons confirmed the magnitude of the rescuing effect *(31)*. Mature red nucleus neurons have high-affinity receptors for BDNF (trk B) and to a lesser degree for NT-3 (trk C) *(33–36)*, suggesting that these neurons may be influenced by these neurotrophins. Recent studies have shown that similar principles of trophic dependence may also operate in the adult CNS after injury (*see* Subheading 7.).

3. IMMATURE AXOTOMIZED NEURONS THAT ARE RESCUED BY FETAL SPINAL CORD TRANSPLANTS REGENERATE AXONS

After SCI during development, transplants of fetal spinal cord tissue at the lesion site not only rescue the immature axotomized neurons but also support the regrowth of their axons into the transplant and back into the host spinal cord caudal to the transplant *(37–42)*. Studies using retrograde double-labeling techniques have shown that the immature axotomized neurons rescued by the transplants regenerate*(37,38)*, and the regrowth after neonatal injury is truly regeneration, not solely axonal plasticity by late developing pathways. Quantification of the double labeling was surprising *(37)*. In each of the brainstem-spinal nuclei examined, approximately 30% of the neurons that were axotomized regenerated to reach spinal cord levels immediately caudal to the transplant, and 10% of the axotomized neurons regrew axons to reach lumbar spinal cord levels *(37)*. This indicates that contrary to the prevalent belief that diffusely projecting pathways have a greater intrinsic capacity for regeneration, both diffusely projecting pathways such as raphe spinal neurons and more point-to-point projecting pathways such as rubrospinal neurons have a similar capacity for regrowth. The specific cellular characteristics that permit particular neurons to survive neonatal axotomy and to regenerate their axons are not clear. It is possible that those neurons that regenerate are those that have established sustaining collateral branches to rostral targets such as the cervical spinal cord or the cerebellum. One reason that cell death after axotomy in the developing animal is more severe than in the adult animal may be that axonal collaterals to divergent targets have not yet developed at birth, but rather are established in the early postnatal period. The existence of axon collaterals to the brainstem, the cerebellum, or the rostral spinal cord may sustain axotomized neurons that have lost their projection to distal targets.

Recent studies have used retrograde double labeling to examine whether the rubrospinal neurons that are rescued by fetal transplants are those that have established collateral projections to either the cerebellum or the cervical spinal cord *(43)*. Double labeling was observed in the axotomized neurons of the red nucleus after injections of the tracer into the cervical spinal cord, but was not seen after injections into the cerebellum. There was a high proportion of double-labeled neurons in the red nucleus, indicating that a significant population of axotomized rubrospinal neurons that regenerate axons through a transplant to innervate the caudal host spinal cord also maintain axon collaterals at cervical spinal cord levels. The cerebellar collaterals, however, do not appear to play a role in the survival and regrowth of the red nucleus neurons after SCI. These observations support the hypothesis that development of a system of axonal collaterals supports neuronal survival and axonal elongation after CNS injury.

4. REGENERATING AND SPROUTING AXONS DIFFER IN THEIR REQUIREMENTS FOR GROWTH

After neonatal SCI, axotomized brainstem-spinal (raphe spinal, rubrospinal, coeruleospinal) and corticospinal neurons regenerate axons into and through a fetal spinal cord transplant placed into the site of either a spinal cord hemisection or transection *(37,38)*. In contrast, if fetal tissue that is not a normal target of the axotomized neurons (embryonic hippocampus or cortex) is placed into a neonatal spinal cord *hemisection,* brainstem-spinal serotonergic axons transiently innervate the transplant but subsequently withdraw *(43–46)*. Experiments were designed to test the hypothesis that after spinal cord *transection,* brainstem spinal axons would cross the non-target transplant, reach normal spinal cord targets caudal to the transection, and gain access to requisite target-derived influences permitting permanent maintenance. Surprisingly, after a complete spinal cord *transection,* brainstem-spinal axons failed to grow into an inappropriate target even transiently *(43,44)*. These observations suggest that the transient axonal ingrowth into non-target transplants observed previously may represent lesion-induced axonal sprouting by contralateral uninjured axons. Retrograde double labeling with fluorescent dyes was used to test directly whether axonal sprouting of neurons that maintain collaterals to uninjured spinal cord targets (1) provide the transient ingrowth of brainstem-spinal axons into an inappropriate transplant; and (2) contribute to permanent ingrowth into target-specific transplants *(43,44)*. Axons of intact red nucleus, raphe nucleus, and locus coeruleus extend into the non-target transplant while maintaining collaterals to the host spinal cord caudal to the transplant. The lesion-induced sprouting by uninjured axons was also observed with a target-specific transplant. Taken together, these studies indicate that sprouting and regenerating axons differ in their requirements for growth after injury. Although the primary form of axonal growth into target-specific transplants after either hemisection or transection is regenerative growth, cues contained in non-target tissues are unable to support regeneration after transection but can support collateral axonal growth from neurons that also send axons to normal targets elsewhere in the cord *(43,44)*.

5. TRANSPLANT-MEDIATED RECOVERY OF FUNCTION AFTER EARLY SPINAL CORD LESIONS

Anatomic regrowth supported by the transplants mediates recovery of function. After spinal cord hemisection lesions at birth, the anatomic remodeling (sprouting and regeneration) elicited by the transplants mediates recovery of locomotor function *(47)*. After spinal cord lesion and transplantation, the extent of recovery of both reflexes and goal-directed overground locomotion is significantly greater in the transplant groups than in the lesion-only groups *(47)*. The transplants support recovery of function after spinal cord transection lesions as well *(48–51)*. Based on a series of anatomic studies from our laboratory and others, it is clear that after SCI at birth, transplants serve both as a bridge and as a relay to restore some of the suprasapinal input to spinal cord levels below the injury site (*see* refs. *52* and *53* for reviews). More recently, we have asked whether transplants are able to influence the recovery of skilled forelimb movement, as well as the more stereotyped movement involved in locomotion. Skilled forelimb movement such as target-directed reaching is not under the influence of pattern generators and may require somewhat greater restoration of supraspinal control for any recovery of function to occur. The neural substrates underlying skilled forelimb movement such as reaching may be far more complex than those underlying rhythmic alternating movements such as locomotion. Recovery of skilled forelimb movement may require more precise restoration of supraspinal control than does recovery of locomotion. Both goal-directed reaching and many of the anatomic projections that contribute to target reaching are either absent or immature at birth. Spinal cord overhemisection lesions were made at a high cervical level in newborn rat pups, and anatomic regrowth and recovery of function in lesion-only and transplant animals were examined *(40,41)*. After cervical SCI, the transplants support recovery not only of rhythmic alternating movements involved in locomotion, but also of skilled forelimb movement *(40)*. The recovery of function is accompanied by extensive anatomic regrowth of supraspinal axons into the transplants *(41)*. There is also restoration of supraspinal input to propriospinal neurons, which may also contribute to the greater recovery of function in these animals *(41)*. Thus, after early spinal cord lesions, transplant-mediated anatomic plasticity contributes to recovery not only of stereotyped movement but also of skilled forelimb movement.

6. AXONAL REGROWTH AFTER INJURY DECREASES AS THE AGE OF THE ANIMAL AT TIME OF INJURY INCREASES

The capacity of CNS neurons for axonal regrowth after injury decreases as the age of the animal at time of injury increases *(29,39,54,55)*. One striking difference between axonal plasticity in the developing and mature CNS is that after lesions in the adult CNS, axonal growth is spatially restricted. It is likely that both intrinsic neuronal and extrinsic environmental influences contribute to this attenuation of the growth capacity in the mature CNS, for example, the decrease in the expression of growth-associated proteins, appearance of myelin-associated inhibitors of axonal growth, alterations in astrocytes and extracellular matrix molecules restricting growth, and so forth *(56–65)*.

Although some axonal regrowth occurs after SCI very early in development *(66–68)*, this capacity for regrowth decreases as the age of the animal at the time of injury increases. For example, some axonal growth across the site of a spinal cord transection is made early in development (at times equivalent to embryonic development) in chicks *(66,67,69)* and opossum *(68)*. This regrowth occurs only during a critical window early in development, during the period of initial

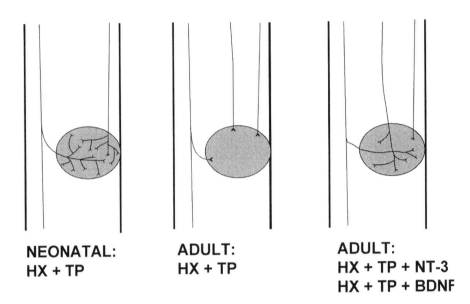

NEONATAL: **ADULT:**   **ADULT:**
HX + TP   **HX + TP**   **HX + TP + NT-3**
                                           HX + TP + BDNF

Fig. 2. Schematic diagram of the changes in axonal growth with increasing age at time of injury. After neonatal spinal cord lesion (HX), supraspinal axons grow into transplants (TP) of fetal spinal cord tissue and project densely throughout the territory provided by the transplant. In addition, after neonatal lesions, supraspinal axons also grow through the transplant and reenter the host spinal cord caudal to the transplant (not shown on diagram). After lesions in the adult, although host axons grow into the transplant, they terminate near the host-transplant border and do not extend throughout the transplant. After lesions in the adult, the exogenous administration of neurotrophic factors NT-3 or BDNF in addition to the transplant increases the extent of axonal regrowth of the injured axons.

axonal elongation and prior to the onset of inhibitory influences in the CNS environment. In contrast, after complete spinal cord transection in the newborn rat or kitten, supraspinal projections to the spinal cord caudal to the lesion are abolished permanently *(41,51)*.

After spinal cord lesions and transplants of fetal spinal cord tissue at birth, however, there is extensive growth of host descending axons into the transplants (Fig. 2). The host axons extend throughout the territory provided by the transplant, and both regenerating axons and sprouting axons contribute to this growth *(37,38,42,44,70)*. The capacity for growth into the transplant decreases within the early postnatal period *(39)*. Although descending axons project into a transplant of fetal spinal cord tissue after injury in the adult, despite the fetal CNS environment within the transplant, host axons terminate within a few hundred microns of the host/transplant border *(39,70–74)* (Fig. 2). It is likely that alterations in both the intrinsic properties of the neurons and the molecular characteristics of the CNS environment contribute to this decrease in the capacity of the mature CNS for regrowth after injury as the animal matures.

In contrast to the relatively restricted growth of supraspinal axons into the transplant after injury in the mature spinal cord described above, regenerative growth of dorsal root axons into transplants of fetal spinal cord tissue is quite robust after lesions in both developing and adult animals. After central rhizotomy and transplant of fetal spinal cord tissue, mature dorsal root axons regrow within the fetal CNS environment and extend throughout the transplant territory *(75–77)*. The conditions that restrict growth are not clear, nor are the reasons for the differences in the capacity of descending and dorsal root axons for growth. Environmental signals must play a role in restricting axonal growth within a mature CNS environment, since mature neurons can extend axons long distances within a graft of peripheral nerve *(78–83)*, but as soon as the axons reenter a CNS environment the axons terminate and do not extend long distances within the CNS environment. Mature dorsal root axons, in contrast to axons that are contained entirely within the CNS, are more robust in their capacity for growth. One difference between the DRG neurons and those contained entirely within the CNS is that the DRG neurons maintain axonal projections to peripheral targets. These targets are known to provide neurotrophic support for the DRG neurons. The availability of the neurotrophic support may render the mature DRG neurons more robust in their regrowth response after central rhizotomy.

7. LESIONS AND TRANSPLANTS IN THE ADULT

After spinal cord lesions and transplants in the adult, neurotrophins increase the density and extent of axonal elongation into transplants of fetal spinal cord tissue.

It is clear that developing neurons are dependent on target-derived trophic support *(15,84–86)*. Recent studies have demonstrated that *mature* CNS neurons are also dependent on trophic support from their targets for their continued survival and function. In the septohippocampal pathway, exogenous NGF, BNDF, or CNTF rescues axotomized basal forebrain neurons and supports the regrowth of their axons *(71–91)*. NGF and BDNF are not required for the survival of mature DRG neurons in culture, but these neurotrophins do increase the rate and extent of neurite elongation in these neurons *(92–94)*. In the adult, in vivo, the exogenous application of NGF increases the growth of dorsal root axons into spinal cord *(95–98)*. A combination of NT-3 and BDNF has been shown to support the regrowth of brainstem fibers into Schwann cell grafts placed into thoracic level lesions in the adult rat; without neurotrophin treatment, brainstem axons do not enter the graft *(99–102)*. Similarly, cells genetically modified to secrete particular neurotrophic factors transplanted to the spinal cord have also been shown to influence the axonal growth of particular populations of spinally projecting neurons *(102,103)*. Taken together. these studies support a role for neurotrophic factors in the mature CNS.

After spinal cord lesion and transplants in adult rats, the administration of neurotrophic factors increases the capacity of mature CNS neurons for regrowth after injury *(74)* (Fig. 2). Spinal cord overhemisection lesions were made at T6 in adult rats. Transplants of embryonic d (E) 14 fetal spinal cord tissue were placed into the lesion site. BDNF, NT-3, NT-4, CNTF, or vehicle alone was administered at the site of the transplant at the time of injury and transplantation. After 1–2 months of survival, neuroanatomic tracing and immunocytochemical methods were used to examine the growth of host axons within the transplants. BDNF, NT-3, and NT-4 *each* increased both the density and extent of serotonergic axonal ingrowth into the transplants dramatically. In controls (transplant only, no neurotrophin), serotonergic axons terminate at the immediate host-transplant border. In the presence of exogenous neurotrophic support, there is an increase in the distance and density of ingrowth within the transplant tissue. In animals treated with either BDNF or NT-3, there was also an increase in corticospinal axonal ingrowth within the transplants, similar to that observed for the serotonergic axons. Qualitatively, the extent of corticospinal ingrowth was similar with each of the neurotrophins examined (BDNF, NT-3, and NT-4). This was surprising in light of previous observations *(104,105)* indicating that NT-3 but not BDNF increased the sprouting of corticospinal axons after spinal cord lesion. The influence of neurotrophins on the growth of injured CNS axons was not a generalized effect of growth factors *per se,* since the administration of CNTF had *no effect* on the growth of any of the descending CNS axons tested (unlike the positive effect of CNTF on the survival of immature axotomized neurons) (*see* above and ref. 106). These results are important because they indicate that in addition to influencing the survival of developing CNS and PNS neurons, neurotrophic factors are able to exert a neurotropic influence on injured mature CNS neurons by increasing their axonal growth within a favorable environment (transplant). In other studies (*see* Section 8.), we have demonstrated that these same interventions that increase the extent of axonal regrowth after spinal cord injury in the adult are also associated with changes in the cell body of the axotomized neurons *(107,108).*

8. INTRINSIC NEURONAL DETERMINANTS OF AXONAL REGROWTH

There are differences in the cellular responses of CNS and PNS neurons to injury. Injury to peripherally projecting neurons (DRG neurons and motorneurons) leads to a rapid and prolonged upregulation of both GAP-43 mRNA and protein. During PNS regeneration there is also an increase in c-Jun expression and in cytoskeletal proteins such as tubulins, while neurofilament expression decreases *(109–111).* Central rhizotomy of DRG neurons, however, fails to induce either c-Jun or growth-associated protein 43 (GAP–43) expression in the injured neurons *(57,112–115).* Although peripheral nerve grafts constitute an extremely favorable environment for axonal regeneration *(78,79,81,82,116)*, only a few of the injured central axons of DRG cells regenerate into peripheral nerves grafted into the dorsal columns of the spinal cord. If, however, the sciatic nerve is also transected as a conditioning stimulus, the alterations in the expression of regeneration-associated genes in the DRG neurons lead to greater axonal growth of the central axons *(81,97,98,117)*. Similarly, axotomy of CNS pathways fails to induce genetic programs associated with regenerative growth. Although the initial pattern of gene expression in axotomized central and peripheral neurons is similar, peripheral injury is often associated with successful axonal regeneration and central lesions with regenerative failure. It is likely that CNS neurons require trophic support from their target not only during development but also in the adult, or that a lesion makes them dependent again in an immature fashion. Thus, the application of neurotrophic factors may increase axonal growth after injury in the adult by acting at the level of the cell body to upregulate cellular programs associated with a regenerative response.

9. TRANSPLANTS INCREASE C-JUN EXPRESSION IN DORSAL ROOT GANGLION NEURONS AFTER CENTRAL AXOTOMY

Although DRG neurons regenerate their axons after peripheral axotomy, they fail to do so after central rhizotomy. It is likely that both intrinsic neuronal and extrinsic environmental factors contribute to successful regeneration after peripheral nerve lesion. We compared the response of DRG neurons with peripheral (Fig. 3A) or central (Fig. 3B,C) axotomy. The response of DRG neurons after central rhizotomy of the L4–L6 DRG (Fig. 3B) was compared with that of DRG neurons in which central rhizotomy was combined with a spinal cord hemisection and transplant (Fig. 3C). As described previously *(52,75,77,118)*, calcitonin-gene–related peptide (CGRP)-positive dorsal root axons are able to regrow into the fetal CNS environment but fail to do so in a mature CNS environment *(119)*. We examined c-Jun expression in DRG neurons after PNS injury and that after central rhizotomy with and without transplants. After peripheral injury, almost all the DRG neurons upregulate c-Jun expression. After central axotomy (unlike peripheral axotomy), there is little c-Jun expression in DRG neurons (Fig. 4A). In contrast, central rhizotomy and transplant lead to the upregulation of c-Jun expression in the DRG neurons (Fig. 4B,C), particularly in the small and medium-sized neurons. Quantitative comparison of the percentage of DRG neurons expressing c-Jun (Fig. 5) after central rhizotomy indicates that under conditions of regrowth 63% of DRG neurons express c-Jun whereas only 18% do so without transplants (Fig. 5). This modest upregulation of c-Jun expression after central rhizotomy alone may correspond to the regrowth of these axons within the Schwann cell environment prior to reaching the spinal cord. In contrast, there is a marked increase in the proportion of DRG neurons expressing c-Jun under conditions of successful axonal regrowth into the CNS.

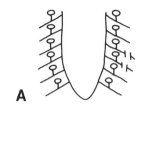

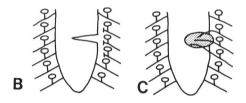

Fig. 3. Schematic diagrams of the surgical paradigm for dorsal root ganglion studies. **(A)** Peripheral axotomy: the right sciatic nerve was axotomized and a 1-mm-long segment of nerve was removed to prevent contact of the cut ends. **(B)**. Central lesion: a lumbar spinal cord hemisection was made, interrupting both dorsal funiculi and the right side of the spinal cord, the central processes of three consecutive dorsal root ganglia (L4–L6) were axotomized, and 1-mm-long segments of the roots were removed. **(C)** In the third group of animals, a hemisection was made, and the central processes of three consecutive dorsal roots (L4–L6) were axotomized, as above. In addition, the cut ends of the dorsal roots were sandwiched between two pieces of embryonic (E14) lumbar spinal cord transplants placed within the lumbar hemisection cavity. (Reproduced with permission from ref. *107*.)

10. EXPRESSION OF C-JUN IN MATURE CNS NEURONS AFTER SPINAL CORD INJURY

Transplants and exogenous neurotrophic support also alter the cell body response to injury in neurons contained entirely within a CNS environment by increasing c-Jun expression in axotomized brainstem-spinal neurons. The protein product of the immediate early gene *c-jun* is highly expressed in axotomized peripheral neurons during regeneration and in central neurons after cervical, but not thoracic spinal cord injury *(120)*. This suggests that the distance that the axotomy occurs from the cell body may play a critical role in determining the regenerative response of the neuron. A recent series of experiments sought to determine whether interventions (transplants and neurotrophic factors) shown to increase the amount of axonal growth also upregulate expression of c-jun in the axotomized neurons *(107,108,121)*. The following experimental interventions were used: hemisection only, hemisection plus transplant, and hemisection plus neurotrophins (BDNF or NT-3) with or without transplant.

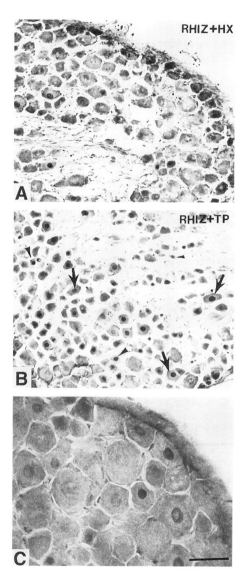

Fig. 4. Expression of c-Jun in dorsal root ganglia (DRG) 7 days after central axotomy. **(A)** There is little c-Jun expression in DRG after dorsal root rhizotomy and lumbar hemisection (RHIZ + HX). Only occasional neurons (arrows) exhibit c-Jun immunoreactivity following central rhizotomy. **(B).** Intense c-Jun immunostaining is observed in the nuclei of both medium (arrows) and small (arrowheads) DRG neurons after dorsal root rhizotomy plus hemisection and embryonic spinal cord transplant (RHIZ + HX + TP). **(C)** c-Jun immunoreactivity appears restricted to the small and medium-sized neurons; little c-Jun immunoreactivity is seen in large DRG cells after rhizotomy and the addition of transplant. Bar = 50 μm. (Reproduced with permission from ref. *107.*)

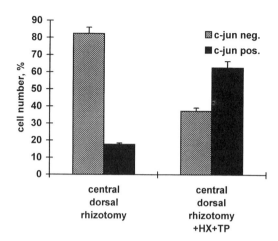

Fig. 5. Increase in the percentage of c-Jun-positive neurons in the DRG after rhizotomy and transplantation. The presence of c-Jun immunoreactivity in DRG neuronal nuclei is expressed as a ratio of the number of c-Jun-positive (solid bars) or c-Jun-negative (hatched bars) neurons to the total number of neurons counted, following either central rhizotomy alone or rhizotomy plus transplant. In each DRG, a total of 120–150 neuronal nuclei were counted. In animals that received a rhizotomy alone, only 18% of the DRG neurons were c-Jun positive. However, following rhizotomy plus hemisection plus transplant, the percentage of c-Jun-positive neurons increased to 63%. Values are expressed as the mean ± standard error. $n = 3–4$ for each group. There was a significant increase in the percentage of c-Jun-positive neurons under conditions that support the regrowth of dorsal root axons (central rhizotomy plus transplant), $*p < 0.001$ (Student's *t*-test). (Reproduced with permission from ref. *107.*)

Axotomized neurons were labeled retrogradely by the application of Fluoro-Gold at the injury site. Animals were sacrificed at 7 and 30 d after surgery. By 7 d after lesion, immunohistochemistry and quantitative image analysis showed increased c-Jun expression in axotomized red nucleus, locus coeruleus, lateral vestibular nucleus, and raphe nuclei of animals from all experimental groups, compared with normal and hemisected-only rats. Figure 6 shows the quantitative analysis of red nucleus neurons expressing c-Jun. Axotomy at the thoracic spinal cord level (HX only) fails to upregulate the expression of c-Jun. Fetal spinal cord transplants, exogenous neurotrophins, and the combination of both increased significantly the number of axotomized neurons expressing c-Jun protein. There was a significant increase in the number of c-Jun-positive neurons in the HX + transplant (TP), HX + BDNF, and HX + NT-3 treatment groups. Surprisingly, the *combination* of spinal cord transplants and exogenous neurotrophic factor treatment was *additive,* leading to a substantially larger increase in the number of axotomized neurons expressing c-Jun (Fig. 6). These results

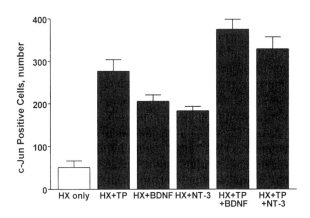

Fig. 6. Number of c-Jun-positive cells in the red nucleus. The expression of c-Jun-inducible transcription factor was assessed at 7 and 28 d after the lesion and addition of embryonic spinal cord transplant and neurotrophins. A total of 300–500 neuronal nuclei were counted in each group. A one-way ANOVA was performed on the means in each treatment group, and these means are expressed ± standard error. Note that the number of c-Jun-positive cells increased approximately four- to five-fold with a hemisection plus transplant or neurotrophic factor. However, in animals that received a hemisection plus a transplant plus neurotrophic factor, the number of c-Jun-positive cells increased seven- to eightfold. Thus, combinations of interventions appear to have an additive influence on c-Jun expression in the injured neurons at short survival times (7 d), and this expression was sustained for a long time after injury (28 d, data not shown). $n = 5–6$ animals for each treatment group. HX, hemisection; TP, transplant; NT-3, neurotrophin-3; BDNF, brain-derived neurotrophic factor. (Reproduced with permission from ref. *121.*)

suggest that expression of c-Jun protein in axotomized mature CNS neurons is elicited by interventions that increase the extent of axonal growth after injury and may play an important role in induction of genes specific for axonal regrowth of the mature CNS.

11. TRANSPLANTS AND NEUROTROPHIC FACTORS PREVENT THE ATROPHY OF MATURE AXOTOMIZED NEURONS

After midthoracic spinal cord lesion at birth, counts of red nucleus neurons indicate a significant loss of such neurons in animals with lesions *(37,122–124).* After the same injury in the adult CNS, however, most axotomized neurons survive, although they typically exhibit cell body, nuclear, and nucleolar atrophy by the first week following the injury *(70,124–128).* Axotomy in the adult CNS, particularly if the lesion is far from the cell body, usually does not result in retrograde cell death of the injured neurons as it does in the newborn. Mature injured neurons, however, often atrophy. We have used quantitative image analysis to measure the size of red nucleus neurons after spinal cord hemisection, transplantation, and neurotrophin treatment *(107)*. Axotomy in the adult leads to a decrease in the number of the largest red nucleus neurons (cross-sectional area greater than 500 μm). At 7 postnatal d, each of the treatments (TP, BDNF, NT-3, TP + BDNF, and TP + NT-3) prevented the atrophy of axotomized red nucleus neurons. The effect of transplants combined with neurotrophin treatment again had an *additive* effect compared with the individual interventions. Transplants and either BDNF or NT-3 led to a preservation of the largest diameter neurons (>500-μm cross-sectional area). Quantitative analysis at 30 postnatal d indicates a further atrophy of the red nucleus neurons in the lesion-only animals *(107)*. This atrophy was attenuated by the transplant by itself and prevented completely by the addition of either BDNF or NT-3 in addition to the transplant *(107)*. Thus, transplants and exogenous neurotrophic support not only rescue immature axotomized neurons from retrograde cell death *(31)*, but they are also able to prevent the atrophy of mature CNS neurons after axotomy *(107)*.

After SCI in the adult, the exogenous application of neurotrophic factors, in combination with fetal spinal cord transplants, upregulates cellular programs associated with regrowth and increases the extent of axonal regrowth within a favorable environment. It is likely that increases in axonal growth in the presence of neurotrophic factors are not only a consequence of increasing the regenerative capacity of the neurons at the level of the cell body but that the extent of growth may also be influenced by a direct effect of the neurotrophin on the cellular and molecular composition of the transplanted tissue. Taken together, these studies support a role for neurotrophic factors in regeneration after CNS injury.

12. CONCLUSIONS

The current optimism in SCI research and treatment derives from the results of fundamental basic science research, which one would not necessarily have predicted *a priori* would have an impact on spinal cord regeneration research. Work from a number of laboratories indicates that the dogma that mature CNS neurons are inherently incapable of regeneration of axons after injury is no longer tenable. The issues, rather, are to identify and reverse the conditions that limit regeneration after SCI. Some of the questions of current interest are: (1) what do CNS neurons require to initiate and maintain a regenerative response? (2) what are the conditions that restrict growth after CNS injury? and (3) what intervention strategies may be required to increase the amount of anatomic plasticity and recovery of function after spinal cord (and other CNS) injury?

The consequences of SCI are complex. It seems unlikely that any single intervention strategy will be sufficient to ensure regeneration of damaged pathways and recovery of function after spinal cord injury. It is likely that after SCI, a hierarchy of "intervention strategies" will be required to restore suprasegmental control leading to recovery of function. The hierarchy may be both temporal and absolute. For example, early interventions (such as the administration of methylprednisolone within hours of the injury) may be required to interrupt the secondary injury cascade and restrict the extent of damage after SCI. At the injury site itself, interventions to minimize the secondary injury effects may be followed by interventions to alter the environment at the site of injury to provide a terrain conducive to axonal elongation. For example, one might envision strategies to downregulate the expression of molecules that limit growth and upregulate the expression of those that support growth. Early after the injury, axotomized neurons may require neurotrophic support either for their survival or to initiate and maintain a cell body response supporting axonal elongation. There may be an absolute hierarchy as well. Particular populations of neurons may have very specific requirements for regenerative growth. For example, the conditions that enhance the regenerative growth of descending motor pathways may differ from those required by ascending sensory systems. One may also want to design strategies to restrict the plasticity of some pathways (e.g., nociceptive) and enhance the growth in other pathways.

It is clear that there are a number of windows of opportunity for intervention strategies to increase regeneration and recovery of function after SCI. The focus must continue to be on understanding the biologic principles and mechanisms that regulate the response of the CNS to injury and identifying strategies to alter that response.

It is possible that the demands on the CNS for anatomic reorganization after SCI may be far less formidable than one might at first imagine. If one assumes that recovery of function will require regenerative growth of large numbers of axotomized axons over long distances in a point-to-point topographically specific fashion, the idea of recovery of function becomes quite daunting. On the other hand, it has been shown in many studies and in many areas of the CNS that as little as 10% of a particular pathway can often subserve substantial function. Furthermore, regrowth over relatively short distances can have major functional consequences. For example, relatively modest changes in the level of SCI can have relatively profound effects on the functional consequences of injury. This is particularly true in cervical SCI; an individual with a C5–C6 spinal cord injury is dramatically more impaired than one with a C7–C8 injury. One might envision relatively short-distance growth across the injury site to reestablish suprasegmental control. Coupled with strategies to enhance the anatomic and functional reorganization of spinal cord circuitry caudal to the level of the injury, even modest long-distance growth may have sufficient functional impact. Given the segmental organization of the spinal cord, the adult spinal cord may have the ability to learn to use even modest quantities of novel inputs in functionally useful, appropriate ways.

ACKNOWLEDGMENTS

Research in my laboratory described in this review has been supported by grants from the National Institutes of Health (NINDS) (NIH NS 19259 and NIH NS 27054) and a grant from the International Spinal Research Trust. I am indebted to Marietta McAtee and Hai Ning Dai for their ongoing outstanding contributions to the work in my laboratory.

REFERENCES

1. Grillner, S., Ekeberg, E., Manira, A., et al. (1998) Intrinsic function of a neuronal network—a vertebrate central pattern generator. *Brain Res. Rev.* **26,** 184–197.
2. Grillner, S. and Georgopoulos, A. P. (1996) Neural control. *Cur. Opin. Neurobiol.* **6,** 741–743.
3. Davies, S. J. A., Fitch, M. T., Memberg, S. P., Hall, A. K., Raisman, G., and Silver, J. (1997) Regeneration of adult axons in white matter tracts of the central nervous system. *Nature* **390,** 680–683.
4. Fitch M. T. and Silver J. (1997) Glial cell extracellular matrix: boundaries for axon growth in development and regeneration. *Cell Tissue Res.* **290,** 379–384.
5. Bregman B. S. and Reier P. J. (1986) Neural tissue transplants rescue axotomized rubrospinal cells from retrograde death. *J. Comp. Neurol.* **244,** 86–95.
6. Bregman, B. S. and Kunkel-Bagden, E. (1988) Effect of target and non-target transplants on neuronal survival and axonal elongation after injury to the developing spinal cord. *Prog. Brain Res.* **78,** 205–212.
7. Bregman, B. S. (1988) Requirements of immature axotomized CNS neurons for survival and axonal elongation after injury, in *Neurology and Neurobiology: Current Issues in Neural Regeneration Research* (Reier, P. J., Bunge, R. P., and Seil, F. J., eds.), Alan R. Liss, New York, pp. 75–87.
8. Snider, W. D., Elliott, J. L., and Yan, Q. (1992) Axotomy-induced neuronal death during development. *J. Neurobiol.* **23,** 1231–1246.
9. Zhang, L., Schmidt, R. E., Yan, Q., and Snider, W. D. (1994) NGF and NT-3 have differing effects on the growth of dorsal root axons in developing mammalian spinal cord. *J. Neurosci.* **14** 5187–5201.
10. Snider, W. D. and Johnson, E. J. (1989) Neurotrophic molecules. *Ann. Neurol.* **26,** 489–506.
11. Yan, Q., Elliott, J. L., Matheson, C., et al. (1993) Influences of neruotrophins on mammalian motoneurons in vivo. *J. Neurobiol.* **24,** 1555–1577.
12. Mu, X., Silos-Santiago, I., Carroll, S. L., and Snider, W. D. (1993) Neurotrophin receptor genes are expressed in distinct patterns in developing dorsal root ganglia. *J. Neurosci.* **13,** 4029–4041.
13. Snider, W. D., Zhang, L., Yusoof, S., Gorukanti, N., and Tsering, C. (1992) Interactions between dorsal root axons and their target motor neurons in developing mammalian spinal cord. *J. Neurosci.* **12,** 3494–3508.
14. Ruit, K. G., Elliott, J. L., Osborne, P. A., Yan, Q., and Snider, W. D. (1992) Selective dependence of mamalian dorsal root ganglion neurons on nerve growth factor during embryonic development. *Neuron* **8,** 573–587.
15. Lindsay, R. M. (1996) Role of neurotropins and trk receptors in the development and maintenance of sensory neurons: an overview. *Philos. Tran. R. Soc. Lond. [Biol.]* **351,** 365–373.

16. Zhang, L., Schmidt, R. E., Yan, Q., and Snider, W. D. (1994) NGF and NT-3 have differing effects on the growth of dorsal root axons in developing mamalian spinal cord. *J. Neurosci.* **14,** 5187–5201.

17. Korshing, S. (1993) The neurotrophic factor concept: a reexamination. *J. Neurosci.* **13,** 2739–2748.

18. Snider, W. D. (1994) Functions of the neurotrophins during nervous system development: what the knockouts are teaching us. *Cell* **77,** 627–638.

19. Lefcort, F., Clary, D. O., Rusoff, A. C., and Reichardt, L. F. (1996) Inhibition of the NT-3 receptor TrkC, early in chick embryogenesis, results in severe reductions in multiple neuronal subpopulations in the dorsal root ganglia. *J. Neurosci.* **16,** 3704–3713.

20. Ernfors, P., Lee, K. F., Kucera, J., and Jaenisch, R. (1994) Lack of neurotrophin-3 leads to deficiencies in the peripheral nervous sytem and loss of limb proprioceptive afferents. *Cell* **77,** 503–512.

21. Sendtner, M., Kreutzberg, G. W., and Thoenen, H. (1990) Ciliary neurotrophic factor prevents the degeneration of motor neurons after axotomy. *Nature* **345,** 440–441.

22. Sendtner, M., Holtmann, B., Kolbeck, R., Thoenen, H., and Barde, Y. A. (1992) Brain-derived neurotrophic factor prevents the death of motoneurons in newborn rats after nerve section. *Nature* **360,** 757–759.

23. Yan, Q., Elliot, J., and Snider, W. D. (1992) Brain-derived neurotrophic factor rescues spinal motor neurons from axotomy-induced cell death. *Nature* **360,** 753–755.

24. Yan, Q., Snider, W. D., Pinzone, J. J., and Johnson, E. M., Jr. (1998) Retrograde transport of nerve growth factor (NGF) in motoneurons of developing rats: assessment of potential neurotrophic effects. *Neuron* **1,** 335–343.

25. Oppenheim, R. W., Qin-Wei, Y., Prevette, D., and Yan, Q. (1992) Brain-derived neurotrophic factor rescues developing avian motoneurons from cell death. *Nature* **360,** 755–757.

26. Yan, Q., Matheson, C., and Lopez, O. T. (1995) In vivo neurotrophic effects of GDNF on neonatal and adult facial motor neurons. *Nature* **373,** 341–344.

27. Oppenheim, R. W., Houenou, L. J., Johnson, J. E., Lin, L. F., and Li, L. (1995) Developing motor neurons rescued from programmed and axotomy-induced cell death by GDNF. *Nature* **373,** 344–346.

28. Bregman, B. S. and Goldberger, M. E. (1983) Infant lesion effect: III. Anatomical correlates of sparing and recovery of function after spinal cord damage in newborn and adult cats. *Dev. Brain Res.* **9,** 137–154.

29. Bregman, B. S. and Goldberger, M. E. (1982) Anatomical plasticity and sparing of function after spinal cord damage in neonatal cats. *Science* **217,** 553–555.

30. Merline, M. and Kalil, K. (1990) Cell death of corticospinal neurons is induced by axotomy before but not after innervation of spinal targets. *J. Comp. Neurol.* **296,** 506–516.

31. Diener, P. S. and Bregman, B. S. (1994) Neurotrophic factors prevent the death of CNS neurons after spinal cord lesions in newborn rats. *Neuroreport* **5,** 1913–1917.

32. Diener, P. S., DiStephano, P. S., and Bregman, B. S. (1993) Tropic and tropic influence of exogenous NT-3 and BDNF after neonatal spinal cord lesions. *Soc. Neurosci. Abstr.* **19,** 1105.

33. Tetzlaff, W., Kobayashi, N. R., and Bedard, A. M. (1993) BDNF prevents atrophy of rat rubrospinal neurons after axotomy. *Soc. Neurosci. Abstr.* **19,** 1164.

34. Tetzlaff, W., Leonard, C. A., and Harrington, K. C. (1992) Expression of neurotrophin receptor mRNAs in axotomized facial and rubrospinal neurons. *Soc. Neurosci. Abstr.* **18,** 1294.

35. Tetzlaff, W., Kobayashi, N. R., Giehl, K. M. G., Tsui, B. J., Cassar, S. L., and Bedard, A. M. (1994) Response of rubrospinal and corticospinal neurons to injury and neurotrophins. *Prog. Brain Res.* **103,** 271–286.

36. Merlio, J. P., Ernfors, P., Jaber, M., and Persson, H. (1992) Molecular cloning of rat trkC and distribution of cells expressing messenger RNAs for members of the trk family in the rat central nervous system. *Neuroscience* **51,** 513–532.

37. Bernstein-Goral, H. and Bregman, B. S. (1993) Spinal cord transplants support the regeneration of axotomized neurons after spinal cord lesions at birth: a quantitative double-labeling study. *Exp. Neurol.* **123,** 118–132.

38. Bregman, B. S. and Bernstein-Goral, H. (1991) Both regenerating and late-developing pathways contribute to transplant-induced anatomical plasticity after spinal cord lesions at birth. *Exp. Neurol.* **112,** 49–63.

39. Bregman, B. S., Kunkel-Bagden, E., McAtee, M., and O'Neill, A. (1989) Extension of the critical period for developmental plasticity of the corticospinal pathway. *J. Comp. Neurol.* **282,** 355–370.

40. Diener, P. S. and Bregman, B. S. (1998) Fetal spinal cord transplants support the development of target reaching and coordinated postural adjustments after neonatal cervical spinal cord injury. *J. Neurosci.* **18,** 763–778.

41. Diener, P. S. and Bregman, B. S. (1998) Fetal spinal cord transplants support growth of supraspinal and segmental projections after cervical spinal cord hemisection in the neonatal rat. *J. Neurosci.* **18,** 779–793.

42. Bregman, B. S. (1987) Spinal cord transplants permit the growth of serotonergic axons across the site of neonatal spinal cord transection. *Dev. Brain Res.* **34,** 265–279.

43. Bernstein-Goral, H. and Bregman, B. S. (1997) Axotomized rubrospinal neurons rescued by fetal spinal cord transplants maintain axon collaterals to rostral CNS targets. *Exp. Neurol.* **148,** 13–25.

44. Bernstein-Goral, H., Diener, P. S., and Bregman, B. S. (1997) Regenerating and sprouting axons differ in their requirements for axonal growth after injury. *Exp. Neurol.* **148,** 51–72.

45. Bernstein-Goral, H. and Bregman, B. S. (1995) Withdrawal of transient serotonergic projections from non-target transplants is regulated by the availability of neurotrophic factors. *Soc. Neurosci. Abstr.* **21,** 1057.

46. Bernstein-Goral, H., Leland, R., and Bregman, B. S. Remodeling of axonal projections after neonatal rat spinal cord injury and transplantation is regulated by the availability of neurotrophic factors, submitted.

47. Kunkel-Bagden, E. and Bregman, B. S. (1990) Spinal cord transplants enhance the recovery of locomotor function after spinal cord injury at birth. *Exp. Brain Res.* **81,** 25–34.

48. Howland, D. R., Bregman, B. S., and Goldberger, M. E. (1995) The development of quadrupedal locomotion in the kitten. *Exp. Neurol.* **135,** 93–107.

49. Howland, D. R., Bregman, B. S., Tessler, A., and Goldberger, M. E. (1995) Development of locomotor behavior in the spinal kitten. *Exp. Neurol.* **135,** 108–122.

50. Howland, D. R., Bregman, B. S., Tessler, A., and Goldberger, M.E. (1995) Transplants enhance locomotion in neonatal kittens whose spinal cords are transected: a behavioral and anatomical study. *Exp. Neurol.* **135,** 123–145.

51. Miya, D., Giszter, S., Mori, F., Adipudi, V., Tessler, A., and Murray, M. (1997) Fetal transplants alter the development of function after spinal cord transection in newborn rats. *J. Neurosci.* **17,** 4856–4872.

52. Bregman, B. S. and Kunkel-Bagden, E. (1994) Potential mechanisms underlying transplant mediated recovery of function after spinal cord injury, in *Neural Transplantation, CNS Neuronal Injury and Regeneration* Marwah, J., Teitelbaum, H., and Prasad, K., eds.), CRC Press, Boca Raton, pp. 81–102.

53. Keifer, J. and Kalil, K. (1991) Effects of infant versus adult pyramidal tract lesions on locomotor behavior in hamsters. *Exp. Neurol.* **111,** 98–105.

54. Bregman, B. S. and Goldberger, M. E. (1983) Infant lesion effect: II. Sparing and recovery of function after spinal cord damage in newborn and adult cats. *Dev. Brain Res.* **9,** 119–135.

55. Bregman, B. S., Kunkel-Bagden, E., Reier, P. J., Dai, H. N., McAtee, M., and Gao, D. (1993) Recovery of function after spinal cord injury: mechanisms underlying transplant-mediated recovery of function differ after spinal cord injury in newborn and adult rats. *Exp. Neurol.* **123,** 3–16.

56. Mamounas, L. A., Blue, M. E., Siuciak, J. A., and Altar, C. A. (1996) Brain-derived neurotrophic factor promotes the survival and sprouting of serotonergic axons in rat brain. *J. Neurosci.* **15,** 7929–7939.

57. Chong, M. S., Fitzgerald, M., Winter, J., et al. (1992) GAP-43 mRNA in rat spinal cord and dorsal root ganglia neurons: developmental changes and re-expression following peripheral nerve injury. *Eur. J. Neurosci.* **4,** 883–895.

58. Benowitz, L. I. and Perrone-Bizzozero, N. I. (1991) The expression of GAP-43 in relation to neuronal growth and plasticity: when, where, how, and why? *Prog. Brain Res.* **89,** 69–87.

59. Smith, G. M. and Silver, J. (1998) Transplantation of immature and mature astrocytes and their effect on scar formation in the lesioned central nervous system, in *Progress in Brain Research Edited by* (Gash, D. M. and Sladek, J. R. Jr., eds.), Elsevier Science Publishers, amsterdam, pp. 353–361.

60. McKeon, R., Schreiber, R. C., Rudge, J. S., and Silver, J. (1991) Reduction of neurite outgrowth in a model of glial scarring following CNS injury is correlated with the expression of inhibitory molecules on reactive astrocytes. *J. Neurosci.* **11,** 3398–3411.

61. Pindzola, R. R., Doller, C., and Silver, J. (1993) Putative inhibitory extracellular matrix molecules at the dorsal root entry zone of the spinal cord during development and after root and sciatic nerve lesions. *Dev. Biol.* **156,** 34–48.

62. Callazo, D., Takahashi, H., and McKay, R. D. G. (1992) Cellular targets and trophic functions of neurotrophin-3 in the developing rat hippocampus. *Neuron* **9,** 643–656.

63. Caroni, P. and Schwab, M. E. (1998) Antibody against myelin-associated inhibitor of neurite growth neutralizes nonpermissive substrate properties of CNS white matter. *Neuron* **1,** 85–96.

64. Savio, T. and Schwab, M. E. (1990) Lesioned corticospinal tract axons regenerate in myelin-free rat spinal cord. *Proc. Natl. Acad. Sci. USA* **87,** 4130–4133.

65. Schwab, M. E. and Schnell, L. (1991) Channeling of developing rat corticospinal tract axons by myelin-associated neurite growth inhibitors. *J. Neurosci.* **11,** 709–721.

66. Keirstead, H. S., Dyer, J. K., Sholomenko, G. N., McGraw, J., Delaney, K. R., and Steeves, J. D. (1995) Axonal regeneration and physiological activity following transection and immunological disruption of myelin within the hatchling chick spinal cord. *J. Neurosci.* **15(10),** 6963–6974.

67. Keirstead, H. S., Pataky, D. M., McGraw, J., and Steeves, J. D. (1997) In vivo immunological suppression of spinal cord myelin development. *Brain Res. Bull.* **44,** 727–734.

68. Nicholls, J. G. and Saunders, N. R. (1996) Regeneration of immature mammalian spinal cord after injury. *Trends Neurosci.* **19,** 229–234.

69. Hasan, S. J., Kierstead, H. S., Muir, G. D., and Steeves, J. D. (1993) Axonal regeneration contributes to repair of injured brainstem-spinal neurons in embryonic chick. *J. Neurosci.* **13,** 492–507.

70. Bregman, B. S. (1987) Development of serotonin immunoreactivity in the rat spinal cord and its plasticity after neonatal spinal cord lesions. *Dev. Brain Res.* **34,** 245–263.

71. Reier, P. J., Bregman, B. S., and Wujek, J. R. (1986) Intraspinal transplantation of embryonic spinal cord tissue in neonatal and adult rats. *J. Comp. Neurol.* **247,** 275–296.

72. Jakeman, L. B. and Reier, P. J. (1991) Axonal projections between fetal spinal cord transplants and the adult rat spinal cord: a neuroanatomical tracing study of local interactions. *J. Comp. Neurol.* **307,** 311–334.

73. Bregman, B. S., Diener, P. S., McAtee, M., Dai, H. N., and James, C. V. (1997) Intervention strategies to enhance anatomical plasticity and recovery of function after spinal cord injury. *Adv. Neurol.* **72,** 257–275.

74. Bregman, B. S., McAtee, M., Dai, H. N., and Kuhn, P. L. (1997) Neurotrophic factors increase axonal growth after spinal cord injury and transplantation in the adult rat. *Exp. Neurol.* **148,** 475–494.

75. Tessler, A., Himes, B. T., Houle, J., and Reier, P. J. (1988) Regeneration of adult dorsal root axons into transplants of embryonic spinal cord. *J. Comp. Neurol.* **270,** 537–548.

76. Houle, J. D. and Reier, P. J. (1989) Regrowth of calcitonin gene-related peptide (CGRP) immunoreactive axons from the chronically injured rat spinal cord into fetal spinal cord tissue transplants. *Neurosci. Lett.* **103,** 253–258.

77. Bregman, B. S. (1994) Recovery of function after spinal cord injury: transplantation strategies, in *Functional Neural Transplantation* (Dunnett, S. B. and Bjorklund, A., eds.), Raven, New York, pp. 489–529.

78. Aguayo, A. J., David, S., Richardson, P., and Bray, G. M. (1979) Axonal elongation in peripheral and central nervous system transplants. *Adv. Cell. Neurobiol.* **3,** 215–234.

79. Aguayo, A. J., David, S., and Bray, G. (1981) Influences of the glial environment on the elongation of axons after injury: transplantation studies in adult rodents. *J. Exp. Biol.* **95,** 231–240.

80. Richardson, P. M., McGuinness, U. M., and Aguayo, A. J. (1980) Axons from CNS neurones regenerate into PNS grafts. *Nature* **284,** 264–265.

81. Richardson, P. M., Issa, V. M. K., and Aguayo, A. J. (1984) Regeneration of long spinal axons in the rat. *J. Neurocytol* **13,** 165–182.

82. David, S. and Aguayo, A. J. (1981) Axonal elongation into peripheral nervous system "bridges" after central nervous system injury in adult rats. *Science* **214,** 931–933.

83. David, S. and Aguayo, A. J. (1985) Axonal regeneration after crush injury of rat central nervous system fibres innervating peripheral nerve grafts. *J. Neurocytol.* **14,** 1–12.

84. Lindsay, R. M., Wiegand, S. J., Altar, C. S., and DiStefano, P. (1994) Neurotrophic factors: from molecules to man. *TINS* **17,** 182–190.

85. Barde, Y. A. (1989) Trophic factors and neuronal survival. *Neuron* **2,** 1525–1534.

86. Barde, Y. A. (1994) Neurotrophic factors: an evolutionary perspective. *J. Neurobiol.* **25,** 1329–1333

87. Kromer, L. F. (1987) Nerve growth factor treatment after brain injury prevents neuronal death. *Science* **235,** 214–216.

88. Knusel, B., Beck, K. D., Winslow, J. W., Rosenthal, A., et al. (1992) Brain-derived neurotrophic factor administration protects basal forebrain cholinergic but not nigral dopaminergic neurons from degenerative changes after axotomy in the adult rat brain. *J. Neurosci.* **12,** 4391–4402.

89. Knusel, B., Winslow, J. W., Rosenthal, A., et al. (1991) Promotion of central cholinergic and dopaminergic neuron differentiation by brain-derived neurotrophic factor but not neurotrophin-3. *Proc. Natl. Acad. Sci. USA* **88,** 961–965.

90. Hagg, T., Quon, D., Higaki, J., and Varon, S. (1992) Ciliary neurotrophic factor prevents neuronal degeneration and promotes low affinity NGF receptor expression in the adult rat CNS. *Neuron* **8,** 145–158.

91. Hagg, T. and Varon, S. (1993) Ciliary neurotrophic factor prevents degeneration of adult rat substantia nigra dopaminergic neurons in vivo. *Proc. Natl. Acad. Sci. USA* **90,** 6315–6319.

92. Lindsay, R. M. (1988) Nerve growth factors (NGF, BDNF) enhance axonal regeneration but are not required for survival of adult sensory neurons. *J. Neurosci.* **8,** 2394–2405.

93. Thanos, S., Bahr, M., Barde, Y. A., and Vanselow, J. (1991) Survival and axonal elongation of adult retinal ganglion cells. In vitro effects of lesioned sciatic nerve and brain-derived neurotrophic factor. *Eur. J. Neurosci.* **1,** 19–26.

94. Mansour-Robaey, S., Clarke, D. B., Wang, Y. C., Bray, G. M., and Aguayo, A. J. (1994) Effects of ocular injury and administration of brain-derived neurotrophic factor on survival and regrowth of axotomized retinal ganglion cells. *Proc. Natl. Acad. Sci. USA* **91,** 1632–1636.

95. Li, Y. and Raisman, G. (1995) Sprouts from cut corticospinal axons persist in the presence of astrocytic scarring in long-term lesions of the adult rat spinal cord. *Exp. Neurol.* **134,** 102–111.

96. Houle, J. D., Wright, J. W., and Ziegler, M. K. (1994) After spinal cord injury chronically injured neurons retain the potential for axonal regeneration, in *Neural Transplantation , CNS Injury and Regeneration* (Marwah, J., Teitelbaum, H., and Prasad, K., eds.), CRC Press, Boca Raton, FL, pp. 103–118.

97. Oudega, M. and Hagg, T. (1996) Nerve growth factor promotes regeneration of sensory axons into adult rat spinal cord. *Exp. Neurol.* **140,** 218–229.

98. Oudega, M., Varon, S., and Hagg, T. (1994) Regeneration of adult rat sensory axons into intraspinal nerve grafts: promoting effects of conditioning lesion and graft predegeneration. *Exp. Neurol.* **129,** 194–206.

99. Xu, X. M., Guenard, V., Kleitman, N., and Bunge, M. B. (1995) Axonal regeneration into Schwann cell-seeded guidance channels grafted into transected adult rat spinal cord. *J. Comp. Neurol.* **351,** 145–160.

100. Xu, X. M., Guenard, V., Kleitman, N., Aebischer, P., and Bunge, M. B. (1995) A combination of BDNF and NT-3 promotes supraspinal axonal regeneration into Schwann cell grafts in adult rat thoracic spinal cord. *Exp. Neurol.* **134,** 261–272.

101. Ye, J. H. and Houle, J. D. (1997) Treatment of the chronically injured spinal cord with neurotrophic factors can promote axonal regeneration from supraspinal neurons. *Exp. Neurol.* **143,** 70–81.

102. Tuszynski, M. H., Gabriel, K., Gage, F. H., Suhr, S., Meyer, S., and Rosetti, A. (1996) Nerve growth factor delivery by gene transfer induces differential outgrowth of sensory, motor, and noradrenergic neurities after adult spinal cord injury. *Exp. Neurol.* **137,** 157–173.

103. Tuszynski, M. H., Murai, K., Blesch, A., Grill, R., and Miller, I. (1997) Functional characterization of NGF secreting cell grafts to the acutely injured spinal cord. *Cell Transplant.* **6,** 361–368.

104. Schnell, L., Schneider, R., Kolbeck, R., Barde, Y.-A., and Schwab, M. E. (1994) Neurotrophin-3 enhances sprouting of corticospinal tract during development and after adult spinal cord lesion. *Nature* **367,** 170–173.

105. Schnell, L. and Schwab, M. E. (1993) Sprouting and regeneration of lesioned corticospinal tract fibres in the adult rat spinal cord. *Eur. J. Neurosci.* **5,** 1156–1171.

106. DiStefano, P. S. and Curtis, R. (1994) Receptor mediated retrograde axonal transport of neurotrophic factors is increased after peripheral nerve injury. *Prog. Brain Res.* **103,** 35–42.

107. Bregman, B. S., Broude, E., McAtee, M., and Kelley, M. S. (1998) Transplants and neurotrophic factors prevent atrophy of mature CNS neurons after spinal cord injury. *Exp. Neurol.* **149,** 13–27.

108. Broude, E., McAtee, M., Kelley, M. S., and Bregman, B. S. (1997) c-Jun expression in adult rat dorsal root ganglion neurons: differential response after central or peripheral axotomy. *Exp. Neurol.* **148,** 367–377.

109. Haas, C. A., Donath, C., and Kreutzberg, G. W. (1993) Differential expression of immediate early genes after transection of the facial nerve. *Neuroscience* **53,** 91–99.

110. Herdegen, T., Fiallos-Estrada, C. E., Schmid, W., Bravo, R., and Zimmermann, M. (1992) The transcription factors c-JUN, JUN D and CREB, but not FOS and KROX-24, are differentially regulated in axotomized neurons following transection of rat sciatic nerve. *Mol. Brain Res.* **14,** 155–165.

111. Robinson, G. A. (1994) Immediate early gene expression in axotomized and regenerating retinal ganglion cells of the adult rat. *Mol. Brain Res.* **24,** 43–54.

112. Chong, M. S., Woolf, C. J., Turmaine, M., Emson, P. C., and Anderson, P. N. (1996) Intrinsic versus extrinsic factors in determining the regeneration of the central processes of rat dorsal root ganglion neurons: the influence of a peripheral nerve graft. *J. Comp. Neurol.* **370,** 97–104.

113. Schreyer, D. J. and Skene, J. H. P. (1994) Injury-associated induction of GAP-43 expression displays axon branch specificity in rat dorsal root ganglion neurons. *J. Neurobiol.* **24,** 959–970.

114. Vaudano, E., Campbell, G., Anderson, P. N., et al. (1995) The effects of a lesion or a peripheral nerve graft on GAP-43 upregulation in the adult rat brain: an in situ hybridization and immunocytochemical study. *J. Neurosci.* **15,** 3594–3611.

115. Chong, M. S., Reynolds, M. L., Irwin, N., et al. (1994) GAP-43 expression in primary sensory neurons following central axotomy. *J. Neurosci.* **14,** 4375–4384.

116. Richardson, P. M., McGuiness, U. M., and Aguayo, A. J. (1982) Peripheral nerve autografts to the rat spinal cord: studies with axonal tracing methods. *Brain Res.* **237,** 147–162.

117. Wiesendanger, M. (1972) Effects of electrical stimulation of peripheral nerves to the hand and forearm on pyramidal tract neurones of the baboon and monkey. *Brain Res.* **40,** 193–197.

118. Johnson, E. M. Jr., Gorin, P. D., Brandeis, L. D., and Pearson, J. (1980) Dorsal root ganglion neurons are destroyed by exposure in utero to maternal antibody to nerve growth factor. *Science* **210,** 916–918.

119. Siegal, J. D., Kliot, M., Smith, G. M., and Silver, J. (1990) A comparison of the regenerating potential of dorsal root fibers into grey or white matter of the adult rat spinal cord. *Exp. Neurol.* **109,** 90–97.

120. Jenkins, R., Tetzlaff, W., and Hunt, S. P. (1993) Differential expression of immediate early genes in rubrospinal neurons following axotomy in rat. *Eur. J. Neurosci.* **5,** 203–209.

121. Broude, E., McAtee, M., Kelley, M. S., and Bregman, B. S. (1999) Fetal spinal cord transplants and exogenous neurotrophic support enhanced cJun expression in axotomized neurons after spinal cord injury. *Exp. Neurol.* **155,** 65–78.

122. Goldberger, M. E. (1988) Partial and complete deafferentation of cat hindlimb: the contribution of behavioral substitution to recovery of motor function. *Exp. Brain Res.* **73,** 343–353.

123. Goldberger, M. E. (1988) Spared-root deafferentation of a cat's hindlimb: hierarchical regulation of pathways mediating recovery of motor behavior. *Exp. Brain Res.* **73,** 329–342.

124. Reynolds, P. J., Talbott, R. E., and Brookhart, J. M. (1972) Control of postural reactions in the dog: the role of the dorsal column feedback pathway. *Brain Res.* **40,** 159–164.

125. Barth, T. M. and Stanfield, B. B. (1990) The recovery of forelimb-placing behavior in rats with neonatal unilateral cortical damage involves the remaining hemisphere. *J. Neurosci.* **10,** 3449–3459.

126. Barbeau, H. and Rossignol, S. (1991) Initiation and modulation of the locomotor pattern in the adult chronic spinal cat by noradrenergic, serotonergic and dopaminergic drugs. *Brain Res.* **546,** 250–260.

127. Bregman, B. S. (1988) Target-specific requirements of immature axotomized CNS neurons for survival and axonal elongation after injury, in *Current Issues in Neural Regeneration Research,* Alan R. Liss, New York, pp. 75–87.

128. Himes, B. T., Goldberger, M. E., and Tessler, A. (1994) Grafts of fetal central nervous system tissue rescue axotomized Clarke's nucleus neurons in adult and neonatal operates. *J. Comp. Neurol.* **339,** 117–131.

9

Cell Transplantation
for Spinal Cord Injury Repair

Juan C. Bartolomei and Charles A. Greer

1. INTRODUCTION

The annual incidence of spinal cord injury (SCI) is estimated at 30–70 cases per million in the United States (Krauss et al., 1995; Braken et al., 1981; Griffin et al., 1985). In addition to the devastating personal consequences of SCI for the individual and family, the postinjury financial costs are estimated to be in excess of 4 billion dollars (Reier et al., 1994). Postinjury intervention strategies, such as the use of steroids (Braken et al., 1990, 1997), are helping to decrease the magnitude of the deficit, but injury to the spinal cord has remained largely irreversible (Reier et al., 1994; Waxman and Kocsis, 1997; Bregman, 1998). However, recent advances in transplant strategies suggest that there may be a window of opportunity following SCI during which the introduction of heterologous cells into the injury site will promote regeneration of the injured axons.

Efforts to promote recovery after SCI have been broadly divided into four areas: (1) cell survival; (2) axon regeneration (growth); (3) correct targeting by growing axons; and (4) establishment of correct and functional synaptic appositions. These areas clearly recapitulate the events occurring during normal development of the nervous system. However, in the mature central nervous system (CNS), many of the molecular substrates that promote early development are downregulated, while others less favorable for these events are upregulated. Understanding the interactions between these different sets of molecules is crucial for the development of strategies that will successfully promote repair and recovery of function in the injured CNS.

In the adult CNS, regeneration of axons following injury is influenced by both glial substrates and the extracellular matrix. Generally, following transection of axons in the CNS, there is a profound proliferation of reactive astrocytes that contribute to the establishment of a dense glial scar that can be a barrier to axonal extension (Hatten et al., 1991). Moreover, oligodendrocytes within the region of injury release inhibitory factors that cause collapse of the growth cone and cessation of axonal growth (Schwab, 1993; Cadelli et al., 1995). A lack of

From: *Neurobiology of Spinal Cord Injury*
Edited by: R. G. Kalb and S. M. Strittmatter © Humana Press Inc., Totowa, NJ

positive growth factors may also reduce axonal regeneration since infusion of nerve growth factor (NGF) into the scar can facilitate axon regeneration (Hatten et al., 1991). This is in sharp contrast to the peripheral nervous system (PNS), in which Schwann cells continue to provide a favorable substrate for the regeneration of axons following injury.

Thus, the potential for recovery from injury to the CNS via regenerative mechanisms may be broadly defined by two strategies (Fig. 1). First, the use of trophic substances may support axon extension or block the effects of inhibitory factors. Such strategies may involve the infusion of specific trophic agents directly into the lesion site or the use of engineered cell transplants that secrete specific trophic factors. Second, cell transplants may be used either to bridge the site of the lesion with an intermediary neuron or to provide a framework through which regenerating axons can traverse the lesion site more effectively.

The potential for the regeneration and extension of CNS axons was recognized by the turn of the century (Ramón y Cajal, 1928). Following a CNS lesion, axons are able to grow for long distances if a permissive environment is provided (Richardson et al., 1980; Davies et al., 1996, 1997; Bregman, 1998). Local environmental and intrinsic neuronal mechanisms are responsible for the degree of axonal sprouting and regeneration. Unlike the developing CNS, in the mature CNS the molecular substrates that promote axonal growth and neuroregeneration are suppressed while those that are less favorable for growth are upregulated and enhanced. Understanding the dynamic interactions between these molecular signals is crucial for developing strategies for regenerative and restorative treatment following SCI.

Rhythmic alternating movements such as stepping and locomotion are controlled by central pattern generators that are intrinsic to the spinal cord and are found mostly in the lumbar region of rodents and primates. Following SCI, the descending and ascending pathways that modulate the central pattern generators as well as the intrinsic circuitry of the cord are physiologically or anatomically disrupted (Murray, 1997; Nicol et al., 1995). This results in loss of voluntary movements and inability to make adaptive postural adjustment or support the body's weight (Murray, 1997). Several reports have revealed that following SCI there is a population of anatomically intact axons that are physiologically dysfunctional as a consequence of demyelination or other pathologic conditions (Bernstein and Goldberg, 1995; Griffith and McCulloch, 1983; Blight, 1983; Bunge et al., 1997). The observation that following SCI undisrupted axons are present opens a window for therapeutic manipulations that could promote the axonal function of supraspinal descending pathways, the establishment of functional synaptic connections with central pattern generators, and the subsequent restitution of locomotion.

Over the last 20 years neural transplants have been successfully employed for physiologic or behavioral restoration and repair in human models of neurodegenerative diseases (Tessler, 1991). The resolution of symptoms following fetal cell transplantation in Parkinson's disease has been a stimulus for the imple-

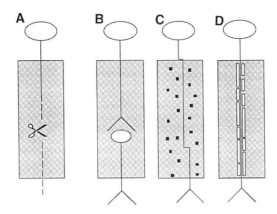

Fig. 1. Strategies for spinal cord injury repair using transplants. **(A)** Following axon transection, the distal axon segments degenerate (broken line). In addition, there is reactive gliosis (gray box) within the injury site. **(B)** One strategy for traversing the injury site is the introduction of embryonic neurons that can act as a functional, synaptic, bridge between the descending axons and their distal spinal cord targets. **(C)** The introduction of trophic/growth substances (squares), by either direct injection or the introduction of transfected cell lines, may provide support for axon extension or may block inhibitory factors intrinsic to the lesion site. **(D)** The introduction of physical channels, supportive populations of glia (open bars), may provide a supportive microenvironment that will be effective in promoting axon regeneration.

mentation of neural intraspinal transplants as a mechanism for SCI repair. Transplanted rat fetal tissue, either brain or spinal cord, has been shown to survive in the spinal cord of adult and newborn hosts (Reier et al., 1994; Diener and Bregman, 1998a,b). Transplanted fetal tissue has also been shown to serve as a substrate for lost neurons as well as axons that could serve as synaptic relays (Bregman, 1998; Diener and Bregman, 1998a,b; Tessler, 1991; Fawcett, 1995). Also, fetal tissue may promote axonal growth by ameliorating the inhibitory effects of the gliotic scar that typically forms following injury (Reier et al., 1986, 1992; Houle and Reier, 1988).

In addition to fetal tissue, remyelinating glial substrates have been employed in intraspinal transplants. Two primary strategies have emerged in this area of inquiry. First, the use of remyelinating glia to rescue preserved demyelinated axons (*see* Chap. 10). In this context, the functional status following SCI is addressed by restoring as much as possible the normal conduction parameters of spinal cord axons. In addition, in light of the effective regeneration of peripheral axons accompanied by Schwann cells, it has seemed a reasonable strategy to determine their effectiveness in promoting regeneration of central axons (Xu et al., 1995a,b). More recently, a unique population of glial cells, the olfactory ensheathing cells, has emerged as a particularly promising heterologous cell transplant for promoting axon extension following SCI (Ramón-Cueto and Avila, 1998). We review these strategies with particular emphasis on the use of transplanted olfactory ensheathing glial cells as a strategy for promoting axon regeneration following SCI.

2. FETAL CELL TRANSPLANTS

Fetal transplants have been extensively employed over the last 20 years in animal models of human neurodegenerative disorders (Tessler, 1991). The rationale for fetal cell intraspinal transplants in SCI is as follows: (1) to restore the damaged neurons that form local intersegmental synaptic circuits; (2) to serve as a bridge through which severed axons can regenerate through the lesion; and (3) to serve as a relay where supraspinal inputs synapse with donor neurons within the graft, which in turn establish synaptic contact with caudal host neurons (Tessler, 1991; Bregman et al., 1993; Reier et al., 1992). Intraspinal fetal cell grafts have been shown to have 80–90% survival rates when transplanted into acutely and chronically injured adult animals as well as when they are used in complete cord transections (Bregman et al., 1993; Houle and Reier, 1988; Stokes and Reier, 1992; Pallini et al., 1989; Itoh et al., 1992; Wirth et al., 1992; Miya et al., 1997).

Fetal cell intraspinal transplants have been shown to prevent retrograde death of red nucleus neurons following acute transection of the rubrospinal tract in the thoracic region (Bregman and Reier, 1986). This appears partially attributable to the endogenous secretion of neurotrophic factors by the fetal cells (Diener and Bregman, 1994). Interestingly, following spinal fetal cell transplants, the grafts develop characteristic areas resembling substantia gelatinosa. Because there were no specific or well-established spinal cord neuronal elements in the transplanted tissue, these data suggest that fetal tissue may contain pluripotent cells with an inherent capacity to establish local segmental circuits (Jakeman et al., 1989). Of particular importance if regenerated axons are to establish normal connections with distal segments of the spinal cord, at the host-graft interface in fetal transplants there is minimal gliotic reaction, and axonal processes appear capable of crossing through the interface (Reier et al., 1986, 1992; Houle and Reier, 1988).

Careful and meticulous behavioral testing has been performed in adult cats and rodents following intraspinal fetal transplants. Following contussive SCI and transplantation in the thoracic region of the adult rat, hindlimb gait support and stride length improved when compared with controls, whereas more complex behaviors, such as inclined plane and grid walking, showed no evidence of improvement following transplants (Stokes and Reier, 1992; Reier et al., 1992;). Reier et al. (1992) studied the locomotive behavior of adult cats after a contussive SCI and a fetal transplant that was delayed for 2–10 weeks following SCI. They transplanted either fetal neocortical (NCx), fetal brainstem (BSt), fetal spinal cord (FSC), or a combination of either FSC/NCx or FSC/BSt. Relative to controls, cats receiving FCS or BSt transplants showed improvements of 10–20% and 58–75% after 3 and 24 weeks, respectively. In contrast, improvement in the NCx group was limited to 20% over the period studied. Moreover, neurophysiologic testing in rats revealed evidence that ascending host projections made synaptic connections with donor neurons (Reier et al., 1992). Recent studies suggest that improvement in both behavioral and physiologic measures of spinal cord function following fetal transplants is likely to reflect an improvement in axon regeneration following SCI (Miya et al., 1997).

Similar analyses of fetal tissue transplants into neonatal animals have suggested better outcomes than in adults (Bregman et al., 1993). These observations led to the hypothesis that differential compensatory mechanisms may be present in adult and neonatal rats and that the latter may also benefit from the relative immaturity or plasticity of developing nervous tissue (Tessler et al., 1997; Bregman et al., 1993; Miya et al., 1997; Diener and Bregman, 1998a,b). In general, this observation is likely to be consistent with the suggestion that one operative mechanism for fetal transplants is the provision of neurotrophic factors and growth factors not normally present in the adult (Bregman, 1998).

A limiting factor in using fetal tissue as the source of transplant has been the availability of tissue, not to mention the ethical issues surrounding its use in clinical settings (Reier et al., 1994; Bartolomei and Spencer, 1999). Another potential problem that may apply to any form of transplantation is the inability to control target-specific synaptic interactions that develop following the transplant. Thus, effort in the neuroregenerative community is devoted not only to elucidating the mechanisms of axonal growth, but also to the establishment of appropriate target-specific synaptic interactions.

3. SCHWANN CELL TRANSPLANTS

The regenerative capacity of the PNS has been long known. In part, this capacity may be due to the absence of inhibitory factors characteristic of central glia (Schwab et al., 1993; Cadelli et al., 1995). In addition, the myelinating glial cells of the PNS, the Schwann cells, promote axonal regeneration (Fawcett and Keynes, 1990). Several laboratories have clearly shown that Schwann cells implanted into the site of SCI will support axon extension/regeneration (Richardson et al., 1980; Xu et al., 1995a,b; Guest et al., 1997). Schwann cells produce several trophic factors including NGF (Bandtlow et al., 1987), brain-derived nerve growth factor (BDNF) (Acheson et al., 1991), and ciliary-derived nerve growth factor (CDNF) (Friedman et al., 1992). In addition, Schwann cells express cell adhesion molecules, as well as laminin, which are known to promote and facilitate axonal guidance and extension. Transplantation of Schwann cells in animal models of demyelinating disease promotes remyelination that is effective in restoring action potential conduction properties in axons (Duncan et al., 1988; Blakemore, 1976; Blakemore and Crang, 1985; Felts and Smith, 1992; Duncan et al., 1988; Itoyama et al., 1983; Honmou et al., 1996; *see* also Chap. 10). Because a subset of axons in SCI may be intact but demyelinated, the transplantation of Schwann cells may be useful in contributing to the restoration of functional properties in such axons through remyelination (Blight, 1983; Bunge et al., 1997; Murray, 1997).

Although several laboratories have successfully demonstrated robust axonal extension into Schwann cell bridges and/or purified populations of Schwann cells injected into the site of SCI, progress has been somewhat limited by the observation that very few axons are able to traverse the lesion site and enter appropriately into the distal segments of the cord (Xu et al., 1995a,b; Paino and Bunge, 1991; Martin et al., 1991; Li and Raisman, 1997; Chen et al., 1996; Cheng et al., 1996; Ramon-Cueto et al., 1998). The precise mechanisms that account for this failure are unknown but seem likely to reflect in part the gliosis present at the host-transplant interface.

Some of the advantages of developing Schwann cells as sources of intraspinal transplants include the ease with which they can be maintained for long intervals in tissue banks (Martin et al., 1991) and their availability from the host (animal or human) without the use of immunosuppression (Honmou et al., 1996; Martin et al., 1991). A potential disadvantage of Schwann cell transplants is the failure of axons to reenter the distal spinal cord robustly. One possibility is that this reflects the failure of the Schwann cells to migrate from the injury site or the site of Schwann cell injection. Several studies have shown that, when iso-lated Schwann cells are intraspinally transplanted following a lesion, such cells remyelinated only within the injury site (Xu et al., 1995a,b; Honmou et al., 1996; Imaizumi et al., 1998). Xu et al. (1995a) produced a three-level cordec-tomy and implanted Schwann cell-seeded guidance channels. The Schwann cells showed no invasion into the host's spinal cord. In contrast, it is notable that Li and Raisman (1997) found evidence for extensive Schwann cell migration following microinjections of purified Schwann cells into the intact spinal cord. Cotransplantation of both Schwann cells and astrocytes has been shown to improve migration and remyelination (Blakemore and Crang, 1985; Honmou et al., 1996; Imaizumi et al., 1998), although the mechanisms underlying this out-come are not yet understood.

4. OLFACTORY ENSHEATHING GLIA TRANSPLANTS

From a neurologic perspective, the olfactory system has not received exten-sive clinical attention. However, the olfactory system has served as an excellent model in which to study molecular and cellular mechanisms of axonal growth, synaptic specificity, and regeneration (Shepherd and Greer, 1998). In studying the principles of organization and plasticity in the olfactory pathway, it has become increasingly clear that the mechanisms supporting plasticity in this part of the CNS may have the potential to contribute to regenerative events else-where in the brain. In particular, recent reports suggest that a population of glia found in the olfactory pathway, the olfactory ensheathing glia, may provide an environment that promotes axon extension following SCI (Li et al., 1997, 1998;

Ramón-Cueto et al., 1998) as well as axon remyelination (Franklin et al., 1996; Imaizumi et al., 1998).

The olfactory system is unique in the CNS because throughout life there is continual turnover of the primary afferent input (Farbman, 1994). The olfactory epithelium contains olfactory receptor neurons whose axons project through the cribriform plate and into the olfactory bulb, where they establish first-order synapses in spherical areas of the neuropil called glomeruli. The primary projection neurons receiving this synaptic input then project to the piriform cortex without a thalamic relay. It is of particular interest that the lifetime of any single olfactory receptor neuron is brief, from 4 to 8 weeks depending on the species and environmental circumstances (Graziadei and Monti-Graziadei, 1979; Mackay-Sim and Kittel, 1991; Hinds et al., 1984). Following the loss of any one olfactory receptor neuron, replacements are derived from basal cells deep in the olfactory epithelium (Huard et al., 1998). As these new cells differentiate, they extend their axons into the CNS, where they correctly target and then terminate in glomeruli (Graziadei and Monti-Graziadei, 1980). The specific glomeruli targeted by an axon appear to be influenced, in part, by the specific odor receptor expressed by the receptor neuron (Singer et al., 1995; Mombaerts et al., 1996). Although the cell bodies of the ensheathing glial are parallel in their major axis with the fascicles of axons (Valverde and Lopez-Mascaraque, 1991; Doucette, 1993a,b), it is not clear whether a single glial cell is responsible for a small subset of axons targeted for a specific glomerulus, or whether a single glial cell can contribute to the fasciculation of axons targeted for many divergent glomeruli. Thus, although the environment of the ensheathing glial process is supportive of growth, it is not known to what extent glial cells contribute to the specificity of axonal targeting. In general, the ability of the olfactory receptor cell axons to traverse the transition zone between the periphery and the CNS repeatedly appears to be unique.

As the axons from the olfactory receptor neurons travel from the olfactory epithelium to their olfactory bulb targets, they are accompanied by the olfactory ensheathing glia. Unlike Schwann cells, which do not cross the PNS/CNS border, olfactory ensheathing glia readily penetrate the pia mater, after which they may contribute to the formation of the glia limitans (Doucette, 1984, 1991; Ramón-Cueto and Valverde, 1995). The recognition that the olfactory ensheathing glia may be the operative mechanism underlying the ability of the olfactory receptor cell axons to traverse the PNS/CNS border throughout life has fueled a rapidly growing interest in understanding the properties of these cells and the degree to which they may be applicable to SCI. The remainder of this review focuses on the biologic and functional properties of the olfactory ensheathing glia and their potential role in the treatment of CNS injury.

4.1. Origin

Ensheathing cells are found along the full length of the olfactory nerve, from the basal lamina of the epithelium to the olfactory bulb. In the olfactory bulb, they are clearly present in the olfactory nerve layer; whether they extend into the glomerular layer as well remains controversial (Ramón-Cueto et al., 1998; Chiu and Greer, 1996; Cushieri and Bannister, 1975; Bailey and Shipley, 1993; Gonzalez et al., 1993; Valverde et al., 1992). Recent studies have established that the olfactory ensheathing glia are derived from progenitor cells found in the olfactory epithelium (Chuah and Au, 1991) and that they migrate toward the olfactory bulb using trophic signals (Liu et al., 1995).

4.2. Ensheathing Glia vs. Schwann Cells and Astrocytes

Ensheathing glia share phenotypic properties with both astrocytes and Schwann cells. Like astrocytes, ensheathing glia express glial fibrillary acidic protein, (GFAP), form endfeet junctions at the blood vessels, and contribute to the glia limitans, which delineates the junction between the olfactory nerve axons and the glomeruli in the olfactory bulb (Ramón-Cueto and Valverde, 1995; Doucette, 1990, 1993a,b, 1995; Raisman, 1985; Barber and Lindsay, 1982). However, they differ from astrocytes in immunocytochemical and ultrastructural characteristics as well as their functional associations with axons. In contrast to astrocytes, ensheathing glia are immunoreactive to the nerve growth factor receptor P75NGR, laminin, the cell adhesion molecule L1, and the neurofilament vimentin (Barber and Lindsay, 1982; Liesi, 1985; Miragall et al., 1988; Barnett et al., 1993; Gong et al., 1994, Vickland et al., 1991). In vitro, ensheathing glia lack immunoreactivity to A2B5, Ran2, or microtubule-associated protein-2, (MAP-2), which are markers for astrocytes (Charriere-Bertran et al., 1991; Miller et al., 1989; Ramón-Cueto and Valverde, 1995). At the level of the electron microscope, ensheathing glia exhibit indented nuclei with patchy chromatin and electron-dense cytoplasm with scattered filaments throughout (Doucette, 1991). On the other hand, astrocytes have ovoid nuclei with an electron-lucent cytoplasm and intermediate filaments arranged in bundles (Ramón-Cueto and Valverde 1995; Doucette, 1984, 1990; Mori and Leblond, 1969). In the olfactory pathway, ensheathing glia wrap bundles of receptor cell axons and guide them through the PNS/CNS transitional zone and the subarachnoid space into the olfactory bulb. In vitro ensheathing cells cocultured with neurons enfold, myelinate, and support neurite outgrowth. Whereas astrocytes may generally be trophic for neurite extension or neuron survival in vitro, they do not develop the multilamellar wraps characteristic of the esheathing glia (Devon and Doucette, 1992; Ramón-Cueto et al., 1993). Ensheathing cells also do not share immunocytochemical properties with oligodendrocytes. Interestingly, although olfactory ensheathing glia do express O-2A, a marker of oligodendrocyte type-2 astrocyte progenitor cells, other markers usually associated are absent, including Rip, anti-galactocerebroside, HNK-1, GD-3, and A2B5, and do not immunolabel ensheathing cells (Barnett et al., 1993. Miller et al., 1989; Ramón-Cueto and Nieto-Sampedro, 1992).

Although olfactory ensheathing glia are sometimes referred to as olfactory Schwann cells, they differ from Schwann cells in both their development and immunocytochemical properties. Schwann cells are derived from neuronal crest cells, and ensheathing glia are derived from the olfactory placode (Ramón-Cueto and Valverde, 1995; Ramón-Cueto and Avila, 1998; Doucette, 1995). Both peripheral Schwann cells and ensheathing glia are immunoreactive for L1, low-affinity NGF receptors, 192-IgG (which also binds to low-affinity NGF), and A5E3, which labels nonmyelinating Schwann cells and has cells with vimentin in their intermediate filaments (Barnett et al., 1993, Ramon-Cueto and Valverde, 1995; Jessen and Mirsky, 1990; Jessen et al., 1990; Liesi, 1985). In contrast, ensheathing glia express the central form of GFAP and Schwann cells do not. Ensheathing glia do not express HNK-1 or galactocerebroside, both of which are seen in nonmyelinating and myelinating Schwann cells (Barnett et al., 1993; Ramón-Cueto and Nieto-Sampedro, 1992; Ramón-Cueto and Valverde, 1995; Ramón-Cueto and Avila, 1998; Goodman et al., 1993; Jessen and Mirsky, 1990).

At the level of the electron microscope, Schwann cells exhibit an electron-dense nucleus with chromatin clumped throughout and a highly electron-dense cytoplasm relative to that seen in the ensheathing glia (Doucette, 1990). Ensheathing glia are able to intermingle among astrocytes, whereas Schwann cells have a sharply distinct basal lamina that separates them from the CNS (Doucette, 1990, 1991; Franklin and Blakemore, 1993; Ramón-Cueto and Valverde, 1995). As noted earlier, ensheathing glia can form endfeet junctions with blood vessels whereas Schwann cells do not. One common feature among ensheathing glia and Schwann cells, discussed further below, is their functional role in axonal growth, regeneration, and guidance.

4.3. Functional Properties

As noted above, the olfactory system is unique in that it supports the continuous regeneration of axons from the olfactory epithelium into the CNS during the lifetime of the individual. The olfactory system's capacity for axonal extension and target-specific synaptic interaction have been attributed to the ensheathing glia's growth-promoting properties. Ensheathing glia have been implicated in the secretion of neurotrophic factors, axonal growth-promoting substrates, and the ability to migrate and myelinate regenerating axons (Ramón-Cueto and Nieto-Sampedro, 1994; Ramón-Cueto et al., 1998; Devon and Doucette, 1992). Immunocytochemical and *in situ* studies in the olfactory system suggest that ensheathing cells are responsible for secreting BDNF, NGF, neurotrophin factor 3 (NT-3), and NT-4 (Ronnet et al., 1991; Roskams et al., 1996; Williams and Rush, 1988). As mentioned above, ensheathing cells are immunoreactive against the low-affinity NGF receptor (P75), which has also been shown to be expressed in the tips of regenerating olfactory receptor cell axons (Deckner et al., 1993, Roskams et al., 1996). In addition to neurotrophic factors, ensheathing cells have been shown to express laminin, L1, fibronectin, S100, glial-derived nexin and neural cell adhesion molecule (N-CAM), all of which are known to support neuronal axon elongation (Chuah and Au, 1994;

Franceschini and Barnett, 1996; Miragall et al., 1988, Ramón Cueto and Avila, 1998; Doucette, 1996; Kafitz and Greer, 1997, 1998; Liesi, 1985; Ramón-Cueto and Nieto-Sampedro, 1992; Van Eldik et al., 1991; Zurn et al., 1988).

In vitro, ensheathing cell glia derived through various purification processes are supportive of axonal extension. Chuah and Au (1994) showed increased survival in primary neuronal cultures plated on ensheathing cell glia. Similarly, Kafitz and Greer (1998) showed longer neurite extension on ensheathing glia substrate than on olfactory bulb or hippocampal derived glia. Similar results were reported by Sonigra et al. (1999) for retinal ganglion cells and by Bartolomei and Greer (1998) for dorsal root ganglion neurons on purified ensheathing cells derived from the olfactory nerve. Using a transformed immortalized line of glia derived from olfactory ensheathing glia, Goodman et al. (1993) reported that chick retinal ganglion cells extended longer neurites on ensheathing derived glia than on alternative glial substrates derived from the cortex or olfactory bulb. Kafitz and Greer (1997) suggested that laminin may be an important component of the axon extension mediated by ensheathing glia because of the presence of high levels of laminin in the olfactory nerve layer, since blocking with antilaminin will decrease neurite extension in vitro, and since laminin is upregulated in the olfactory nerve during periods of extensive axon growth (Liesi, 1985). It is of particular interest that Devon and Doucette (1992) demonstrated that ensheathing glia adopted a Schwann cell-like myelinating phenotype when plated with dorsal root ganglion neurons and that they would establish stereotypic myelin sheaths around the axons.

In vivo, transplanted ensheathing glia have been shown to remyelinate axons in the spinal cord (Franklin et al., 1996), and the remyelination supports at least a partial restoration of function as measured by action potential conduction (Imaizumi et al., 1998; *see* Chap. 10). The pluripotential properties of transplanted ensheathing glia are not yet understood. In the olfactory nerve in vivo, ensheathing glia encompass larger populations of axons, mesaxons, or fascicles. However, when transplanted they appear to establish a 1:1 myelinating phenotype with axons. This suggests that the effectiveness of transplanted glia in restoring function to demyelinated tracts may be dependent on the number of cells that can be effectively transplanted.

A stunning report in late 1997 from the Raisman laboratory (Li et al., 1997) suggested that olfactory ensheathing glia were an unusually effective intervention for promoting axon regeneration in the injured spinal cord, with an accompanying recovery of function. Following transections of the corticospinal tract in rats, they demonstrated that suspensions of ensheathing glia injected into the site of the lesion promoted growth/regeneration of transected axons as well as a marked improvement in the use of the forelimb. A subsequent report (Li et al., 1998) showed clearly the significant extension of regenerating axons into distal regions of the spinal cord accompanied by cells with the ensheathing glia-like phenotype. Of particular importance, the transplanted ensheathing glia migrated from the lesion site into more distal regions of the spinal cord. This clearly sets them apart from Schwann cell transplants that exhibit less extensive migration (*see* Section 3. above). The transplanted ensheathing glia adopted two distinct phenotypes, one in which single regenerating axons were closely accompanied, and a second in which larger bundles were encompassed. The two phenotypes may work in concert to shepherd growing axons through the site of the lesion and into healthy tissue.

Ramón-Cueto et al. (1998) also reported significant axon regeneration in SCI following injections of ensheathing glia suspensions, although their data also suggested that fibroblasts present in the transplant bridge may also contribute to the regeneration of some subpopulations of axons. Finally, their data also demonstrated that both ascending and descending spinal axons traversed the site of injury following ensheathing glia transplants.

In both reports, a critical feature appears to be the interface between the transplant/graft placed into the lesion site and the healthy tissue surrounding the lesion site. As noted above, the ensheathing glia readily migrated from the lesion, both longitudinally or in parallel with the spinal cord axons as well as radially out toward the pial surface. The ability to migrate into the spinal cord had previously been noted when ensheathing glia facilitated the reinnervation of the dorsal root entry zone following a rhizotomy (Ramón-Cueto and Nieto-Sampedro, 1994; *see* also Navarro et al., 1999). The transplanted ensheathing glia appeared to intermingle with reactive glia; there appeared to be no clear boundaries between these populations of cells (Ramón-Cueto et al., 1998). Indeed, Li et al. (1998) suggested that the ensheathing glia may contribute to a suppression of the astrocytic scar that would normally accompany SCI. However, this seems unlikely to be the sole mechanism because there were no individual axons that traversed the site of the lesion with a 1:1 accompaniment by a transplanted ensheathing glia.

The ensheathing glia appear to offer one of the most promising emerging strategies for SCI. However, procedures for generating large numbers of the ensheathing glia must first be developed. Also, a more complete understanding of the relationship, both structural and functional, between ensheathing glia and regenerating axons is required if we are to identify the underlying mechanisms. Despite these limitations, it seems certain that the further study of this unique population of glia will improve our understanding of the capacity of the CNS for regenerative events and those mechanisms that may be most effective in realizing that capacity.

5. SUMMARY

SCI remains an insidious and challenging problem in neurobiology. The use of transplants for promoting regeneration in the spinal cord was long thought to be limited by a minimal ability of CNS axons to regenerate, but we now recognize that the potential for axon growth is extensive, even in the adult CNS. The primary hurdles that appear to inhibit axon regeneration include the presence of secreted inhibitory factors at the site of SCI, the physical barrier of the gliotic scar, and the downregulation of growth factors in the adult. Transplantation of Schwann cells appears to overcome these hurdles partially. However, a more robust response, including long-distance extension of regenerated axons, follows transplants of a unique population of glia from the olfactory nerve, the olfactory ensheathing glia. Although the specific properties of ensheathing glia underlying enhanced axon regeneration remain to be established, it seems certain that they will contribute to the establishment of new horizons in SCI research.

REFERENCES

Acheson, A., Barker, P. A., Alderson, F. D., Miller, F. D., and Murphy, R. A. (1991) Detection of brain-derived neurotrophic factor-like activity in fibroblasts and Schwann cells: inhibition by antibodies to NGF. *Neuron* **7,** 265–275.

Bailey, M. S. and Shipley, M. T. (1993) Astrocyte subtypes in the rat olfactory bulb: morphological heterogeneity and differential laminar distribution. *J. Comp. Neurol.* **328,** 501–526.

Bandtlow, C. E., Heumann, R., Schwabb, M. E., and Thoenen H. (1987) Cellular localization of nerve growth factor synthesis by in situ hybridization. *EMBO J.* **6,** 891–899.

Barber, P. C. and Lindsay, R. M. (1982) Schwann cells of the olfactory nerves contain glial fibrillary acidic protein and resemble astrocytes. *Neuroscience* **7,** 3077–3090.

Barnett, S. C., Hutchins, A. M., and Noble M. (1993) Purification of olfactory nerve ensheathing cells from the olfactory bulb. *Dev. Biol.* **155,** 337–350.

Bartolomei, J. C. and Greer C. A. (1998) Differential axon extension from DRG cells on poly-L-lysine, laminin, ensheathing cell and cortical astrocyte substrates. *Soc. Neurosci. Abst.* **24,** 1054.

Bartolomei, J. C. and Spencer D. D. (1999) Fetal mesencephalic tissue implantation. *Techniques Neurosurg,* **5,** 73–78.

Bernstein, J. J. and Goldberg W. J. (1995) Experimental spinal cord transplantation as a mechanism of spinal cord regeneration. *Paraplegia* **33,** 250–253.

Blakemore, W. F. (1976) Invasion of Schwann cells into the spinal cord of the rat following local injection of lysolecithin. *Neuropathol. Appl. Neurobiol.* **2,** 21–39.

Blakemore, W. F. and Crang A. J. (1985) The use of cultured autologous Schwann cells to remyelinate areas of persistent demyelination in the central nervous system. *J. Neurol. Sci.* **70,** 207–223.

Blight A. R. (1983) Cellular morphology of chronic spinal cord injury in the cat: analysis of myelinated axons by line sampling. *Neuroscience* **10,** 521–543.

Bracken M. B., Shephard, M. J., Holford, T. R., Leo-Summers, L., Aldrich, E. F., Fazl, M., Fehlings, M., Herr, D. L., Hitchon, P. W., Marshall, L. F., Nockels, R. P., Pascale, V., Perot, P. L., Piepmeier, J., Sonntag, V. K. H., Wagner, F., Wilberger, J. E., Winn, H. R., and Young W. (1997) Administration of methylprednisolone for 24 or 48 hours or trilizad mesylate for 48 hours in the treatment of acute spinal cord injury. *JAMA* **277,** 1597–1604.

Braken M. B., Shepard, M. J., Collins, W. F., et al. (1990) A randomized controlled trial of methylprednisolone or naloxone in the treatment of acute spinal cord injury; results of the second National Acute Spinal Cord Injury Study. *N. Engl. J. Med.* **322,** 1405–1411.

Bracken M. B., Freeman, D. H., and Hellebrand K. (1981) Incidence of acute traumatic hospitalized spinal cord injury in the United States, 1970–1977. *Am. J. Epidemiol.* **3,** 615–622.[???]

Bregman, B.S. (1998) Regeneration in the spinal cord. *Curr. Opin. Neurobiol.* **8,** 800–807.

Bregman, B. S., Kunkel-Bagden, E., Reier, P. J., Dai, H. N., McAtee, M., and D. Gao (1993) Recovery of function after spinal cord injury: mechanisms underlying transplant-mediated recovery of function differ after spinal cord injury in newborn and adult rats. *Exp. Neurol.* **123,** 3–16.

Bregman B. S., and Reier P. J. (1986) Neural tissue transplant rescue axotomized rubrospinal cells from retrograde death. *J. Comp. Neurol.* **244,** 86–95.

Bunge R P, Puckett, W. R., and Hiester E. D. (1997) Observations on the pathology of several types of human spinal cord injury, with emphasis on the astrocyte response to penetrating injuries. *Adv. Neurol.* **72,** 305–315.

Cadelli, D.S., Bandtlow, C. E., and Schwab M. E. (1995) Oligodendrocyte-and myelin-associated inhibitors of neurite outgrowth: their involvement in the lack of CNS regeneration. *Exp. Neurol.* **115,** 189–192.

Charriere-Bertrand C., Garner, C., Tardy, M., and Nunez J. (1991) Expression of various microtubule-associated protein 2 forms in the developing mouse brain and in cultured neurons and astrocytes. *J. Neurochem.* **56,** 385–391.

Chen, A., Xu, X. M., Kleitman, N., and Bunge M. B. (1996) Methylprednisolone administration improves axonal regeneration into Schwann cell grafts in transected adult rat thoracic spinal cord. *Exp. Neurol.* **138,** 261–276.

Cheng, H., Cao, Y., and Olson L. (1996) Spinal cord repair in adult paraplegic rats: partial restoration of hind limb function. *Science* **273,** 510–513.

Chiu, K. and Greer, C. A. (1996) Immunocytochemical analyses of astrocyte development in the olfactory bulb. *Dev. Brain Res.* **95,** 28–37.

Chuah M. I. and Au, C. (1991) Olfactory Schwann cells are derived from precursors cells in the olfactory bulb epithelium. *J. Neurosci. Res.* **29,** 172–180.

Chuah, M. I. and Au, C. (1994) Olfactory cell cultures on ensheathing cell monolayers. *Chem. Senses* **19,** 25–34.

Cuschieri, A. and Bannister L. H. (1975) The development of the olfactory mucosa in the mouse: light microscopy. *J. Anat.* **119,** 277–286.

Davies, S. J., Field, P. M., and Raisman G. (1996) Regeneration of cut adult axons fails even in the presence of continuous aligned glial pathways. *Exp. Neurol.* **142,** 203–216.

Davies, S. J., Field, P. M., and Raisman G. (1997) Embryonic tissue induces growth of adult axons from myelinated fiber tracts. *Exp. Neurol.* **145,** 471–476.

Deckner M. L., Frisen, J. Verge, V. M. K., Hokfelt, T., and Risling M. (1993) Localization of neurotrophin receptors in the olfactory epithelium and bulb. *Neuroreport* **5,** 301–304.

Devon, R. and Doucette R. (1992) Olfactory ensheathing cells myelinate dorsal root ganglion neurites. *Brain Res.* **589**, 175–179.

Diener P. S. and Bregman B. S. (1994) Neurotrophic factors prevent the death of CNS neurons after spinal cord lesions in newborn rat. *Neuroreport* **5**, 1913–1917.

Diener, P. S. and Bregman B. S. (1998a) Fetal spinal cord transplants support the development of target reaching and coordinated postural adjustments after neonatal cervical spinal cord injury. *J. Neurosci.* **18**, 763–778.

Diener, P. S. and Bregman B. S. (1998b) Fetal spinal cord transplants support growth of supraspinal and segmental projections after cervical spinal cord hemisection in the neonatal rat. *J. Neurosci.* **18**, 779–793.

Doucette R. (1984) The glial cells in the nerve fiber layer of the rat olfactory bulb. *Anat. Rec.* **210**, 385–391.

Doucette, R. (1990) Glial influences on axonal growth in the primary olfactory system. *Glia* **3**, 433–449.

Doucette, R. (1991) PNS-CNS transitional zone of the first cranial nerve. *J. Comp. Neurol.* **312**, 451–466.

Doucette, R. (1993a) Glial progenitor cells of the nerve fiber layer of the olfactory bulb: effect of astrocyte growth media. *J. Neurosci. Res.* **35**, 274–287.

Doucette, R. (1993b) Glial cells in the nerve fiber layer of the main olfactory bulb of embryonic and adult mammals. *Microsc. Res. Tech.* **24**, 113–130.

Doucette, R. (1995) Olfactory ensheathing cells: potential for glial cell transplantation into areas of CNS injury. *Histol. Histopathol.* **10**, 503–507.

Doucette R. (1996) Immunocytochemical localization of laminin, fibronectin and collagen type IV in the nerve layer of the olfactory bulb. *Int. J. Dev. Neurosci.* **14**, 945–959.

Duncan, I. D., Hammang, J. P., and Gilmore S. A. (1988) Schwann cell myelination of the myelin deficient rat spinal cord following X-irradiation. *Glia* **1**, 233–239.

Farbman, A. I. (1994) Developmental biology of olfactory sensory neurons. *Semin. Cell Biol.* **5**, 3–10.

Fawcett, J. (1995) Spinal cord transplants: a future treatment for spinal injury? [Editorial]. *Paraplegia* **33**, 491–492.

Fawcett J. and Keynes R. J. (1990) Peripheral nerve regeneration. *Annu. Rev. Neurosci.* **13**, 43–60.

Felts, P. A. and Smith K. J. (1992) Conduction properties of central nerve fibers remyelinated by Schwann cells. *Brain Res.* **574**, 178–192.

Franceschini, I. A. and Barnett S. C. (1996) Low-affinity NGF-receptor and E-N-CAM expression define two types of olfactory nerve ensheathing cells that share a common lineage. *Dev. Biol.* **173**, 327–343.

Franklin, R. J., Gilson, J. M., Franceschini, I. A., and Barnett S. C. (1996) Schwann cell-like myelination following transplantation of an olfactory bulb-ensheathing cell line into areas of demyelination in the adult CNS. *Glia* **17**, 217–224.

Franklin R. J. M. and Blakemore W. F. (1993) Migration of Schwann cells. Requirements for Schwann cell migration within CNS environments: A viewpoint. *Int. J. Dev. Neurosci.* **11**, 641–649.

Friedman D., Scherer, S. S., Rudge, J. S., et al. (1992) Regulation of cilliary neurotrophic factor expression in myelin-related Schwann cells in vivo. *Neuron* **9**, 295–305.

Gong Q. Z., Bailey, M. S., Pixley, S. K., Ennis, M. Liu, W., and Shipley M. T. (1994) Localization an deregulation of the low affinity nerve growth factor receptor expression in the rat olfactory system during development and regeneration. *J. Comp. Neurol.* **344**, 336–348.

Goodman, M. N., Silver, J., and Jacobberger J. W. (1993) Establishment and neurite outgrowth properties of neonatal and adult rat olfactory bulb glial cell lines. *Brain Res.* **619**, 199–213.

Gonzalez, M. L., Malemud, C. J., and Silver J. (1993) Role of astroglial extracellular matrix in the formation of rat olfactory bulb glomeruli. *Exp. Neurol.* **123**, 91–105.

Graziadei P. P. C. and Monti-Graziadei G. A. (1980) Neurogenesis and neuron regeneration in the olfactory system of mammals. III Deafferentation and reinnervation of the olfactory bulb following section of fila olfactoria in rat. *J. Neurocytol.* **9**, 145–162.

Graziadei, P. P. C. and Monti-Graziadei G. A. (1979) Neurogenesis and neuron regeneration in the olfactory system of mammals. I. Morphological aspects of differentiation and structural organization of the olfactory sensory neurons. *J. Neurocytol.* **8**, 1–18.

Griffin M. R., Opitz, J. L., Kurland, L. T., Ebersold M. J., and O'Fallon W. M. (1985) Traumatic spinal cord injury in Olmsted County, Minnesota, 1935–1981. *Am. J. Epidemiol.* **121**, 884–895.

Griffith I. R. and McCulloch. M. C. (1983) Nerve fibers in spinal cord impact injuries. 1. Changes in the myelin sheath during the initial five weeks. *J. Neurol. Sci.* **58**, 335–345.

Guest, J. D., Rao, A., Olson, L., Bunge, M. B., and Bunge R. P. (1997) The ability of human Schwann cell grafts to promote regeneration in the transected nude rat spinal cord. *Exp. Neurol.* **148**, 502–522.

Hatten, M. E., Liem, R. K., Shelanski, M. L., and Mason C. A. (1991) Astroglia in CNS injury. *Glia* **4**, 233–243.

Hinds, J. W., Hinds, P. L., and McNelly N. A. (1984) An autoradiographic study of the mouse olfactory epithelium: evidence for long-lived receptors. *Anat. Rec.* **210**, 375–383.

Honmou, O., Felts, P. A., Waxman, S. G., and Kocsis J. D. (1996) Restoration of normal conduction properties in demyelinated spinal cord axons in the adult rat by transplantation of exogenous Schwann cells. *J. Neurosci.* **16**, 3199–3208.

Hotz, M. A., Gong, J., Traganos, F., and Darzynkiewicz Z. (1994) Flow cytometric detection of apoptosis: comparison of the assays of in situ DNA degradation and chromatin changes. *Cytometry* **15**, 237–244.

Houle J. D. and Reier P. J. (1988) Transplantation of fetal spinal cord tissue into chronically injured adult rat spinal cord. *J. Comp. Neurol.* **269**, 535–547.

Huard, J. M. T., Youngentob, S. L., Goldstein, B. J., Luskin, M. B., and Schwob J. E. (1998) Adult olfactory epithelium contains multipotent progenitors that give rise to neurons and non-neural cells. *J. Comp. Neurol.* **400**, 469–486.

Imaizumi, T., Lankford, K. L., Waxman, S. G., Greer, C. A., and Kocsis J. D. (1998) Transplanted olfactory ensheathing cells remyelinate and enhance axonal conduction in the demyelinated dorsal columns of the rat spinal cord. *J. Neurosci.* **18,** 6176–6185.

Itoh, Y., Sugawara, T., Kowada, M., and Tessler A. (1992) Time course of dorsal root axon regeneration into transplants of fetal spinal cord: I. A light microscopic study. *J. Comp. Neurol.* **323,** 198–208.

Itoyama, Y., Webster, H. D., Richardson, E. P. J., and Trapp B. D. (1983) Schwann cell remyelination of demyelinated axons in spinal cord multiple sclerosis lesions. *Ann. Neurol.* **14,** 339–346.

Jakeman L. B., Reier, P. J., Bregman B. S., et al. (1989) Differentiation of substantia gelatinosa-like regions in intraspinal and intracerebral transplantation of embryonic spinal cord tissue in the rat. *Exp. Neurol.* **103,** 17–33.

Jessen K. R. and Mirsky R. (1991) Schwann cell precursor and their development. *Glia* **4,** 185–194.

Jessen, K. R., Morgan, L., Stewart, H. J. S., and Mirsky, R. (1990) Three markers of adult non-myelin-forming Schwann cell, 217c(Ran-1), A5E3 and GFAP. Development and regulation by neuron-Schwann cell interactions. *Development* **109,** 91–103.

Kafitz, K. W. and Greer, C. A. (1997) Role of laminin and axonal extension from olfactory receptor cells. *J. Neurobiol.* **32,** 298–310.

Kafitz, K. W. and Greer, C. A. (1998) Differential expression of extracellular matrix and cell adhesion molecules in the olfactory nerve and glomerular layers of adult rats. *J. Neurobiol.* **34,** 271–282.

Kraus, J., Silberman, T. A., and McArthur, D. L. (1995) Epidemiology of spinal cord injury, in *Principles of Spinal Surgery* (Sonntag, V. K. H. and Menezes, A. H., eds.) McGraw-Hill, New York, pp. 41–58,

Li, Y., Field, P. M., and Raisman G. (1998) Regeneration of adult rat corticospinal axons induced by transplanted olfactory ensheathing cells. *J. Neurosci.* **18,** 10514–10524.

Li, Y. and Raisman, G. (1995) Sprouts from cut corticospinal axons persist in the presence of astrocytic scarring in long-term lesions of the adult rat spinal cord. *Exp. Neurol.* **134,** 102–111.

Li, Y., Field, P. M., and Raisman, G. (1997a) Repair of adult rat corticospinal tract by transplants of olfactory ensheathing cells. *Science* **277,** 2000–2002.

Li, Y. and Raisman, G. (1997b) Integration of transplanted cultured Schwann cells into the long myelinated fiber tracts of the adult spinal cord. *Exp. Neurol.* **145,** 397–411.

Liesi, P. (1985) Laminin-immunoreactive glia distinguish regenerative adult CNS systems from non-regenerative ones. *EMBO J.* **4,** 2505–2511.

Liu, K. L., Chuah, M. I., and Lee, K. K. (1995) Soluble factors from the olfactory bulb attract olfactory Schwann cells. *J. Neurosci.* **15,** 990–1000.

Mackay-Sim, A. and Kittel, P. (1991) Cell dynamics in the adult mouse olfactory epithelium: a quantitative autoradiographic study. *J. Neurosci.* **11,** 979–984.

Martin, D., Schoenen, J., Delree, P., Leprince, P., Rogister, B., and Moonen, G. (1991) Grafts of syngenic cultured, adult dorsal root ganglion-derived Schwann cells to the injured spinal cord of adult rats: preliminary morphological studies. *Neurosci. Lett.* **124,** 44–48.

Miller, R. H., French-Constant, C., and Raff, M. C. (1989) The macroglial cells of the rat optic nerve. *Annu. Rev. Neurosci.* **12,** 517–534.

Miragall, F., Kadmon, G., Husmann, M., and Schachner, M. (1988) Expression of cell adhesion molecules in the olfactory system of the adult mouse. Presence of the embryonic form of N-CAM. *Dev. Biol.* **129,** 516–531.

Miya, D., Giszter, S., Mori, F., Adipudi, V., Tessler, A., and Murray, M. (1997) Fetal transplants alter the development of function after spinal cord transection in newborn rats. *J. Neurosci.* **17,** 4856–4872.

Mori, S. and Leblond, C. P. (1969) Electron microscopic features and proliferation of astrocytes in the corpus callosum of the rat. *J. Comp. Neurol.* **137,** 197–226.

Mombaerts, P., Wang, F., Dulac, C., et al. (1996) The molecular biology of olfactory perception. *Cold Spring Harbor Symp. Quant. Biol.* **LXI,** 135–145.

Murray, M. (1997) Strategies and mechanisms of recovery after spinal cord injury. *Adv. Neurol.* **72,** 219–225.

Navarro, X., Valero, A., Gudino, G., et al. (1999) Ensheathing glia transplants promote dorsal root regeneration and spinal reflex restitution after multiple lumbar rhizotomy. *Ann. Neurol.* **45,** 207–215.

Nicol, D. J., Granat, M. H., Baxendale, R. H., and Tuson, S. J. M. (1995) Evidence for a human spinal stepping generator. *Brain Res.* **684,** 230–232.

Paino, C. L. and Bunge, M. B. (1991) Induction of axon growth into Schwann cell implants grafted into lesioned adult rat spinal cord. *Exp. Neurol.* **114,** 254–257.

Pallini, R., Fernandez, E., Gangitano, C., Del Fa, A., Olivieri-Sangiacomo, C., and Sbricoli, A. (1989) Studies on embryonic transplants to the transected spinal cord of adult rats. *J. Neurosurg.* **70,** 454–462.

Raisman, G. (1985) Specialized neuroglial arrangement may explain the capacity of vomeronasal axons to reinnervate central neurons. *Neuroscience* **14,** 237–254.

Ramón-Cueto, A. and Avila, J. (1998) Olfactory ensheathing glia: properties and function. *Brain Res. Bull.* **46,** 175–187.

Ramón-Cueto, A. and Nieto-Sampedro, M. (1992) Glial cells from adult rat olfactory bulb: immunocytochemical properties of pure cultures of ensheathing cells. *Neuroscience* **47,** 213–220.

Ramón-Cueto, A. and Nieto-Sampedro, M. (1994) Regeneration into the spinal cord of transected dorsal root axons is promoted by ensheathing glia transplants. *Exp. Neurol.* **127,** 232–244.

Ramón-Cueto, A., Perez, J., and Nieto-Sampedro, M. (1993) In vitro enfolding of olfactory neurites by p75 NGF receptor positive ensheathing cells from adult rat olfactory bulb. *Eur. J. Neurosci.* **5,** 1172–1180.

Ramón-Cueto, A., Plant, G. W., Avila, J., and Bunge, M. B. (1998) Long-distance axonal regeneration in the transected adult rat spinal cord is promoted by olfactory ensheathing glia transplants. *J. Neurosci.* **18,** 3803–3815.

Ramón-Cueto, A. and Valverde, F. (1995) Olfactory bulb ensheathing glia: a unique cell type with axonal growth-promoting properties. *Glia* **14,** 163–173.

Ramón y Cajal, S. (1928) *Studies on Degeneration and Regeneration of the Nervous System.* Oxford Press, London.

Reier, P. J., Anderson, D. K., Young, W., Michel, M. E., and Fessler, R. (1994) Workshop on intraspinal transplantation and clinical application. *J. Neurotrauma* **11,** 369–377.

Reier, P. J., Bregman, B. S., [???]B.S,[???] and Wujek, J. R. (1986) Intraspinal transplantation of embryonic spinal cord tissue in neonatal and adult rats. *J. Comp. Neurol.* **247,** 275–296.

Reier, P. J., Strokes, B. T., Thompson, F. J., and Anderson, D. K. (1992) Fetal cell grafts into resection and contusion/compression injuries of the rat and cat spinal cord. *Exp. Neurol.* **115,** 177–188.

Richardson, P. M., McGuinness, U. M., and Aguayo, A. J. (1980) Axons from CNS neurons regenerate into PNS grafts. *Nature* **284,** 264–265.

Ronnett, G. V., Hester, L. D., and Snyder, S. H. (1991) Primary culture of neonatal rat olfactory neurons. *J. Neurosci.* **11,** 1243–1255.

Roskams, A. J. I., Bethel, M. A., Hurt, K. J., and Ronnet, G. V. (1996) Sequential expressions of Trks A,B, and C in the regenerating olfactory neuroepithelium. *J. Neurosci.* **16,** 1294–1307.

Schwab, M. E., Kapfhammer, J. P., and Bandtlow, C. E. (1993) Inhibitors of neurite growth. *Annu. Rev. Neurosci.* **16,** 565–595.

Shepherd, G. M. and Greer, C. A. (1998) Olfactory bulb, in *The Synaptic Organization of the Brain.* (Shepherd, G. M., ed.), Oxford University Press, New York, pp. 159–204.

Singer, M. S., Shepherd, G. M., and Greer, C. A. (1995) Olfactory receptors guide axons. *Nature* **377,** 19–20.

Sonigra, R. J., Brighton, P. C. Jacoby, J. Hall, S., and Wigley, C. B. (1999) Adult rat olfactory nerve ensheathing cells are effective promoters of adult central nervous system neurite outgrowth in coculture. *Glia.* **25,** 256–269.

Stokes, B. T. and Reier, P. J. (1992) Fetal grafts alter chronic behavioral outcome after contusion damage to the adult rat spinal cord. *Exp. Neurol.* **116,** 1–12.

Tessler, A. (1991) Intraspinal transplants. *Ann. Neurol.* **29,** 115–123.

Tessler, A., Fischer, I., Giszter, S., et al. (1997) Embryonic spinal cord transplants enhance locomotor performance in spinalized newborn rats. *Adv. Neurol.* **72,** 291–303.

Valverde, F. and Lopez-Mascaraque, L. (1991) Neuroglial arrangements in the olfactory glomeruli of the hedgehog. *J. Comp. Neurol.* **307,** 658–674.

Valverde, F., Santacana, M., and Heredia, M. (1992) Formation of an olfactory glomerulus: morphological aspects of development and organization. *Neuroscience* **49,** 255–275.

Van Eldik, L. J., Christie-Pope, B., Bolin, L. M., Shooter, E. M., and Whetsell, W. O. (1991) Neurotrophic activity of S100β in cultures of dorsal root ganglia from embryonic chick and fetal rat. *Brain Res.* **542,** 280–285.

Vickland, H., Westrum, L. E., Kott, J. N., Patterson, S. L., and Bothwell, M. A. (1991) Nerve growth factor receptor expression in the young and adult rat olfactory system. *Brain Res.* **565,** 269–279.

Waxman, S. G. and Kocsis, J. D. (1997) Spinal cord repair progress towards a daunting goal. *Neuroscientist* **3,** 263–269.

Williams, R. and Rush, R. A. (1988) Electron microscopic immunocytochemical localization of nerve growth factor in developing mouse olfactory neurons. *Brain Res.* **463,** 21–27.

Wirth, E. D., Theele, D. P., Mareci, T. H., Anderson, D. K., Brown, S. A., and Reier, P. J. (1992) In vivo magnetic resonance imaging of fetal cat neural tissue transplants in the adult cat spinal cord. *J. Neurosurg.* **76,** 261–274.

Xu, X. M., Guenard, V., Kleitman, N., Aebischer, P., and Bunge, M. B. (1995a) A combination of BDNF and NT-3 promotes supraspinal axonal regeneration into Schwann cell grafts in adult rat thoracic spinal cord. *Exp. Neurol.* **134,** 261–272.

Xu, X. M., Guenard, V., Kleitman, N., and Bunge, M. B. (1995b) Axonal regeneration into Schwann cell-seeded guidance channels grafted into transected adult rat spinal cord. *J. Comp. Neurol.* **351,** 145–160.

Zurn, A. D., Nick, H., and Monrad, D. (1988) A glial-derived nexin promotes neurite outgrowth in cultured chick sympathetic neurons. *Dev. Neurosci.* **10,** 17–24.

10

Experimental Approaches to Restoration of Function of Ascending and Descending Axons in Spinal Cord Injury

Stephen G. Waxman and J. D. Kocsis

1. INTRODUCTION

In many cases of spinal cord injury (SCI) (even in some clinically "complete" cases of nonpenetrating SCI), there is a subpopulation of axons that survive. These residual axons are not transected, but fail to conduct normal action potentials as a result of demyelination; they display conduction block or slowing. Recognition of the existence of these spared axons and identification of demyelination as a significant factor contributing to their dysfunction have triggered an important shift in our thinking about restoration of function in SCI. In addition to examining the regrowth of severed axons within the spinal cord, contemporary SCI research is thus focusing on the restoration of axonal conduction in demyelinated spinal cord axons. This chapter reviews the background of this approach as well as recent developments that suggest new strategies for restoration of function following spinal cord trauma.

2. DEMYELINATION IN SPINAL CORD INJURY

Numerous histologic and electron microscopic studies have demonstrated the presence of demyelination within ascending and descending tracts of the spinal cord white matter in experimental models of compressive and contusive SCI (Gledhill et al., 1973; Harrison and McDonald, 1977; Griffiths and McCulloch, 1983). Although the mechanism of demyelination is not definitively understood, electron microscopic studies in experimental animals have been interpreted as suggesting that loss of myelin in SCI may result from invasion by inflammatory cells that strip the myelin from spinal cord axons following contusive injury (Blight, 1985). Alternatively, it has been suggested that apoptosis of oligodendrocytes may occur (possibly due to loss of trophic signals from injured axons) in SCI, with a resultant loss of myelin sheaths (Shuman et al., 1997; Beattie et al., 1998).

From: *Neurobiology of Spinal Cord Injury*
Edited by: R. G. Kalb and S. M. Strittmatter © Humana Press Inc., Totowa, NJ

Postmortem studies in humans have also provided evidence for demyelination in spinal cord white matter following SCI and spinal cord compression (Byrne and Waxman, 1990; Bunge et al., 1993). A functional parallel exists in the observation, from some patients in whom SCI has been judged on the basis of clinical criteria to be "complete," of residual descending influences on spinal reflex activity, a finding that has led to the concept of "dyscomplete" spinal cord injury (Sherwood et al., 1992). Interestingly, in some patients who were initially judged to have complete SCI, careful serial clinical assessments have demonstrated improvements in neurologic status that can occur over periods extending at least as long as 1 year (Young, 1989). This delayed clinical improvement, together with the evidence for demyelination cited above, suggests the possibility that restoration of conduction in previously demyelinated axons may provide at least a partial basis for functional recovery following SCI. Indeed, in at least some experimental models of traumatic SCI in rodents, there is evidence for a degree of spontaneous remyelination (Harrison and McDonald, 1977; Salgato-Ceballos et al., 1998). In humans, however, the degree of spontaneous remyelination (if any) is not known.

The clinical deficits associated with demyelination are due in large part to conduction failure in demyelinated axons. Several approaches have been taken in an attempt to restore normal action potential conduction in demyelinated spinal cord axons. One strategy involves the pharmacologic manipulation of ion channels in demyelinated axons, and the other focuses on spinal cord repair via the transplantation of myelin-forming cells.

3. IONIC CHANNEL ARCHITECTURE
OF DEMYELINATED AXONS

The response of axons to demyelination reflects not only the loss of the high-resistance, low-capacitance myelin shielding, but also the distribution of voltage-sensitive ion channels, which are distributed within the axon membrane of myelinated axons in a nonuniform pattern. Figure 1 shows a simplified model of the mammalian myelinated fiber. Na^+ channels are clustered in high density ($\sim 1000/\mu m^2$) in the axon membrane at the node of Ranvier, where action potentials are generated in normal myelinated axons (Ritchie and Rogart, 1977; Waxman, 1977). The Na^+ channel density is much lower ($< 25/\mu m^2$) within the paranodal and internodal axon membrane (i.e., the axon membrane under the myelin sheath). At least three types of K^+ channels, a "fast" channel, a "slow" K^+ channel, and an inward rectifier, are present in myelinated axons (for reviews, *see* Kocsis et al., 1993; Waxman and Ritchie, 1993). Fast K^+ channels are expressed in highest density in the axon membrane under the myelin, but only in low densities in the axon membrane at the node (Chiu and Ritchie, 1981; Foster et al., 1982; Kocsis et al., 1982; Ritchie, 1982). Measurements made using voltage clamp techniques indicate that the density of fast K^+ channels decreases to less than one-sixth of its peak value (which is reached in the paranode) in the node and internode (Röper and Schwartz, 1989).

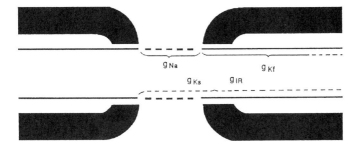

Fig. 1. Model showing putative organization of ion channels in the myelinated fiber. g_{Na}, Na^+ channels; g_{Kf}, fast K^+ channels; g_{KS}, slow K^+ channels; g_{IR}, inward rectifier. Sodium channels are clustered in high density in the axon membrane at the node of Ranvier and are present in much lower densities in the internode. Fast K^+ channels, in the axon membrane under the myelin, are unmasked by demyelination.

This nonuniform distribution of fast K^+ channels has permitted a pharmacologic approach to understanding, and altering the behavior of, the demyelinated axons. 4-Aminopyridine (4-AP), when applied on the extracellular side of the axon membrane, blocks fast K^+ channels (Kocsis et al., 1986; Baker et al., 1987). As a result, 4-AP leads to delayed repolarization and prolongation of the action potential in normal nonmyelinated fibers, in both the peripheral nervous system (PNS) (Sherratt et al., 1980; Bostock et al., 1981) and the central nervous system (CNS) (Malenka et al., 1981; Preston et al., 1983); these results are consistent with the conclusion that fast K^+ channels contribute to action potential repolarization in these fibers, which are not covered by myelin. In contrast, in normal myelinated axons within the adult mammalian PNS and CNS, fast K^+ channels at the node of Ranvier do not appear to play a significant role in action potential repolarization, and as shown in Figure 2B (Kocsis et al., 1982), the action potential is not prolonged following exposure to externally supplied 4-AP (Bostock et al., 1981; Kocsis and Waxman, 1980). This is not due to an absence of fast K^+ channels in mature myelinated fibers. As noted above, these channels are expressed in the paranodal part of the axon membrane and are masked by the overlying myelin. Consistent with this finding, developmental studies show a relatively large effect of 4-AP on action potential duration in premyelinated axons, but this effect is attenuated during maturation of the myelin and associated axoglial junctions (Fig. 2A) (Eng et al., 1988; Foster et al., 1982; Kocsis et al., 1982; Ritchie, 1982).

The contribution of fast K^+ channels to action potential electrogenesis can also be seen following acute demyelination (Targ and Kocsis, 1985, 1986). When 4-AP is applied to demyelinated fibers (where it has access to the exposed, formerly internodal axon membrane), it blocks fast K^+ channels and thereby produces a significant delay in repolarization. This is manifested by a broadened action potential (Fig. 2C), which delivers more inward current to the fiber.

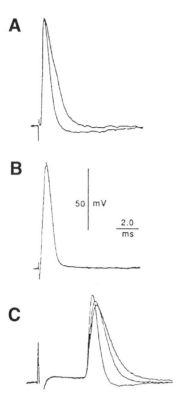

Fig. 2. Effect of 4-aminoyridine (0.5 m*M*) on action potentials in **(A)** premyelinated axon in regenerating sciatic nerve after crush injury, **(B)** myelinated axon in
adult rat sciatic nerve, and **(C)** acutely demyelinated (lysophosphatidyl-choline)
ventral root axon. Conduction slows after demyelination, as evidenced by the long
latency to spike onset. Extracellular application of 4-AP results in prolongation of
the action potential in premyelinated and demyelinated axons but not in mature
myelinated axons, where fast K⁺ channels are masked by the myelin. (A and B,
reproduced with permission from Kocsis et al., 1982; C, reproduced with permission
from Bowe et al., 1987.)

4. PHARMACOLOGIC REVERSAL OF CONDUCTION BLOCK IN DEMYELINATED AXONS

Safety factor (defined as "current produced at a given node of Ranvier ÷ current needed to depolarize the next node to threshold") is decreased in demyelinated axons. On theoretical grounds, it might be expected that maneuvers
prolonging the action potential would increase the time integral of inward current and thus might improve conduction (Schauf and Davis, 1974). Bostock and
co-workers observed that 4-AP in fact increases the temperature at which con-

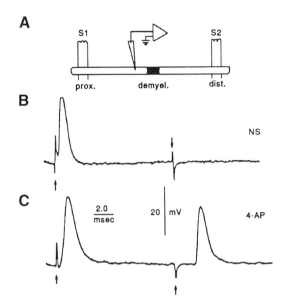

Fig. 3. Restoration of conduction in demyelinated sciatic nerve axon with 4-AP (1.0 mM). The axon was focally demyelinated with lysophosphatidyl choline. **(A)** Experimental design permits examination of conduction through normal (stimulation at S[1]) or demyelinated (stimulation at S[2]) regions. **(B)** Stimulation at S[1] evokes an action potential, but conduction block occurs following S[2] stimulation. **(C)** Following application of 4-AP, conduction block is reversed and the action potential propagates, with increased latency, through the demyelinated zone. Action potential duration is increased, due to blockade of fast K^+ channels. NS, normal solution. (Adapted from Targ and Kocsis, 1985.)

duction failure occurs in demyelinated ventral root axons, in some cases reversing conduction block at physiologic temperatures (Sherratt et al., 1980; Bostock et al., 1981). Reversal of conduction block has also been demonstrated at the single axon level with restoration of secure impulse conduction, following treatment of experimentally demyelinated sciatic nerve axons with 4-AP (Targ and Kocsis, 1985, 1986). Reversal of conduction block in an experimentally demyelinated (lysophosphatidyl choline-treated) axon following application of 4-AP is illustrated in Figure 3 (Targ and Kocsis, 1985). In this experiment, stimulating electrodes were positioned on both ends of the nerve, and recordings from single axons were obtained with microelectrodes positioned intraaxonally on one side of the lesion (Fig. 3A). Following stimulation proximal to the lesion (i.e., with a conduction path that did not include the demyelinated region), reliable conduction occurred and a propagated action potential could be recorded in normal Ringer's solution. In contrast, following stimulation on the contralateral side of the lesion (so that conduction had to cross the zone of demyelination), an action potential could not be recorded because conduction block had occurred (Fig. 3B). The effect of 4-AP in the axon is shown in Figure 3C. Following block of the fast K$^+$ channels with 4-AP, action potential duration was increased and reliable conduction through the lesion was restored. Subsequent studies showed reversal of conduction block in vivo with 4-AP in experimental models of chronic SCI (Blight, 1989). Clinical studies have been carried out with 4-AP, and with the related drug 3,4-diaminopyridine (3,4-DAP), to examine the effects of these agents on neurologic status in patients with multiple sclerosis (see, e.g., Jones et al., 1983; Stefoski et al., 1987; Davis et al., 1990). Improvements in motor function, improved brainstem function (e.g., improvement in extraocular movements), reduction in the size of scotomata, and improved critical flicker fusion have been observed in such patients. Clinical studies are under way in humans with SCI. Hopefully these studies will determine whether 4-AP and related drugs provide symptomatic improvement in humans with SCI.

5. REMYELINATION IN SPINAL CORD INJURY

Remyelination, even with relatively thin or short myelin segments, can support the restoration of action potential propagation through previously demyelinated fibers, if the remyelinated nodes of Ranvier develop membrane properties similar to those in normal fibers so that they can support action potential electrogenesis (Koles and Rasminsky, 1972; Waxman and Brill, 1978). The development of relatively normal Na$^+$ channel densities at remyelinated nodes of Ranvier in the spinal cord is, in fact, suggested by cytochemical studies showing that newly formed nodes along remyelinated axons develop normal properties (Weiner et al., 1980). Saxitoxin-binding experiments in remyelinated sciatic nerve have demonstrated an increase in the number of Na$^+$ channels that is proportional to the increase in nodal membrane area imposed by the shorter internode spacing seen in remyelinated fibers, again suggesting that the newly formed nodes acquire a relatively normal number of sodium channels (Ritchie, 1982). Increased conduction velocity and restoration of the ability to conduct high-frequency impulse trains have been observed in association with remyelination in peripheral nerves demyelinated with lysophophatidyl choline (Smith and Hall, 1980). Microelectrode studies have also provided evidence for recovery of conduction following remyelination in spinal cord (dorsal column) axons, with refractory periods returning to normal levels following remyelination even with abnormally thin and short myelin segments (Smith et al., 1983). There is also evidence for a link between clinical recovery and remyelination in experimental allergic encephalomyelitis (Stanley and Pender, 1991). Both oligodendrocytes and Schwann cells can spontaneously remyelinate CNS axons, and action potential conduction can be facilitated by remyelination by both endogenous oligodendrocytes and endogenous Schwann cells (Blight and Young, 1989; Felts and Smith, 1992). These observations raise the questions of whether remyelination can be induced by transplantation of *exogenous* myelin-forming cells and if so, whether this can enhance axonal conduction within the spinal cord.

6. TRANSPLANTATION OF MYELIN-FORMING CELLS INTO THE INJURED SPINAL CORD

Recognition that demyelination is an important component of the pathology of SCI and that remyelination can enhance conduction in demyelinated axons has suggested a new strategy of restoration of function, i.e., the transplantation of myelin-forming glial cells to demyelinated parts of the injured spinal cord. (Utzschneider et al., 1994; Honmou et al., 1994; Imaizumi et al., 1998a,b).

Electron microscopic studies have demonstrated that transplanted Schwann cells and oligodendrocytes can form myelin around demyelinated axons within the spinal cord (Blakemore, 1976; Groves et al., 1993; Duncan et al., 1988a; Gout et al., 1988; Rosenbluth et al., 1990). It should be noted, however, that the formation of morphologically intact myelin around demyelinated axons by transplanted glial cells does not, *per se,* guarantee the restoration of secure impulse conduction; action potential electrogenesis and secure impulse conduction depend not only on the formation of myelin around axons, but also on the presence of appropriate numbers and types of ion channels at the newly formed nodes of Ranvier. There is also a need for mature paranodal axoglial junctions to be established, following myelin formation by transplanted cells, between the myelin-forming cells and axons; otherwise there will be a current shunt under the myelin that may interfere with conduction (Hirano and Dembitzer, 1978). Moreover, remyelination should optimally meet the requirements for impedance matching. In this regard, patchy or incomplete remyelination can lead to impedance mismatch at the junction between myelinated and demyelinated axon zones, which can cause or exacerbate conduction failure (Sears et al., 1978; Waxman, 1978).

7. GLIAL CELL TRANSPLANTATION IN THE *MYELIN-DEFICIENT (MD)* SPINAL CORD

Our first electrophysiologic studies, aimed at providing proof of principle that action potential conduction could be improved following transplant-induced myelination, used the dorsal columns axons in the *myelin-deficient (md)* rat as a model (Utzschneider et al., 1994). This mutant provides an excellent experimental model because, as a result of a point mutation in the proteolipid protein gene (Boison and Stoffel, 1989; Simons and Riordan, 1990; Zeller et al., 1989; Hudson et al., 1989), it shows a total lack of CNS myelin; this facilitates definitive confirmation that physiologic changes are due to myelination by exogenous, transplanted glial cells. Thus, in this model system we can rule out the possibility of background myelination by endogenous cells (Duncan et al., 1988a,b).

Fig. 4. *(Right)* Conduction velocity is increased in dorsal column axons of the *md* rat following transplantation of myelin-forming cells. **(A)** Field potentials recorded from transplant region (~3 mm in length) and nontransplant region 16 d following glial cell transplantation. Stimulation at two sites (S_1 and S_2) provided recording tracks within and outside the transplant region. **(B)** Conduction latencies are shown for nontransplant (upper graphs) and transplant regions (lower graph) in the *md* dorsal columns. The upper graph displays the latency of the main N negatively (from 100 recording sites from 17 recording tracks) outside the transplant region. The slope of the linear regression indicates an average conduction velocity of 0.9 ± 0.03 m/s. The lower graph shows a significantly smaller increase in latency with increasing conduction distance within the transplant region, where there is an average conduction velocity of 3.2 ± 0.23 m/s. **(C)** Outside the transplant zone, field potentials typically display a single primary negativity with occasional early or late components. **(D)** Field potentials from transplant regions (same animal) show two negativities (N_1 and N_2) with increasingly distinct latencies as the conduction distance in increased. The N_1 component displays an increased conduction velocity. The stimulus site was outside the transplant region, demonstrating conduction of the impulse across the dysmyelinated-myelinated junction. (Adapted from Utzschneider et al., 1994.)

In preparation for transplantation, glial cell suspensions were prepared from the spinal cords of female littermates of the animals to be transplanted, at 4–5 postnatal d. Following overnight culture, glial cells were concentrated at 50,000/µL. Injections of 1.0µL of cell suspension were made via a dorsal laminectomy at the thoracolumbar junction, using a fine glass micropipete, into two or three sites along the dorsal columns of the spinal cord. Conduction in the transplanted spinal cords was studied, 15–17 d following cell transplantation, on postnatal d 20, 21, or 22 using field potential and single cell recordings as described by Utzschneider et al. (1994).

As viewed from their dorsal aspect, the transplanted spinal cords displayed opaque white patches, which correspond to zones of myelination, extending for several millimeters along the longitudinal axis of the dorsal surface of the spinal cord. Histologic examination demonstrated numerous oligodendrocytes and myelination of most axons within these patches. Rostral and caudal to these areas of myelination there were few or no myelinated axons within the dorsal columns, and the histology was similar to that of nontransplanted *md* spinal cords. Conduction was significantly enhanced as a result of myelination by the transplanted cells. This can be seen clearly in Figure 4, which demonstrates that conduction velocity was increased threefold, to 3.2 ± 0.2 m/s within the zone of myelination, compared with 0.9 ± 0.03 m/s in nontransplanted regions. Notably, in experiments in which the conduction track encompassed both myelinated and nonmyelinated regions, we observed a distinct increase in conduction velocity as the action potential propagated into the region that had been myelinated by the transplanted glial cells. Importantly, we also observed that action potentials

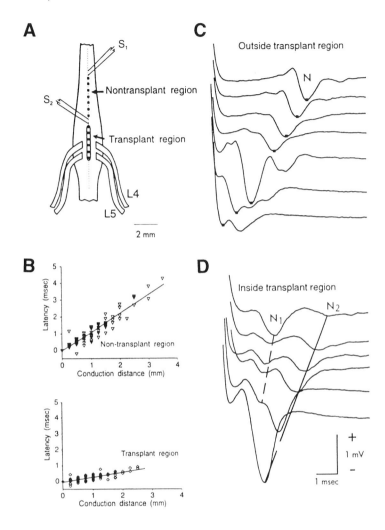

could propagate either into, or out of, the transplant area in all the spinal cords studied. This shows that impedance mismatch, which can block conduction at sites where the pattern of myelination changes abruptly (Waxman and Brill, 1978), did not block conduction at this crucial interface following transplantation. Moreover, the ability of the axons to carry high-frequency impulse trains was enhanced following transplantation (Fig. 5).

To assess the functional consequences of myelination by transplanted cells at the single-cell level, we also obtained intracellular recordings from dorsal root ganglion neurons while the axons of these cells were antidromically stimulated within the dorsal columns (Fig. 6A). These experiments demonstrated that conduction velocity was approx £threefold faster for axons in the transplant region,

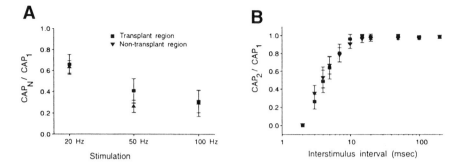

Fig. 5. (A) High-frequency conduction in *md* axons is similar inside and outside the transplant region. Ratios of the amplitudes of the first (CAP_1) and final (CAP_N) compound action potentials (CAPs) for repetitive stimuli at 20 Hz (10 s), 50 Hz (10 s), and 100 Hz (2 s) are shown. **(B)** Double-shock experiments showing the ratio of test CAP (CAP_2) to control CAP (CAP_1) for interstimulus intervals of 2–200 ms. The time-course of recovery is similar for axons inside and outside the transplant region, which are known to conduct high-frequency trains as well as spinal cord axons in control rats. (Adapted from Utzschneider et al., 1994.)

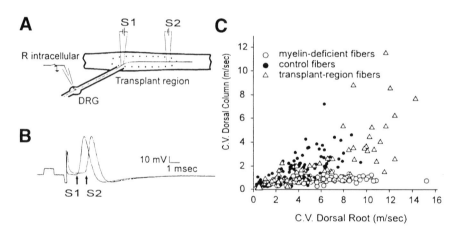

Fig. 6. (A) Single-cell recording showing conduction of action potentials through the transplant region in *md* spinal cord. **(B)** Action potentials recorded from dorsal root ganglion cell in response to stimulation at two sites (S^1 and S^2) in the transplant region. Propagation of the action potential from stimulating electrode S^2 to the dorsal root ganglion cell indicates that conduction was not blocked at the zone of potential impedance mismatch between the transplant zone and nontransplanted parts of the host nervous system. From the latency shift and interstimulus distance, a conduction velocity of 2.6 m/s within the transplant zone can be calculated for this axon. **(C)** Aggregate conduction velocity is shown for axons within the *md* transplant region, the *md* nontransplant region, and control spinal cord; conduction velocity is significantly increased in axons within transplant zone. (Adapted from Utzschneider et al., 1994.)

compared with the nontransplanted *md* dorsal column axons. In fact, conduction velocity in transplanted axons was comparable to that in age-matched controls in which myelination had occurred normally (transplanted, 2.3 ± 0.3 m/s; *n* = 67; *md*, 0.76 ± 0.02 m/s; *n* = 258; age-matched controls 1.9 ± 0.13 m/s; *n* = 95). Conduction velocities for axons in the transplant region were significantly faster than those of axons with comparable dorsal root conduction velocities from nontransplanted *md* rats (Fig. 6C). Moreover, as illustrated in Figure 6B, propagation of action potentials, from the transplant region into nontransplanted regions, was also apparent at the single-axon level in these experiments. These experiments provided additional evidence that conduction can proceed securely through the region of potential impedance mismatch at the junction between myelinated and nonmyelinated axon segments.

These results from the *md* rat spinal cord provided the first demonstration of enhanced conduction along spinal cord axons following myelination by transplanted glial cells. These findings also provided another important piece of information: enhanced axonal conduction was not confined to the cell injection site but was observed throughout the longitudinal extent of myelin formation extending rostrally and caudally from the cell injection site. These results demonstrate that transplanted glial cells can migrate within CNS white matter away from their site of implantation and further suggest that they can myelinate axons at distances as far as 5–10 mm.

As a result of the absence of endogenous CNS myelin in the *md* rat, these results can be interpreted as showing that the enhanced axonal conduction is due to myelination by exogenous, transplanted cells and is not due to background myelination by endogenous glial cells. Despite its advantages as an experimental model, however, the CNS of the *md* rat has some limitations. Because affected *md* rats rarely survive for more than 3–5 postnatal weeks, analysis of this model system is confined to immature hosts. Moreover, long-term effects and stability of glial cell transplantation cannot be studied in this model. To address these issues, we have more recently studied a model of acute demyelination in the spinal cord of the adult rat (Honmou et al., 1996).

8. GLIAL CELL TRANSPLANTATION INTO THE DEMYELINATED DORSAL COLUMNS

To determine whether it is possible to use glial cell transplantation to reconstruct demyelinated white matter and improve electrophysiologic function within the adult CNS, we have more recently transplanted suspensions of various cell types into the demyelinated dorsal columns in the adult rat. In these studies, acute demyelinating lesions that exhibited minimal endogenous remyelination were produced by X-irradiation and injection of ethidium bromide (which chelates nucleic acids and leads to oligodendrocyte death) using a modification of the method of Blakemore and Patterson (1978). The ethidium bromide-X radiation lesion remains glial cell free for more than 5–6 weeks, in contrast to most other models, in which endogenous remyelination can com-

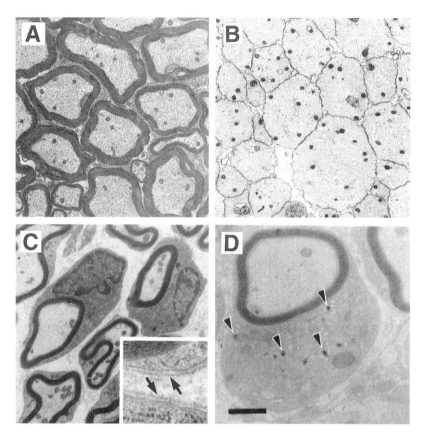

Fig. 7. Electron micrographs showing normal (**A**) and acutely demyelinated (**B**) axons in the dorsal columns of adult rat. Following cell transplantation all of the dorsal column axons were remyelinated (**C**). Higher magnification (inset in **C**) demonstrates basal lamina (arrows) and extracellular collagen fibrils surrounding axons. (**D**) Schwann cells carrying the β-gal reporter gene (arrowheads) are present in the lesion following treatment of the tissue with the substrate X-Gal. Bar = 4 μm for A, B and C; 2 μm for D; 0.6 μm for inset in C. (Adapted from Honmou et al., 1996.)

mence within days of demyelination. Histologic and ultrastructural examination (Fig. 7A,B) confirmed that axons within the lesion were totally demyelinated and showed that the lesion site was essentially free of glial cells (Honmou et al., 1994). Demyelinated lesions encompassed 70–80% of the extent of the dorsal columns and, using this demyelination protocol, measured 7–8 mm along the rostrocaudal axis of the spinal cord.

For these experiments we transplanted Schwann cells transfected with a *Lac Z* reporter gene, together with astrocytes derived from immature rats, into the demyelinated dorsal columns (Honmou et al., 1996). The Schwann cells were cultured from neonatal (P1–P3) rat sciatic nerve (Brockes et al., 1979), and primary astrocyte cultures were established from neonatal rat optic nerve (McCarthy and de Vellis, 1980). Histologic identification of transplanted Schwann cells was facilitated by expression of, β-galactose (β-gal), introduced using a replication-defective BAG retroviral vector (Price et al., 1987) from the φ2 packaging line (Mann et al., 1983) as a reporter. Suspensions of 5×10^4 Schwann cells and astrocytes (in a ratio of approximately 3:2) were injected into ethidium bromide X-irradiation-induced lesions, 3 d following ethidium bromide injection.

A large number of axons within the lesion zone were myelinated by 3 weeks following Schwann cell transplantation, with the exception of the smallest diameter axons, which are normally nonmyelinated (Fig. 7C). Demonstration of β-gal gene product in Schwann cells associated with remyelinated axons confirmed that the new myelin had been formed by exogenous transplanted Schwann cells (Fig. 7D).

As seen in Figure 8, conduction in the demyelinated dorsal columns was severely impaired compared with controls. In control spinal cords (no lesions), as the compound action potential propagated for 5 mm through the dorsal columns, its amplitude was reduced to $13.4 \pm 4.4\%$ (mean $\pm$ SEM; $n = 5$) of the amplitude at 2 mm (Fig. 8B1). In contrast, there was a more rapid fall-off in amplitude of the compound action potential as conduction distance increased within the demyelinated dorsal columns (Fig. 8B2), and at a conduction distance of 5 mm the reduction in amplitude of the compound action potential was so severe that a response could not be detected, demonstrating conduction block in the demyelinated axons.

These abnormalities in axonal conduction were reversed following cell transplantation. Recordings of compound action potentials showed that, following myelination by transplanted Schwann cells, a relatively normal pattern of conduction was restored and the amplitude decrement with distance was indistinguishable from controls (Fig. 8B3), indicating that propagation of action potentials had penetrated substantially further within the lesion without block. There was also restoration of conduction velocities to levels close to, or even above, normal levels following transplantation (11.4 ± 0.7 m/s transplanted; 0.9 ± 0.1 m/s demyelinated; 10.2 ± 0.9 control).

To study conduction in single axons that traversed the lesion site in this model, we used intraaxonal recording methods with arrays of stimulating electrodes that were positioned along the spinal cord within the lesion and at distances up to several millimeters rostral and caudal to the lesion. Axons were impaled between the two arrays of stimulating electrodes within the nondemyelinated region, so that conduction along axon segments both running through, and excluding, the lesion could be studied (Fig. 9A). These experiments confirmed that in demyelinated spinal cords that had not received a transplant, conduction velocity decreased abruptly within the demyelinated zone (Fig. 9B). In axons that had been myelinated following transplantation of

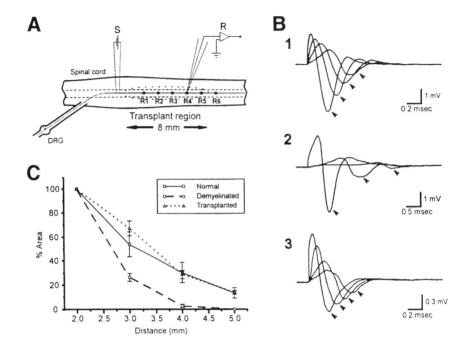

Fig. 8. (A) Schematic drawing illustrating the arrangement of stimulating and recording (R) electrodes for studies on the effect of Schwann cell transplantation. **(B)** Compound action potentials recorded at 1-mm increments along the dorsal columns in control (1), ethidium bromide/X-irradiation demyelinated (2), and transplant-induced remyelinated (3) axons. **(C)** compound action potential area plotted vs. conduction distance for normal, demyelinated, and transplant-induced remyelinated dorsal columns (*n* = 5). (Adapted from Honmou et al., 1996.)

Schwann cells, conduction velocities were restored to close-to-normal levels; in these spinal cords, conduction velocities were the same within the lesion (where axons had been remyelinated by transplanted Schwann cells) and outside of the lesion (where demyelination had not occurred) (Fig. 9C,D). Action potentials, evoked by stimulation within the transplant region, propagated into the nondemyelinated portion of the spinal cord (Figs. 6C,9A), demonstrating that conduction had successfully traversed the junction between remyelinated and normal parts of the host nervous system.

In this model of CNS demyelination, as expected, the refractory period was prolonged, and the ability to conduct action potential impulse trains at high frequencies was impaired. Both of these conduction abnormalities were reversed by remyelination with transplanted Schwann cells. Figure 10A shows that, in axons remyelinated following transplantation, the relative refractory period was reduced so that it approached, or was even less than, control values. The onset of

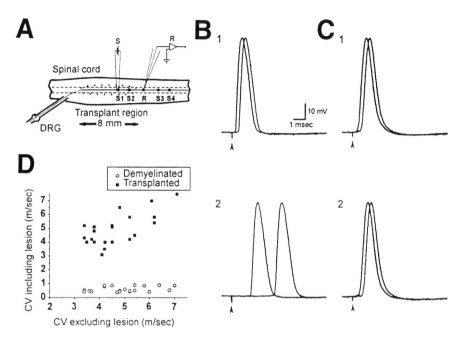

Fig. 9. (A) Schematic drawing showing arrangement of extracellular stimulation and recording sites. Intraxonal impalements permitted recordings to be obtained from dorsal column axons outside the lesion, where the axons were normally myelinated (R). Stimulating electrodes (S1–S4) within and outside the lesion zone permitted conduction velocity to be examined in the demyelinated or remyelinated parts of the axons, and in a normally myelinated segment of the same axon. **(B)** Pairs of action potentials from spinal cord that did not receive a transplant, showing conduction (for comparable conduction distances) along trajectories that included the demyelinated lesion (2) and excluded (1) the lesion. Conduction latency is prolonged through the demyelinated region. **(C)** Similar stimulation-recording protocol for transplant-induced remyelinated axons. Conduction latencies through the remyelinated region are reduced and are similar to those outside of the lesion zone; this increase in conduction velocity is a result of myelin formation. **(D)** Plot showing axonal conduction velocity (CV) in an axon within the lesion vs. conduction velocity outside the lesion, for demyelinated and transplant-induced remyelinated groups. Conduction velocities are increased as a result of myelination by transplanted cells. (Adapted from Honmou et al., 1996.)

recovery occurred sooner, and the slope of recovery was greater in axons remyelinated following demyelination. Figure 10B shows that the ability of dorsal column axons to carry high-frequency impulse trains was impaired following demyelination, and the amplitude of the field potential of the demyelinated axons was reduced compared with controls at frequencies of 50 Hz and higher.

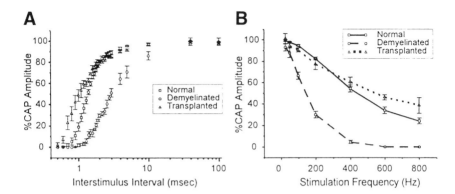

Fig. 10. Refractory period and ability to conduct high-frequency impulse trains are restored to close to normal in the demyelinated dorsal columns as a result of cell transplantation. **(A)** Paired pulse stimuli at varying intervals were used to quantitate the refractory period for transmission in normal, demyelinated, and transplant-induced remyelinated axons. Compound action potentials (CAPs) evoked by the second of two paired stimuli measured at increasing interstimulus intervals. Amplitude recovery was impaired in the demyelinated axon compared with normal axon, but transplant-induced remyelinated axons exhibited faster recovery properties control axons. **(B)** The ratio of the compound action potential amplitudes of the last response of a train (0.5 s, 36°C) over the first response was examined at various frequencies. Axons that had been remyelinated and axons following transplantation showed improved high-frequency conduction capabilities compared with those of the demyelinated axons; at high frequencies the transplanted axons, which had been remyelinated by Schwann cells, displayed a smaller amplitude decrement compared with controls. (Adapted from Honmou et al., 1996.)

Following remyelination by transplanted Schwann cells, axons were able to follow high-frequency stimulation as well as controls, with smaller amplitude decrements at high stimulus frequencies than in control dorsal.

9. TRANSPLANTATION OF OLFACTORY ENSHEATHING CELLS

More recently, we have begun to examine a variety of other cell types, in the search for the optimal source of cells for transplantation into the spinal cord. Our previous studies (Honmou et al., 1996) as well as those of others (Blakemore and Crang, 1985), had demonstrated that, although transplantation of Schwann cells alone results in myelination of demyelinated spinal cord axons, transplantation of Schwann cells together with astrocytes results in more robust migration and more extensive areas of remyelination. On the basis of these results, we reasoned that olfactory ensheathing cells, which are pluripotent and exhibit properties of both Schwann cells and astrocytes (Ramón-Cueto and

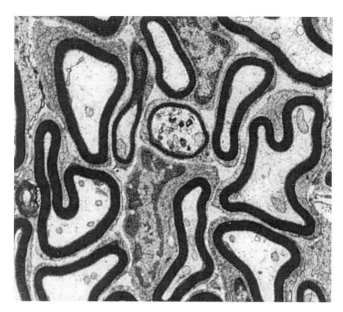

Fig. 11. Electron micrograph showing remyelinated axons within rat dorsal columns following transplantation of olfactory ensheathing cells. (×12,000.) (Adapted from Imaizumi et al., 1998.)

Valverde, 1995), might provide a cell type that would restore axonal conduction properties without cotransplantation of astrocytes. Although olfactory ensheathing cells do not normally produce myelin, experimental studies have shown that, under some circumstances, they can myelinate axons both in vitro (Devon and Doucette, 1992) and in vivo (Franklin et al., 1996). Thus, we have carried out studies (Imaizumi et al., 1998b) to determine whether olfactory ensheathing cells can remyelinate extensive areas within the spinal cord and enhance conduction in demyelinated spinal cord axons.

In these studies, olfactory ensheathing cells were separated from neonatal rats (2 or 3 d old) as described by Chauh and Chau (1993). Spinal cords were studied using electrophysiologic recording methods 21–25 d following cell injection into myelinating lesions produced using the ethidium bromide-X irradiation method. Transplantation of olfactory ensheathing cells into previously demyelinated regions resulted in remyelination of large areas, near the site of cell transplantation at the approximate center of the lesion and for 2–3 mm rostral and caudal to its midpoint.

As seen in Figure 11, olfactory ensheathing cells formed compact myelin around numerous axons within the lesion. Migration of myelin-forming cells was extensive, with at least 200 remyelinated axons being detected across distances of 4–6 mm along the spinal cord. Our results indicate that transplantation

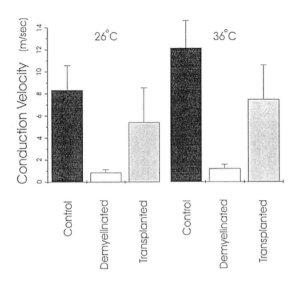

Fig. 12. Conduction velocities in normal, demyelinated, and remyelinated axons following transplantation of olfactory ensheathing cells, at 26°C and 36°C. (Adapted from Imaizumi et al., 1998.)

of olfactory ensheathing cells results in remyelination of approximately 17 ± 2% of demyelinated axons.

Consistent with our morphologic evidence for extensive remyelination by transplanted olfactory ensheathing cells, our electrophysiologic results (Fig. 12) revealed significant enhancement of conduction velocities to 7.48 m/s, compared with the demyelinated spinal cord (1.21 m/s). Moreover, the larger decrement in the compound action potential with increased conduction distance, due to extensive conduction block in the demyelinated dorsal columns, was partially reversed following transplantation of olfactory ensheathing cells (Fig. 13). Within the transplanted spinal cords, the remyelinated axons were able to conduct for at least 5 mm into the lesion, providing clear evidence for a reduction of conduction block.

Transplantation of olfactory ensheathing cells also improved the recovery properties of demyelinated axons within the spinal cord. This is shown in Figure 14. Thus, the conduction block that occurs at high frequencies in demyelinated axons was partially reversed following olfactory ensheathing cell transplantation, and olfactory ensheathing cell-remyelinated axons showed considerable restoration in their ability to follow high-frequency stimulation. Another potential advantage of olfactory ensheathing cell transplantation is the ability of these cells to enhance axonal regeneration in the spinal cord after axonal transection (Li et al., 1997; Ramón-Cueto et al., 1995; Imaizumi et al., 1998). Transplanta-

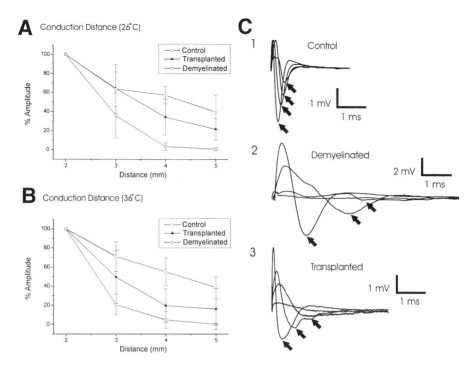

Fig. 13. (A, B) Improvement in amplitude decrement of the dorsal column compound action potential following transplantation of olfactory ensheathing cells. **(C)** Field potential recordings from control (C1), demyelinated (C2), and remyelinated (C3) dorsal columns. Compound action potentials were recorded at 1-mm increments (arrows) along the dorsal columns. There is an increase in latency for recordings obtained from the demyelinated axons, which is corrected in the dorsal columns with olfactory ensheathing cell transplantation. (Adapted from Imaizumi et al., 1998.)

tion of these cells into the injured spinal cord could then both induce remyelination and enhance axonal regeneration.

10. REPAIR OF DEMYELINATED AXONS IN SPINAL CORD INJURY

The demonstration that patients with SCI, including some patients with "clinically complete" lesions, harbor a subpopulation of nontransected axons that runs through the level of injury and fails to conduct as a result of demyelination has resulted in an important paradigm shift. As a result of this change in perspective, one goal of research on SCI now focuses on restoring or enhancing conduction in demyelinated spinal cord axons. As outlined in this chapter, the

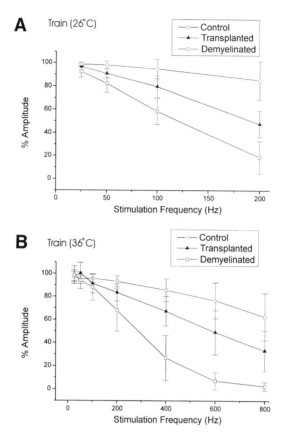

Fig. 14. Olfactory ensheathing cell transplantation improves the ability of dorsal column axons to transmit high-frequency impulse traits. These graphs show amplitude recovery for the second of two stimuli at varying interstimulus intervals for normal, demyelinated, and transplanted dorsal columns at 26°C **(A)** and 36°C **(B)**. There is less amplitude decrement for the remyelinated axons than for the demyelinated axons. (Adapted from Imaizumi et al., 1998.)

first steps have been taken in this direction. We now know that it is possible to enhance action potential conduction in demyelinated fibers with pharmacologic agents and that, at least in acute experiments in model systems, it is possible to improve the conduction of action potentials via the transplantation of myelin-forming cells. There is still much work to do, and many important questions remain unanswered. Nevertheless, proof of principle has now been achieved via pharmacologic manipulation of demyelinated axons, and via their remyelination using cell transplantation. Hopefully, the initial steps, which have been taken in the laboratory, will ultimately be followed by steps taken by humans with SCI.

ACKNOWLEDGMENTS

Research in the authors' laboratories has been supported in part by the Medical Research Service and Rehabilitation Research Service, by the Veterans Administration, and by grants from the National Institutes of Health, National Multiple Sclerosis Society, and The Myelin Project.

REFERENCES

Adorini, L. (1993) Selective inhibition of T cell responses by protein and peptide-based immunotherapy. *Clin. Exp. Rheumatol.* **11 (suppl. 8),** S41–44.

Baker, M., Bostock, P., Grafe, P., and Martins, P. (1987) Function and distribution of three types of rectifying channel in rat spinal root myelinated axons. *J. Physiol. (Lond.)* **383,** 45–67.

Beattie, M. S., Shuman, S. L., and Bresnahan, J. C. (1998) Apoptosis and spinal cord injury. *Neuroscientist* **4,** 163–171.

Blakemore, W. F. (1976) Invasion of Schwann cells into the spinal cord of the rat following local injections of lysolecithin. *Neuropathol. Appl. Neurobiol.* **2,** 21–39.

Blakemore, W. F. and Crang, A. J. (1985) The use of cultured autologous Schwann cells to remyelinate areas of persistent demyelination in the central nervous system. *J. Neurol. Sci.* **70,** 207–223.

Blakemore, W. F. and Patterson, R. C. (1978) Suppression of remyelination in the CNS by X-irradiation. *Acta Neuropathol.* **42,** 105–113.

Blight, A. R. (1989) Effect of 4-AP on axonal conduction block in chronic spinal cord injury. *Brain Res. Bull.* **22,** 47–52.

Blight, A. R. and Young, W. (1989) Central axons in injured cat spinal cord recover electrophysiological function following remyelination by Schwann cells. *J. Neurol. Sci.* **91,** 15–34.

Boison, D. and Stoffel, W. (1989) Myelin-deficient rat: a point mutation in axon III (A—C, Thr75—Pro) of the myelin proteolipid protein causes dysmyelination and oligodendrocyte death. *EMBO J.* **8,** 3295–3302.

Bostock, H., Sears, T. A., and Sherratt, R. M. (1981) The effects of 4-aminopyridine and tetraethylammonium ions on normal and demyelinated mammalian nerve fibers. *J. Physiol. (Lond.)* **313,** 301–315.

Bowe, C. M., Kocsis, J. D., Targ, E. F., and Waxman, S. G. (1987) Physiological effects of 4-aminopyridine on demyelinated mammalian motor and sensory fibers. *Ann. Neurol.* **22,** 264–268.

Brockes, J. P., Fields, K. L., and Raff, M. C. (1979) Studies on cultured rat Schwann cells. I. Establishment of purified populations from cultures of peripheral nerve. *Brain Res.* **165,** 105–118.

Bunge, R. P., Puckett, W. R., Becerra, J. L., Marcillo, A., and Quencer, R. M. (1993) Observations on the pathology of human spinal cord injury. A review and classification of 22 new cases with details from a case of chronic cord compression with extensive focal demyelination, in *Advances in Neurology,* vol. 59. *Neural Injury and Regeneration* (Seil, FJ, ed.), Raven, New York, pp. 75–89.

Byrne, T. N. and Waxman, S. G. (1990) *Spinal Cord Compression,* F. A. Davis, Philadelphia.

Chauh, M. I., Au, C. (1993) Cultures of ensheathing cells from neonatal rat olfactory bulbs. *Brain Res.* **601,** 213–220.

Chiu, S. Y. and Ritchie, J. M. (1981) Evidence for the presence of potassium channels in the paranodal region of acutely demyelinated mammalian nerve fibers. *J. Physiol. (Lond.)* **313,** 415–437.

Davis, F. A., Stefoski, D., and Rush, J. (1990) Orally administered 4-aminopyridine improves clinical signs in multiple sclerosis. *Ann. Neurol.* **27,** 186–192.

Devon, R. and Doucette, R. (1992) Olfactory ensheathing cells myelinate dorsal root ganglion neurites. *Brain Res.* **589,** 175–179.

Duncan, I. D., Hammang, J. P., Jackson, K. F., Wood, P. M., Bunge, R. P., and Langford, L. (1988a) Transplantation of oligodendrocytes and Schwann cells into the spinal cord of the myelin-deficient rat. *J. Neurocytol.* **17,** 351–360.

Duncan, I. D., Hammang, J. P., and Gilmore, S. A. (1988b) Schwann cell myelination of the myelin deficient rat spinal cord following X-irradiation. *Glia* **1,** 233–239.

Felts, P. A. and Smith, K. J. (1992) Conduction properties of central nerve fibers remyelinated by Schwann cells. *Brain Res.* **574,** 178–192.

Foster, R. E., Connors, B. W., and Waxman, S. G. (1982) Rat optic nerve: electrophysiological, and anatomical studies during development. *Dev. Brain Res.* **3,** 361–376.

Franklin, R. J., Gilsson, J. M., Franceschini, I. A., and Barnett, S. C. (1996) Schwann cell-like myelniation following transplantation of an olfactory bulb-ensheathing cell line into areas of demyelination in the adult CNS. *Glia* **17,** 217–224.

Gledhill, R. F., Harrison, B. M., and McDonald, W. I. (1973) Demyelination and remyelination after acute spinal cord compression. *Exp. Neurol.* **38,** 472–487.

Gout, O., Gansmuller, A., Baumann, N., and Gumpel, M. (1988) Remyelination by transplanted oligodendrocytes of a demyelinated lesion in the spinal cord of the adult shiverer mouse. *Neurosci. Lett.* **87,** 195–199.

Griffiths, I. R. and McCulloch, M. C. (1983) Nerve fibers in spinal cord impact injuries. 1. Changes in the myelin sheath during the initial five weeks. *J. Neurol. Sci.* **58,** 335–345.

Groves, A. K., Barnett, S. C., Franklin, R. J. M., et al. (1993) Repair of demyelinated lesions by transplantation of purified O-2A progenitor cells. *Nature* **362,** 453–456.

Harrison, B. M. and McDonald, W. I. (1977) Remyelination after transient experimental compression of the spinal cord. *Ann. Neurol.* **1,** 542–551.

Hirano, A. and Dembitzer, H. M. (1978) Morphology of normal central myelinated axons, in *Physiology and Pathobiology of Axons* (Waxman S. G., ed.), Raven, New York, pp. 68–82.

Honmou, O., Felts, P. A., Waxman, S. G., and Kocsis, J. D. (1996) Restoration of normal conduction in demyelinated spinal cord axons in the adult rat by transplantation of exogenous Schwann cells. *J. Neurosci.* **16,** 3199–3209.

Hudson, L. D., Puckett, C., Berndt, J., Chan, J., and Gencic, S. (1989) Mutation of the proteolipid protein gene PLP in a human X chromosome-linked myelin disorder. *Proc. Natl. Acad. Sci. USA* **86,** 8128–8131.

Imaizumi, T., Lankford, K. L., and Kocsis, J. D. (1998a) Facilitation of axonal conduction across transected rat spinal cord following transplantation of olfactory ensheathing cells or Schwann cells. *Soc. Neurosci. Abstr.* **24,** 2010.

Imaizumi, T., Lankford, K. L., Waxman, S. G., Greer, C. A., and Kocsis, J. D. (1998b) Transplanted olfactory ensheathing cells remyelinate and enhance axonal conduction in the demyelinated dorsal columns of the rat spinal cord. *J. Neurosci.* **18,** 6176–6185.

Jones, R. E., Heron, J. R., Foster, D. H., Snelgar, R. S., and Mason, R. J. (1983) Effects of 4-aminopyridine inpatients with multiple sclerosis. *J. Neurol. Sci.,* **60,** 353–362.

Kocsis, J. D., Black, J. A., and Waxman, S. G. (1993) Pharmacological modification of axon membrane molecules and cell transplantation as approaches to the restoration of conduction in demyelinated axons, in S. G. Waxman (Ed.), *Molecular and Cellular Approaches to the Treatment of Neurological Disease* (Waxman S. G., ed.), Raven, New York, pp. 265–292.

Kocsis, J. D., Gordon, T. R., and Waxman, S. G. (1986) Mammalian optic nerve fibers display two pharmacologically distinct potassium channels. *Brain Res.* **393,** 357–361.

Kocsis, J. D. and Waxman, S. G. (1980) Absence of potassium conductance in central myelinated axons. *Nature* **287,** 348–349.

Kocsis, J. D., Waxman, S. G., Hildebrand, C., and Ruiz, J. A. (1982) Regenerating mammalian nerve fibres: changes in action potential waveform and firing characteristics following blockage of potassium conductance. *Proc. R. Soc. Lond. [Biol.]* **217,** 277–287.

Koles, Z. J. and Rasminsky, M. (1972) A computer simulation of conduction in demyelinated nerve fibres. *J. Physiol. (Lond.)* **227,** 351–364.

Li, Y., Field, P. M., and Raisman G. (1997) Repair of adult rat corticospinal tract by transplants of olfactory ensheathing cells. *Science* **277,** 2000–2002.

Malenka, R. C., Kocsis, J. D., Ransom, B. R., and Waxman, S. G. (1981) Modulation of parallel fiber excitability by postsynaptically mediated changes in extracellular potassium. *Science* **214,** 339–341.

Mann, R., Mulligan, R. C., and Baltimore, D. (1983) Construction of a retrovirus packaging mutant and its use to produce helper-free defective retrovirus. *Cell* **33,** 153–159.

McCarthy, K. D., de Vellis, J. (1980) Preparation of separate astroglial and oligodendroglial cell cultures from rat cerebral cortex. *J. Cell Biol.* **85,** 890–902.

Preston, R. J., Waxman, S. G., and Kocsis, J. D. (1983) Effects of 4-aminopyridine on rapidly and slowly conducting axons of rat corpus allosum. *Exp. Neurol.* **79,** 808–820.

Price, J., Turner, D., and Cepko, C. (1987) Lineage analysis in the vertebrate nervous system by retrovirus-mediated gene transfer. *Proc., Natl. Acad. Sci. USA* **84,** 156–160.

Ramón-Cueto, A. and Valverde, F. (1995) Olfactory ensheathing glia: a unique cell type with axonal growth-promoting properties. *Glia* **14,** 163–173.

Ritchie, J. M. (1982) Sodium and potassium channels in regenerating and developing mammalian myelinated nerves. *Proc. R. Soc. Lond. [Biol.]* **215,** 273–287.

Ritchie, J. M. and Rogart, R. B. (1977) The density of sodium channels in mammalian myelinated nerve fibers and the nature of the axonal membrane under the myelin sheath. *Proc. Natl. Acad. Sci. USA* **74,** 211–215.

Röper, J. and Schwartz, J. R. (1989) Heterogeneous distribution of fast and slow potassium channels in myelinated rat nerve fibers. *J. Physiol. (Lond.)* **416,** 93–110.

Rosenbluth, J., Hasegawa, M., Shirasaki, N., Rosen, C. L., and Liu, Z. (1990) Myelin formation following transplantation of normal fetal glia into myelin-deficient rat spinal cord. *J. Neurocytol.* **19,** 718–730.

Salgado-Ceballos, H., Guizar-Sahagun, G., Feria-Velasco, A., et al. (1998) Spontaneous long-term remyelination after traumatic spinal cord injury in rats. *Brain Res.* **782,** 126–135.

Schauf, C. L. and Davis, F. A. (1974) Impulse conduction in multiple sclerosis: a theoretical basis for modification by temperature and pharmacological agents. *J. Neurol. Neurosurg. Psychiatry* **37,** 152–161.

Sears, T. A., Bostock, H., and Sherratt, M. (1978) The pathophysiology of demyelination and its implications for the symptomatic treatment of multiple sclerosis. *Neurology* **28,** 21–26.

Sherratt, R. M., Bostock, H., and Sears, T. A. (1980) Effects of 4-aminopyridine on normal and demyelinated mammalian nerve fibers. *Nature* **283,** 570–572.

Sherwood, A. M., Dimitrijevic, M. R., and McKay, W. B. (1992) Evidence of subclinical brain influence in clinically complete spinal cord injury: discomplete SCI. *J. Neurol. Sci.* **110,** 90–98.

Shuman, S. L., Bresnahan, J. C., and Beattie, M. S. (1997) Apoptosis of microglia and oligodendrocytes after spinal cord injury in rats. *J. Neurosci. Res.* **50,** 798–808.

Simons, R. and Riordan, J. R. (1990) Single base substitution in codon 74 of the MD rat myelin proteolipid gene. *Ann. NY Acad. Sci.* **605,** 146–154.

Smith, K. J., Blakemore, W. F., and McDonald, W. I. (1983) Central remyelination restores secure conduction. *Nature* **280,** 395–396.

Smith, K. J. and Hall, S. M. (1980) Nerve conduction during peripheral demyelination and remyelination. *J. Neurol. Sci.* **48,** 201–219.

Stanley, G. P. and Pender, M. P. (1991) Pathophysiology of chronic relapsing experimental allergic encephalomyelitis. *Brain* **114,** 1827–1853.

Stefoski, D., Davis, F. A., Faut, M., and Schauf, C. L. (1987) 4-Aminopyridine improves clinical signs in multiple sclerosis. *Ann. Neurol.* **21,** 71–77.

Targ, E. F. and Kocsis, J. D. (1985) 4-Aminopyridine leads to restoration of conduction in demyelinated rat sciatic nerve. *Brain Res.* **328,** 358–361.

Targ, E. F. and Kocsis, J. D. (1986) Action potential characteristics of demyelinated rat sciatic nerve following application of 4-aminopyridine. *Brain Res.* **363,** 1–9.

Utzschneider, D. A., Archer, D. R., Kocsis, J. D., Waxman, S. G., and Duncan, I. D. (1994) Transplantation of glial cells enhances action potential conduction of amyelinated spinal cord axons in the myelin-deficient rat. *Proc. Natl. Acad. Sci. USA* **91,** 53–57.

Waxman, S. G. (1977) Conduction in myelinated, unmyelinated, and demyelinated fibers. *Arch. Neurol.* **34,** 585–590.

Waxman, S. G. (1978) Prerequisites for conduction in demyelinated fibers. *Neurology* **28,** 27–34.

Waxman, S. G., Black, J. A., Duncan, I. D., and Ransom, B. (1990) Macromolecular structure of axon membrane and action potential conduction in myelin deficient and myelin deficient heterozygote rat optic nerves. *J. Neurocytol.* **19,** 11–27.

Waxman, S. G. and Brill, M. H. (1978) Conduction through demyelinated plaques in multiple sclerosis: computer simulations of facilitation by short internodes. *J. Neurol. Neurosurg. Psychiatry* **41,** 408–417.

Waxman, S. G. and Richie, J. M. (1993) Molecular dissection of the myelinated axon. *Ann. Neurol.* **33,** 121–136.

Weiner, L. P., Waxman, S. G., Stohlman, S. A., and Kwan, A. (1980) Remyelination following viral-induced demyelination: ferric ion-ferrocyanide staining of nodes of Ranvier within the CNS. *Ann. Neurol.* **8,** 580–583.

Young, W. (1989) Recovery mechanisms in spinal cord injury: implications for regenerative therapy, in *Neural Regeneration and Transplantation* (Seil F. J., ed.), Alan R. Liss, New York, pp. 157–169.

Zeller, N. K., Dubois-Dalcq, M., and Lazzarini, R. A. (1989) Myelin protein expression in the myelin-deficient rat brain and cultured oligodendrocytes. *J. Mol. Neurosci.* **1,** 139–149.

11
Molecular and Cellular Mechanisms of Spinal Cord Injury Therapies

Wise Young

1. INTRODUCTION

For much of human history, clinicians have considered spinal cord injury (SCI) an incurable condition. Several millennia ago, an anonymous Egyptian physician wrote in the Edwin Smith Papyrus that water should be withheld from spinal-injured warriors *(1)*. This pessimistic view has held sway. Until recently, the paucity of therapies for spinal cord injury did not pose a problem because most people did not survive long after injury. However, the advent of modern antibiotics, the ventilator, intermittent catheterization, and modern intensive care turned this situation around. By 1980, over 90% of people survived SCI.

About 9 years ago, a major discovery markedly altered the prospects for SCI therapies. In 1990, the second National Acute Spinal Cord Injury Study (NASCIS 2) showed that high-dose methylprednisolone (MP) significantly improves neurologic recovery even when given several hours after injury *(2)*. MP was the first neuroprotective therapy for central nervous system (CNS) injury and introduced the concept of therapeutic time window for neuroprotective therapies.

The availability of an effective neuroprotective therapy transformed acute SCI care. Before 1990, people with severe "complete" SCI outnumbered people with milder "incomplete" injuries by a ratio of 6:4. By 1995, however, this ratio had inverted, so that there are four "complete" injuries for every six "incomplete" injuries *(3)*. This inversion is probably not due solely to MP but may be related to faster and better emergency care of SCI victims, a practice that emerged when a treatment became available.

The discovery that MP improves recovery stimulated researchers to look for other SCI therapies. Several dozen therapies besides MP have been reported to improve recovery in animal SCI models. These include naloxone, nalmefene, thyrotropin-releasing hormone, cycloheximide, lipopolysaccharide (LPS), and many others. Indeed, there is now an embarrassing number of therapies that seem to be beneficial, but without clear mechanisms of action.

From: *Neurobiology of Spinal Cord Injury*
Edited by: R. G. Kalb and S. M. Strittmatter © Humana Press Inc., Totowa, NJ

Several investigators have even reported successful attempts to regenerate the spinal cord of rats, overturning the long-held dogma that the adult mammalian spinal cord cannot regrow and reconnect. Perhaps the most interesting of all are several reports that the spinal cord can regenerate spontaneously in contused spinal cords *(4)*. These results suggest strongly that the spinal cord is capable of more repair and growth than we realized and that endogenous factors are present to stimulate such growth.

For years, scientists in the field have given passing nods of acknowledgement to the roles of inflammatory and immunomodulatory mechanisms in neuroprotection, apoptosis, demyelination-remyelination, and axonal growth. Recent studies of molecular and cellular mechanisms of inflammation and immunity suggest that degenerative and regenerative phenomena are linked in a complex web of injury-initiated factors. Such studies may explain why many disparate therapies are beneficial in SCI and why they have unusually narrow dose-response ranges and short therapeutic time windows. These studies and their implications for SCI therapies are the subject of this chapter. Note that the field has enlarged sufficiently that a thorough review or even mention of most of the beneficial therapies for SCI is not possible in one chapter.

2. GLUCORTICOID THERAPY

NASCIS 2 compared MP and the opiate receptor antagonist naloxone (NX) against placebo controls *(5)*. Plegic patients given MP within 8 hr recovered 21% of lost motor function compared with 8% in controls; MP-treated paretic patients recovered 75% compared with 59% in controls. Sensory improvements were similar to motor improvements. Patients who received MP or NX more than 8 hr after injury showed no benefit compared with controls. Patients who received NX within 3 hr showed a trend for better motor and sensory recovery. Although the initial analysis suggested that the NX effect was not significant, subsequent analyses revealed a significant effect of NX that was not as great as MP *(6)*. The NASCIS group consequently recommended rapid treatment of SCI with high-dose MP. The effective clinical dose of MP is 30 mg/kg followed by 5.4 mg/kg/hr for 23 hr, adding up to 10 or more g of MP over a 24-hr period in an average size person. This dose is an order of magnitude greater than the previous so-called megadose steroid therapy of 1 g a day *(7)*. Likewise, the dose of NX given in the trial was 5.4 mg/kg followed by 3.0 mg/kg/hr for 23 hr, much greater than the 0.5 mg of NX given to block opiate receptors in heroin overdose. At such high doses, both drugs have effects beyond their normal mechanisms of action. For example, at 30-mg/kg doses, MP is a free radical scavenger that effectively inhibits lipid peroxidation *(8)*. The high doses of MP are necessary for neuroprotection *(8,9)*, and overdoses are deleterious *(10)*. The dose of MP required for neuroprotection far exceeded the dose necessary to activate glucocorticoid receptors.

Early theories emphasized the ability of MP to scavenge free radicals, to inhibit lipid peroxidation *(11–16)*, to improve blood flow, and to prevent calcium influx into cells *(17)*. Injured tissues produce free radicals that can damage the tissue further *(18)*. Several investigators have reported beneficial effects of antioxidants on spinal cord injury *(19–21)*. In combination with calcium-activated phospholipases, free radicals contribute to membrane breakdown *(22,23)*. To test the hypothesis that MP is acting primarily as a free radical scavenger, Hall et al. created a novel family of molecules called 21-aminosteroids. One member of this family is tirilazad mesylate (TM), a steroid molecule that is a potent antioxidant but does not activate glucocorticoid receptors *(24–27)*. Animal studies had shown that TM significantly improved recovery after spinal cord injury *(25,28)*.

The third National Acute Spinal Cord Injury Study (NASCIS 3) had three treatment groups: a standard 24-hr course of MP (30 mg/kg + 5.4 mg/kg/hr × 23), a 48-hr course of MP (30 mg/kg + 5.4 mg/kg/hr × 47), and a single MP bolus (30 mg/kg) followed by a 48-hr course of TM (6 mg/kg/d). All three treatments had similar beneficial effects on neurologic recovery when started within 3 hr after injury *(3)*. In people treated more than 3 hr after injury, the 48-hr course of MP yielded greater motor and sensory recovery than either the 24-hr MP or 48-hr TM treatments. The data therefore suggest that TM could substitute for MP when treatment was initiated within 3 hr after injury. Because all three treatment groups received a bolus of dose of MP, the trial did not rule out the possibility that a single bolus dose of MP is as effective as a 24- or 48-hr course of MP when started within 3 hr after injury. Nevertheless, the data support a beneficial effect of TM and antioxidation and suggest that MP may be acting through mechanisms in addition to glucocorticoid receptor activation.

MP is a potent antiinflammatory glucocorticoid drug. MP effectively attenuates expression of tumor necrosis factor (TNF) and interleukins-1 and -6 (IL-1 and IL-6) in injured spinal cords *(29,30)*. These cytokines mediate inflammatory and immune responses *(31–33)*, as well as allergic *(34)* responses in tissues. In injured spinal cords, TNF message and protein appear within an hour at the injury site and remain elevated for about 8 hr; they decrease steadily thereafter *(35)*. Both TNF and c-FOS messages are detectable in spinal cord as early as 30 min after injury *(36)*. IL-1β mRNA levels are also elevated at 1 and 6 hr after injury but fall by 12 hr *(29)*. In situ hybridization studies indicate that IL-1, IL-6, and TNF messages increase significantly within an hour *(37)*.

Glucocorticoids suppress cytokines by antagonizing nuclear transcription factors such as AP-1 *(38,39)* and NF-κB *(40–46)*. NF-κB stimulates TNF and interleukin production *(47)* and is associated with apoptosis *(48–51)*. Glucocorticoid receptors directly inhibit NF-κ activity *(52–55)* and NF-κB transcription *(56)*. Glucocorticoids block the NF-κB transcription factor Rel1A *(57–59)* without affecting DNA binding of NF-κB *(60)* or inducing IκBα *(61–63)*, which regulates nuclear transport of NF-κB *(64)*. Injury activates NF-κB in spinal cord *(64a)*.

Glucocorticoids control the inflammatory processes at multiple levels. First, dexamethasone shortens the half-life of mRNA for TNF and interleukins *(65)*. Second, dexamethasone prevents IL-1β induction of several inflammatory enzymes *(66)*, including nitric oxide synthetase *(67)*. Third, glucocorticoids discourage invasion of inflammatory cells by downregulating key cellular adhesion molecules *(68)* such as the intercellular adhesion molecule *(69,70)*, E-selectin *(71,72)*, and neutrophil chemokines *(73,74)*. Fourth, glucocorticoids inhibit neurotoxin production by macrophages *(75)*. Finally, dexamethasone induces glutamine synthetase (GS), an enzyme present in astrocytes *(76–78)*. GS appears in developing central nervous tissues when the organism begins secreting glucocorticoids *(79)*, a response regulated by c-jun *(79–81)*. Blockade of GS induces apoptosis *(82)*. Glucocorticoids protect isolated retinas against anoxia *(83)*, an effect that requires glucocorticoid receptor-induced GS production. Exogenously applied GS protects retinas against anoxia *(83)*. Because GS is too large to penetrate cells, it probably acts from the extracellular space. However, MP does not alter extracellular glutamate levels in injured spinal cords *(84)*.

In summary, MP prevents progressive tissue damage through multiple mechanisms. At the doses that are neuroprotective, MP is a potent antioxidant, a powerful antiinflammatory agent, and an effective immunosuppressant. In addition to suppressing NF-κB, a nuclear transcription factor that induces cytokine production, MP depresses nitric oxide, reduces the half-life of RNA for cytokines and chemokines, and induces GS. glutamine synthetase. These mechanisms act in concert to dampen free radicals, cytokine production, and inflammatory cellular reactions to the injury.

3. OTHER ANTIINFLAMMATORY THERAPIES

Many classes of molecules besides glucocorticoids regulate NF-κB *(58)*. These include cycloheximide *(85)*, the phorbol ester 12-0-tetradecanoyl-phorbol-13-acetate (TPA) *(86)*, TNF *(87–90)*, interleukins *(91)*, morphine *(92)*, tritiprenoids *(93)*, oxidative stress-inducing compounds *(94,95)*, the human immunodeficiency virus (HIV) immunosuppressive gene product HIV-1 Vpr *(96)*, progesterone *(97)* and other sex hormones *(98)*, β-adrenergic agonists *(99)*, cyclic adenosine monophosphate (cAMP) *(100)*, thyrotropin-releasing hormone (TRH) *(101–104)*, and even bacterial toxins such as lipopolysaccharide (LPS) *(45,105,106)*. Antirheumatic treatments such as gold thiolates, penicillamines, and chloroquine act by inhibiting NF-κB gene expression *(107)*. Some of these drugs may be neuroprotective. In fact, several have been reported to be neuroprotective and regenerative in SCI models, including cycloheximide *(108–114)* and LPS *(115–119)*.

Nonsteroidal antiinflammatory drugs, such as the cyclooxygenase (COX) inhibitors ibuprofen and indomethacin, are neuroprotective in SCI *(120)*. COX inhibitors not only inhibit production of prostaglandins and leukotrienes in injured spinal cords *(121–123)* and can directly reduce NF-κB expression *(124)*. Indomethacin reduces edema *(125–128)*, blood-brain barrier breakdown *(129)*, prostaglandin release *(130)*, LPS-induced behavioral changes *(131)*, and neuronal loss *(132)*. However, COX inhibitors do not appear to be as neuroprotective as high-dose MP in SCI *(133)*, perhaps because they may also inhibit potentially beneficial inflammatory responses. For example, indomethacin represses heat-shock proteins *(134)*. Combinations of indomethacin-heparin-prostaglandin-12 *(135)* and LPS-indomethacin-pregnenolone *(136)* have been reported to improve recovery after SCI.

The immune system shares many cytokines with inflammation. Cyclosporin A (CyA) and FK506 block activation of calcineurin, a phosphatase that is found primarily in neurons and lymphocytes. Calcineurin activates the nuclear transcription factor NFAT, which induces IL-2 *(137)* and other cytokines. For example, calcineurin stimulates release of TRANCE, a member of the TNF family of cytokines *(138)*. Immunosuppression delays onset of immune-mediated lesions, such as experimental allergic encephalitis (EAE) in rats *(139)*. However, immunosuppression may directly affect tissue repair and growth. For example, CyA has neurotrophic actions on rats with transected spinal cords *(140,141)* but both CyA and FK506 inhibit growth cone activity *(142)*. The immunophilins, cyclophilin and FK506 binding protein (FKBP), bind respectively to cyclophilin and FKBP to modulate calcineurin activity. FKBP expression is upregulated in regenerating facial nucleus *(143)*. FKBP and cyclophilin colocalize with calcineurin *(144,145)* and glucocorticoid receptors *(146)*. These findings have attracted significant interest in immunophilin analogs as potential therapies for SCI.

Opioid receptors affect the immune system. Both exogenously administered and endogenous opioids are immunosuppressive *(147–155)*. Opiate receptors depress cellular *(156)* and humoral *(157)* immunity in stress *(158)*. Lymphocytes possess opioid receptors *(159–161)*. Opioid receptors are present in lymphocytes from invertebrates and therefore evolutionarily primitive *(162–164)*. Morphine administration reduces microglia and peripheral monocyte expression of TNF *(165,166)*. Morphine suppresses allogeneic lymph node responses in mice *(167)*. However, naloxone did not block exercise-induced increases in immune cell numbers or activity *(168–170)*. Interestingly, morphine increases glucocorticoid secretion *(171)*, and corticotrophin-releasing hormone appears to stimulate opiate release *(172)*. The effects of naloxone on the immune system are clearly complex and indirect and may account for some of the neuroprotective effects of naloxone in SCI.

Multiple feedback loops complicate attempts to manipulate inflammatory processes *(173,174)*. For example, in addition to suppressing TNF *(175)* and interleukin *(176)* expression, glucocorticoids induce expression of cytokine receptors. Glucocorticoids induce low-affinity IL-2 receptors (IL-2R) and thereby facilitate IL-2-initiated apoptosis of lymphocytes *(177)*. However, the apoptotic effects of glucocorticoids on lymphocytes can be countered by IL-2, which inhibits IκBα activity *(178)*. The transcription initiation site for the thromboxane receptor gene has binding sites for AP-1, NF-κB, and glucocorticoids *(91)*. Glucocorticoid receptor and NF-κB activities mutually repress each other *(179)* and can paradoxically enhance interleukin production *(180,181)*. When given

with the protein synthesis blocker cycloheximide, IL-1 can cause paradoxical
"superstimulation" of NF-κB expression *(182)*. Cycloheximide was recently
reported to be neuroprotective in spinal cord injury *(108,109,183)*. Interestingly,
nonsteroidal antiinflammatory drugs promote TNF and interleukin production
in peripheral but not central nervous tissues.

In summary, many factors regulate inflammatory and immune processes in
injured spinal cords, perhaps explaining why such drugs as disparate as glucocorticoids, naloxone, indomethacin, cycloheximide, immunosuppressants, and
immunophilins are beneficial in SCI. The multiplicity of antiinflammatory
treatments for SCI is turning out to be an embarrassment of riches. Many of
the drugs activate nuclear factors such as NF-κB, AT-1, and NF-AT that are not
only expressed and induced at different times after SCI but, in turn, have complex positive and negative feedback loops that can cause contradictory and
paradoxic effects. These findings may account for the unusual narrow dose-
response curves of many neuroprotective drugs, the constraints on the therapeutic time windows, and the importance of treatment duration. For example,
very high doses of some drugs are necessary for neuroprotection. It is possible
that the high doses are necessary to overwhelm some of the compensatory
feedback mechanisms. Likewise, the timing of genetic responses to injury dic-
tates the timing of therapies aimed at these responses. Finally, much evidence
suggests that injury-induced inflammatory and immune processes may not all
be deleterious.

4. PROINFLAMMATORY THERAPIES

Injury-induced cytokines may be beneficial in SCI. For example, whereas
TNF is elevated in EAE and is believed to injure oligodendroglia *(184)*, direct
application of TNF does not appear to have toxic effects on myelinated axons in
cultures *(185)*. Transgenic mice that overexpress TNF-α show little damage,
although EAE seems to be more severe in these mice *(186)*. Transgenic mice
with a dominant negative mutation that reduces IL-1 activity show less neuronal
death associated with growth factor withdrawal *(187)*. Administration of IL-6 to
injured spinal cords reduced tissue damage *(188)*. Likewise, combination therapy with IL-1, IL-6, and TNF reduced spinal cord injury volumes *(37,189)*.
Recently, Bethea et al. *(190)* reported that antiinflammatory IL-10 given systemically 15 min after injury significantly reduced 7-d spinal cord lesion volumes by 55% and improved 6-week locomotor recovery in rats. Paradoxically,
rats that received IL-10 at 15 min and 3 d after injury did not show improved
locomotor recovery (presented at the Neurotrauma Society Meeting, November
14, 1998). NF-κB upregulation may be neuroprotective *(191)*.

Proinflammatory bacterial toxins may be neuroprotective. For example, LPS
may improve locomotion after spinal cord crushes in rats *(115,116)*. Although
LPS itself appeared to be effective, a combination (LIP) of LPS, indomethacin,
and pregnenolone was the most effective. Although these results have not been

confirmed by other laboratories to date, the underlying rationale for using combination pro- and antiinflammatory therapy is interesting. LPS *(192–194)* and sepsis *(192,195)* markedly induced glutamine synthetase expression in tissues. Another bacterial toxin was recently reported to improve locomotor recovery in spinal cord injury *(196)*. CM-101, a bacterial polysaccharide produced by group B streptococci restored locomotion in mice that were rendered hemiplegic or paraplegic by spinal cord crushes. The CM-101 (30–60 µg/kg) was given intravenously shortly after injury and every other day for 5 times. Electrophysiologic studies confirmed recovery of action potential transmission in injured spinal cords. Histologic examinations of the spinal cord suggested reduced gliosis and fibrosis in the cords. The authors suggested that CM-101 improves recovery by suppressing acute and chronic inflammatory responses in the spinal cord. Unfortunately, four of five of the hemiplegic controls and eight of nine paraplegic controls died by the first week after injury. Thus, the study needs to be repeated and confirmed.

Several recent studies suggest that activated macrophages and T-lymphocytes are neuroprotective and stimulate regeneration in injured optic nerves and spinal cords *(197–199)*. Until recently, most scientists regarded inflammatory cellular responses as potential contributors to secondary tissue damage. Here again, the data suggest a complex situation. For example, while macrophages activated by exposure to peripheral nerves improved locomotor recovery and histologic appearance of the spinal cord, similar macrophages activated by exposure to optic nerve were not effective. This finding suggests that the specific activation states of the macrophages and inflammatory cells determine whether the inflammatory cells are beneficial or deleterious. Other investigators have shown that microglial implants are beneficial after SCI *(200)*.

The monosialic ganglioside GM1 antagonized the beneficial effects of MP *(201)* when both drugs were given shortly after injury. Clinical trials have shown that GM1 improved locomotor recovery in human spinal cord injury *(202–206)* and prevented progression of Parkinson's disease *(207,208)*. A recently completed trial suggests that GM1 significantly improves recovery rate during the first 3 months but not recovery extent at 1 year (Geisler, F., 1999, unpublished data). The study randomized nearly 800 patients to MP alone or MP followed by a 2-month course of 100 mg/d GM1, starting 24 hr after spinal cord injury. Much evidence suggests that GM1 is neuroprotective *(209,210)* and can also stimulate axonal growth. For example, GM1 potentiates growth factors *(211)*, prevents apoptosis *(212–215)*, and stimulates axonal growth *(211)*. GM1 facilitates trimerization *(216)* and phosphorylation *(217)* of the neurotrophin trk-A receptors, and suppresses amyloid-induced cytokine release *(218)* but enhances trimethyltin-induced cytokine release *(219)*, activates microtubule-associated protein (MAP)-kinase *(220–222)*, reduces citrulline release and nitric oxide production *(210)*, and regulates opiate *(223,224)* and glutamate receptors *(225,226)*. The plethora of potential GM1 mechanisms is confusing but may explain the puzzling interaction between MP and GM1. GM1 may stimulate cytokine release during the first few hours after injury, blocking the effects of MP, and then suppress longer and slower cytokine release by inhibiting nitric oxide production. At the same time, it facilitates neurotrophin receptors and activates MAP-kinase to stimulate axonal growth.

Thus, these studies suggest three novel therapeutic approaches to SCI. The first is the use of opposing proinflammatory and antiinflammatory drugs to stimulate beneficial injury responses and suppress others that may be deleterious. The second is sequential therapy using a drug like MP to reduce acute tissue damage and a drug like GM1 to stimulate tissue repair and regrowth. The third is to implant inflammatory cells that have been selectively activated to produce appropriate factors that protect cells and facilitate tissue repair. These approaches differ considerably from previous thinking in the field. Injury initiates a cascade of tissue responses. Many researchers assume that the tissue responses are deleterious because they correlate with injury severity and tissue damage. However, some components of the cascade may be beneficial. Although some of these studies need to be validated, the concept of using sequential opposing therapies to enhance desirable components and to suppress deleterious components of the injury response is an important breakthrough.

In summary, much data support potential beneficial effects of inflammatory factors in SCI. For example, several oft-regarded cytokine villains (IL-1, IL-6, and TNF) turn out not to be particularly damaging to cells in culture and may in fact reduce tissue damage when applied to the spinal cord in vivo. Several proinflammatory toxins, such as LPS and CM-101, may be neuroprotective. Even activated macrophages implanted into the injury site appear to improve neurologic recovery in animal SCI models. GM1, a drug reported to improve locomotor recovery in human SCI, paradoxically antagonizes the neuroprotective effects of MP when given early after injury. However, delayed GM1 therapy after MP significantly improves recovery, emphasizing the importance of sequential therapy addressing different phases of injury and recovery.

5. NEUROTRANSMITTER RECEPTORS AND AXOTOXICITY

Several neurotransmitter receptor blockers are neuroprotective in spinal cord injury. The opiate receptor blocker NX was one of the first drugs reported to be neuroprotective when given after SCI, suggesting that opioids play a major role in secondary tissue damage *(227,228)*. NASCIS 2 compared NX and MP. An earlier clinical trial had shown that increasing doses of NX improved somatosensory evoked potential recovery in human SCI *(229,230)*. In NASCIS 2, NX improved motor and sensory recovery *(5)*, but the difference did not reach statistical significance. A subsequent analysis suggested that the NX effect was significant *(6)*. The dose of NX used in NASCIS 2 was less than the optimal 10 mg/kg dose found to be effective in animal studies *(230,231)*. The clinical trial results closely parallel the animal findings *(231)*, suggesting that opioid receptors do play a role in secondary tissue damage in SCI.

The mechanisms by which NX protects the spinal cord are not clear. The dose of NX required for neuroprotection far exceeded doses required to block μ-opiate receptors *(229,232)* and in fact were sufficiently high to scavenge free radicals *(233)*. Faden proposed that the neuroprotective effects of NX are due to κ-opiate receptor blockade *(234)*. Whereas NX binds both μ and κ-receptors, much more NX is necessary to block κ-receptors. Indeed, several κ-selective opiate receptor blockers are neuroprotective in SCI models, including nalmefene *(235–238)*, thyrotropin-releasing hormone (TRH) *(239–243)*, the TRH analog YM-14673 *(26,243,244)*, and others *(245,246)*. However, a κ-receptor agonist is also protective in SCI *(244,247–249)*, possibly by inhibiting glutamate release *(250)*. Faden et al. hypothesized that opiate receptors contribute to secondary injury by modulating glutamate receptors *(251–255)* and vice versa *(237,256)*. If this is the case, however, why not use glutamate receptor blockers?

Excitatory neurotransmitter receptors contribute to secondary tissue damage in brain injury *(257)*. Several investigators have reported that blockers of the *N*-methyl-D-aspartate (NMDA) subtype of glutamate receptors protect injured spinal cords *(237,251,256,258–264)*, although other laboratories have not found similar beneficial effects of MK801 *(265)* or memantine *(266)*. Extracellular glutamate levels do rise in injured cords *(267–273)*. Injection of NMDA agonists produces gray matter lesions in the spinal cord *(274)*. Likewise, the alphaamino-5-methyl-3-hydroxyisoxazolone-4-propionate (AMPA) receptor blockers are neuroprotective *(261,264,275–284)*.

Adult spinal axons do not possess glutamate receptors. Whereas injection of the excitatory glutamatergic neurotransmitters NMDA and quisqualate will produce gray matter lesions in the spinal cord, they typically do not cause white matter damage. Glutamate may affect NMDA receptors on astrocytes or oligodendroglia in white matter *(285–287)*. Glutamate may also amplify the release of other axotoxic neurotransmitters or substances *(288,289)*. Blockade of glutamate receptors reduces neurotransmitter release from gray matter *(268,290–292)*. Gray matter has an abundance of NMDA and AMPA receptors *(293)*, and applied glutamate will release other neurotransmitters *(294–298)*. Spinal axons, however, are sensitive to α-aminobutyric acid (GABA), serotonin *(299–303)*, and norepinephrine *(304)*. GABA is of particular interest because GABA-A receptor blockers protect spinal white matter against profound anoxic injury *(305,306)*. In the presence of bicuculline, a GABA-A receptor blocker, isolated neonatal spinal cords are remarkably resistant to prolonged and profound anoxia.

These findings suggest a new theory of axotoxicity as opposed to neurotoxicity. In neurotoxicity, glutamate damages neuronal cell bodies by activating NMDA and AMPA receptors, depolarizing neuronal cell bodies and allowing calcium influx, whereas GABA tends to hyperpolarize the cell bodies and antagonize the effects of glutamate. When applied to spinal axons, however, glutamate had very little effect while GABA depolarizes. GABA normally depresses axonal excitability. In the presence of anoxia or Na/K ATPase blockade with ouabain, GABA produces irreversible declines in axonal excitability. Other axonal neurotransmitters, such as serotonin (5-HT), may contribute to axotoxicity. For example, spinal axons express both 5-HT$_{1A}$ and 5-HT$_{2A}$ receptors *(301,307,308)*. The former depresses *(309)* whereas the latter increases axonal excitability. Injury releases high levels of glutamate *(310–312)*, GABA *(84,271,313)*, and 5-HT *(314)* in the spinal cord. In addition, injury causes Na/K ATPase depression *(315)* in the spinal cord, and glucocorticoid therapy upregulates Na/K ATPase in spinal neurons *(316)*.

In summary, opiate and glutamate opiate and glutamate receptor blockers are neuroprotective in animal SCI models. The mechanisms by which these receptor blockers protect the spinal cord, particularly white matter, are not clear. Opiate receptor blockers depress lymphocytic activity and immune response to injury. However, direct immunosuppression with CyA does not produce dramatic recoveries. Several investigators have reported that opiate receptors modulate the glutamate receptors. If so, wouldn't glutamate receptors be a better and more direct therapy? Indeed, glutamate receptor blockers do protect injured spinal cords. However, since adult spinal axons do not possess glutamate receptors, it is not clear how glutamate receptor blockers protect white matter. Glutamate receptor blockers may prevent release of other neurotransmitters or neurotoxins that affect spinal axons. An alternative explanation is that glutamate protects astrocytes and oligodendroglia rather than axons. Some evidence suggests that "axotocity" differs from "neurotoxicity" and may involve GABA rather than glutamate receptors.

6. SPINAL CORD REGENERATION

Many investigators have now reported successful efforts to regenerate the adult mammalian spinal cord. These include the use of the IN-1 antibody, which blocks a white matter growth inhibitor *(317–319)*, applied neurotrophins *(320,321)*, fibroblast growth factors *(322)*, X-rays *(323,324)*, cyclosporin *(140)*, and even MP *(325)*. In addition, many investigators have shown that a variety of cells will not only survive implantation into the spinal cord but will provide stimulating environments for axonal growth. These include Schwann cells *(326–329)*, olfactory ensheathing glial cells *(330,331)*, macrophages activated by exposure to peripheral nerves *(198)*, nonactivated macrophage microglial cells *(200)*, neurotrophin-expressing fibroblasts *(332–334)*, and fetal cells *(335–337)*. These studies have decisively overturned the long-held dogma that the spinal cord cannot regenerate; please note that the above is not a comprehensive literature review.

Perhaps most interesting development is an emerging body of work indicating that spontaneous or "primary" regeneration can occur in injured spinal cords without any special treatment. For example, Olby et al. *(338)* observed dorsal column regeneration in rats with photochemical lesions. The Multicenter Animal Spinal Cord Injury Study (MASCIS) found large numbers of axons invading the injury site by 6 weeks after injury *(4)*. The main difference between these studies and previous work is the type of injury. Olby et al. used the photochemical lesion, and MASCIS used a weight drop contusion. In contrast, most previous regenerative studies transected the spinal cord, leaving a gap between the two ends of the cord. Although this gap can be bridged, the surgical procedures are not trivial *(339,340)*. Nevertheless, others have successfully used peripheral nerve grafts to reconnect the phrenic nucleus *(341,342)*. Alternatively, very sharp transections and careful reapposition of the cut ends will allow dorsal column and corticospinal growth across the gap in young rats *(343,344)*. However, in the case of a contusion or photochemical lesion, some tissue remains at the injury site.

Observations of "primary" regenerating fibers in injured spinal cords are often dismissed as dorsal root ingrowth, abortive sprouting, or insignificant data because they may not penetrate into the other side of the injury site. Several arguments can be made against dismissing these observations too cavalierly. First, whereas most of the axons invading the injury site may originate from dorsal roots, many are probably central. As many as 70% of the rats show hundreds or even thousands of axons invading the injury site. Second, the finding of dorsal root fibers growing in the spinal cord is of some interest. After all, dorsal root axons are subject to the same inhibitory constraints as central axons when they are growing in spinal cord white matter. Third, the amount and length of the growth correlate with greater injury severity, suggesting that greater tissue destruction or injury response promotes axonal growth. Fourth, after a severe contusion injury, most of the spinal axons die back a distance of 1 cm or more, while the injury site cavitates. These fibers are absent from the injury at 1 week after injury. For the fibers to invade back to the injury site, these fibers must grow distances of many millimeters. Finally, due to the location of the corticospinal tract close to the central canal, even a mild contusion invariably transects the tract. The corticospinal tracts die back and regrow in the proximal spinal cord back into the injury site. They cannot be surviving fibers.

Studies by Davies et al. *(345,346)* indicate that axons can grow long distances in central white matter, as long as the axons do not come into contact with extracellular matrix proteins produced by reactive astrocytes growing in response to blood-brain barrier breakdown. For example, when adult dorsal root ganglion cells are implanted into the corpus callosum, they can grow remarkably long distances if the transplantation is small and does not interrupt the blood-brain barrier. With larger transplants and blood-brain barrier breakdown, reactive astrocytosis develops with extracellular matrix proteins at the transplantation site. In such cases, the transplanted neurons do not send out axons. A more recent study *(347,347a)* provided striking evidence indicating that the long-distance axonal growth can occur in injured spinal cord white matter. Silver et al. showed that dorsal root ganglion cells transplanted to the spinal dorsal columns will grow long distances rostrally and caudally. A cut of the spinal cord below the transplantation site does not prevent axonal growth above the cut.

However, the growing axons stop at the cut where reactive astrocytosis develops. These data suggest that axons can grow in mature white matter. Furthermore, the data suggest that contact with these proteins will dramatically change the behavior of the axon.

The possibility that "primary" regeneration occurs in the spinal cord is important for several reasons. First, spontaneous recovery occurs in many animals and people after SCI. As a rule, most people recover at least one to two segments even after the most severe SCI. This recovery may take many months or years, consistent with the slow grow of axons in the spinal cord. Second, injured spinal cords must be expressing factors that stimulate axonal regrowth through proximal white matter. Therapies that enhance or prolong these factors may be useful. Third, the regenerating fibers are somehow evading, overcoming, or ignoring white matter growth inhibitors *(319,348)*. One possibility is that axons simply do not express the receptors for the white matter inhibitors until they come into contact with certain extracellular matrix proteins. The possibility that the white matter inhibitors are only expressed in injured nervous systems was ruled out by the above-described studies. Both the case of contused spinal cords where the corticospinal tract grew back to the injury site and the study by Silver et al. *(347)* indicate that the growth will occur in injured white matter with degenerating fiber tracts present.

In summary, the spinal cord clearly has a greater capability for regrowth and repair than we realized. A variety of therapies have been shown to stimulate and facilitate functional regeneration in adult mammalian spinal cords. These include use of antibodies to block white matter growth inhibitors, neurotrophins applied directly or through genetically modified cells, olfactory ensheathing glial cells, and inflammatory cells. Some data suggest that regeneration can occur spontaneously in injured spinal cords, particularly those that have been contused or lesioned without leaving a gap. Even when there are gaps, several studies have now shown that the gaps can be bridged with peripheral nerve grafts or by careful reapposition of sharply cut ends of the spinal cord. Very recent studies have raised questions of whether white matter growth inhibitors play a major role in preventing regeneration since dorsal root ganglion cells implanted into white matter can grow long distances as long as they do not come into contact with certain extracellular matrix proteins secreted by reactive astrocytes. One possibility is that central axons do not express receptors for white matter growth inhibitors until they come into contact with these extracellular matrix proteins. The future for regenerative therapies of the spinal cord is clearly very bright.

REFERENCES

1. Breasted, J. H. (1930) *The Edwin Smith Papyrus,* The University of Chicago Press, Chicago.
2. Bracken, M. B., Shepard, M. J., Collins, W. F., et al. (1990) A randomized controlled trial of methylprednisolone or naloxone in the treatment of acute spinal-cord

injury: results of the Second National Acute Spinal Cord Injury Study. *N. Engl. J. Med.* **322,** 1405–1411.

3. Bracken, M. B., Shepard, M. J., Holford, T. R., et al. (1997) Administration of methylprednisolone for 24 or 48 hours or tirilazad mesylate for 48 hours in the treatment of acute spinal cord injury. Results of the Third National Acute Spinal Cord Injury Randomized Controlled Trial. National Acute Spinal Cord Injury Study. *JAMA* **277,** 1597–1604.

4. Beattie, M. S., Bresnahan, J. C., Komon, J., et al. (1997) Endogenous repair after spinal cord contusion injuries in the rat. *Exp. Neurol.* **148,** 453–463.

5. Bracken, M. B., Shepard, M. J., Collins, W. F., et al. (1990) A randomized, controlled trial of methylprednisolone or naloxone in the treatment of acute spinal-cord injury. Results of the Second National Acute Spinal Cord Injury Study [see comments]. *N. Engl. J. Med.* **322,** 1405–1411.

6. Bracken, M. B. and Holford, T. R. (1993) Effects of timing of methylprednisolone or naloxone administration on recovery of segmental and long-tract neurological function in NASCIS 2. *J. Neurosurg.* **79,** 500–507.

7. Bracken, M. B., Shepard, M. J., Hellenbrand, K. G., et al. (1985) Methylprednisolone and neurological function 1 year after spinal cord injury. Results of the National Acute Spinal Cord Injury Study. *J. Neurosurg.* **63,** 704–713.

8. Hall, E. D. (1992). The neuroprotective pharmacology of methylprednisolone [Review]. *J. Neurosurg.* **76,** 13–22.

9. Holtz, A., Nystrom, B., and Gerdin, B. (1990). Effect of methylprednisolone on motor function and spinal cord blood flow after spinal cord compression in rats. *Acta Neurol. Scand.* **82,** 68–73.

10. Rosenberg, L. J., Jordan, R. S., Gross, G. W., Emery, D. G., and Lucas, J. H. (1996) Effects of methylprednisolone on lesioned and uninjured mammalian spinal neurons: viability, ultrastructure, and network electrophysiology. *J. Neurotrauma* **13,** 417–437.

11. Demopoulos, H. B., Flamm, E. S., Pietronigro, D. D., and Seligman, M. L. (1980) The free radical pathology and the microcirculation in the major central nervous system disorders. *Acta Physiol. Scand. Suppl.* **492,** 91–119.

12. Anderson, D. K., Saunders, R. D., Demediuk, P., et al. (1985) Lipid hydrolysis and peroxidation in injured spinal cord: partial protection with methylprednisolone or vitamin E and selenium. *Cent. Nerv. Syst. Trauma* **2,** 257–267.

13. Braughler, J. M. and Hall, E. D. (1981) Acute enhancement of spinal cord synaptosomal (Na^+-K^+)-ATPase activity in cats following intravenous methylprednisolone. *Brain Res.* **219,** 464–469.

14. Braughler, J. M. and Hall, E. D. (1981) Correlation of methylprednisolone pharmacokinetics in cat spinal cord with its effect on (Na^+-K^+)-ATPase, lipid peroxidation and motor neuron function. *J. Neurosurg.* **56,** 838–844.

15. Hall, E. D. and Braughler, J. M. (1982) Glucocorticoid mechanisms in acute spinal cord injury: a review and therapeutic rationale. *Surg. Neurol.* **18,** 320–327.

16. Hall, E. D. and Braughler, J. M. (1993) Free radicals in CNS injury [Review]. *Res. Publ. Assoc. Res. Nerv. Mental Dis.* **71,** 81–105.

17. Young, W. and Flamm, E. S. (1982) Effect of high dose corticosteroid therapy on blood flow, evoked potentials, and extracellular calcium in experimental spinal injury. *J. Neurosurg.* **57,** 667–673.

18. Seligman, M. L., Flamm, E. S., Goldstein, B. D., Poser, R. G., Demopoulos, H. B., and Ransohoff, J. (1977). Spectrofluorescent detection of malonaldehyde as a measure of lipid free radical damage in response to ethanol potentiation of spinal cord trauma. *Lipids* **12,** 945–950.

19. Anderson, D. K. and Means, E. D. (1983) Free radical-induced lipid peroxidation in spinal cord: $FeCl_2$ induction and protection with antioxidants. *Neurochem. Pathol.* **1,** 249–264.

20. Anderson, D. K. and Means, E. D. (1985) Iron-induced lipid peroxidation in spinal cord: protection with mannitol and methylprednisolone. *Free Radical Biol. Med.* **1,** 59–64.

21. Anderson, D. K. and Means, E. D. (1986) Alpha-tocopherol, mannitol, and methylprednisolone prevention of $FeCl_2$ initiated free radical induced lipid peroxidation in spinal cord, in *Oxygen Free Radicals in Shock* (Novelli, U., ed.), Karger, Basel, pp. 224–230.

22. Young, W. (1992). Role of calcium in central nervous system injuries. *J. Neurotrauma* **9,** S9–25.

23. Braughler, J. M. and Hall, E. D. (1992) Involvement of lipid peroxidation in CNS injury. *J. Neurotrauma* **9,** S1–7.

24. Jacobsen, E. J., McCall, J. M., Ayer, D. E., et al. (1990). Novel 21-aminosteroids that inhibit iron-dependent lipid peroxidation and protect against central nervous system trauma. *J. Med. Chem.* **33,** 1145–1151.

25. Anderson, D. K., Braughler, J. M., Hall, E. D., Waters, T. R., McCall, J. M., and Means, E. D. (1988) Effects of treatment with U74006F on neurological outcome following experimental spinal cord injury. *J. Neurosurg.* **69,** 562–567.

26. Behrmann, D. L., Bresnahan, J. C., and Beattie, M. S. (1994) Modeling of acute spinal cord injury in the rat: neuroprotection and enhanced recovery with methylprednisolone, U-74006F and YM-14673. *Exp. Neurol.* **126,** 61–75.

27. Hall, E. D. (1993) Neuroprotective actions of glucocorticoid and nonglucocorticoid steroids in acute neuronal injury [Review]. *Cell Mol. Neurobiol.* **13,** 415–432.

28. Holtz, A. and Gerdin, B. (1992) Efficacy of the 21-aminosteroid U74006F in improving neurological recovery after spinal cord injury in rats. *Neurol. Res.* **14,** 49–52.

29. Hayashi, M., Ueyama, T., Tamaki, T., and Senba, E. (1997) Expression of neurotrophin and IL-1 beta mRNAs following spinal cord injury and the effects of methylprednisolone treatment. *Kaibogaku Zasshi J. Anat.* **72,** 209–213.

30. Xu, J., Fan, G., Chen, S., Wu, Y., Xu, X. M., and Hsu, C. Y. (1998) Methylprednisolone inhibition of TNF-alpha expression and NF-kB activation after spinal cord injury in rats [In Process Citation]. *Brain Res. Mol. Brain Res.* **59,** 135–142.

31. Ray, A., Zhang, D. H., and Ray, P. (1995) Antagonism of nuclear factor-kappa B functions by steroid hormone receptors. *Biochem. Soc. Trans.* **23,** 952–958.

32. Auphan, N., DiDonato, J. A., Rosette, C., Helmberg, A., and Karin, M. (1995) Immunosuppression by glucocorticoids: inhibition of NF-kappa B activity through induction of I kappa B synthesis [see comments]. *Science* **270,** 286–290.

33. Scheinman, R. I., Cogswell, P. C., Lofquist, A. K., and Baldwin, A. S., Jr. (1995) Role of transcriptional activation of I kappa B alpha in mediation of immunosuppression by glucocorticoids [see comments]. *Science* **270,** 283–286.

34. Hein, H., Schluter, C., Kulke, R., Christophers, E., Schroder, J. M., and Bartels, J. (1997) Genomic organization, sequence, and transcriptional regulation of the human eotaxin gene. *Biochem. Biophys. Res. Commun.* **237,** 537–542.

35. Wang, C. X., Nuttin, B., Heremans, H., Dom, R., and Gybels, J. (1996) Production of tumor necrosis factor in spinal cord following traumatic injury in rats. *J. Neuroimmunol.* **69,** 151–156.

36. Yakovlev, A. G. and Faden, A. I. (1994) Sequential expression of c-fos protoonco-gene, TNF-alpha, and dynorphin genes in spinal cord following experimental traumatic injury. *Mol. Chem. Neuropathol.* **23,** 179–190.

37. Bartholdi, D. and Schwab, M. E. (1997) Expression of pro-inflammatory cytokine and chemokine mRNA upon experimental spinal cord injury in mouse: an in situ hybridization study. *Eur. J. Neurosci.* **9,** 1422–1438.

38. Sakurai, H., Shigemori, N., Hisada, Y., Ishizuka, T., Kawashima, K., and Sugita, T. (1997) Suppression of NF-kappa B and AP-1 activation by glucocorticoids in experimental glomerulonephritis in rats: molecular mechanisms of anti-nephritic action. *Biochim. Biophys. Acta* **1362,** 252–262.

39. Adcock, I. M., Brown, C. R., Gelder, C. M., Shirasaki, H., Peters, M. J., and Barnes, P. J. (1995) Effects of glucocorticoids on transcription factor activation in human peripheral blood mononuclear cells. *Am. J. Physiol.* **268,** C331–338.

40. Baeuerle, P. A. and Baichwal, V. R. (1997) NF-kappa B as a frequent target for immunosuppressive and anti-inflammatory molecules. *Adv. Immunol.* **65,** 111–137.

41. Barnes, P. J. and Karin, M. (1997) Nuclear factor-kappaB: a pivotal transcription factor in chronic inflammatory diseases. *N. Engl. J. Med.* **336,** 1066–1071.

42. Barnes, P. J. (1997) Nuclear factor-kappa B. *Int. J. Biochem. Cell Biol.* **29,** 867–870.

43. Wissink, S., van Heerde, E. C., vand der Burg, B., and van der Saag, P. T. (1998) A dual mechanism mediates repression of NF-kappaB activity by glucocorticoids. *Mol. Endocrinol.* **12,** 355–363.

44. Ramdas, J. and Harmon, J. M. (1998) Glucocorticoid-induced apoptosis and regulation of NF-kappaB activity in human leukemic T cells. *Endocrinology* **139,** 3813–3821.

45. Saura, M., Zaragoza, C., Diaz-Cazorla, M., et al. (1998) Involvement of transcriptional mechanisms in the inhibition of NOS2 expression by dexamethasone in rat mesangial cells. *Kidney Int.* **53,** 38–49.

46. McKay, L. I. and Cidlowski, J. A. (1998) Cross-talk between nuclear factor-kappa B and the steroid hormone receptors: mechanisms of mutual antagonism. *Mol. Endocrinol.* **12,** 45–56.

47. Ray, A., Zhang, D. H., Siegel, M. D., and Ray, P. (1995) Regulation of inter-leukin-6 gene expression by steroids. *Ann. NY Acad. Sci.* **762,** 79–87; discussion 87–78.

48. Choi, J. S., Kim-Yoon, S., and Joo, C. K. (1998) NF-kappa B activation following optic nerve transection [In Process Citation]. *Korean J. Ophthalmol.* **12,** 19–24.

49. Doyle, C. A. and Hunt, S. P. (1997) Reduced nuclear factor kappaB (p65) expression in rat primary sensory neurons after peripheral nerve injury. *Neuroreport* **8,** 2937–2942.

50. Migheli, A., Piva, R., Atzori, C., Troost, D., and Schiffer, D. (1997) c-Jun, JNK/SAPK kinases and transcription factor NF-kappa B are selectively activated in astrocytes, but not motor neurons, in amyotrophic lateral sclerosis. *J. Neuropathol. Exp. Neurol.* **56,** 1314–1322.

51. Ma, W. and Bisby, M. A. (1998) Increased activation of nuclear factor kappa B in rat lumbar dorsal root ganglion neurons following partial sciatic nerve injuries. *Brain Res.* **797,** 243–254.

52. Mukaida, N., Morita, M., Ishikawa, Y., et al. (1994) Novel mechanism of glucocorticoid-mediated gene repression. Nuclear factor-kappa B is target for glucocorticoid-mediated interleukin 8 gene repression. *J. Biol. Chem.* **269,** 13289–13295.

53. Ray, A. and Prefontaine, K. E. (1994) Physical association and functional antagonism between the p65 subunit of transcription factor NF-kappa B and the glucocorticoid receptor. *Proc. Natl. Acad. Sci. USA* **91,** 752–756.

54. Ray, A., Siegel, M. D., Prefontaine, K. E., and Ray, P. (1995) Anti-inflammation: direct physical association and functional antagonism between transcription factor NF-KB and the glucocorticoid receptor. *Chest* **107,** 139S.

55. Scheinman, R. I., Gualberto, A., Jewell, C. M., Cidlowski, J. A., and Baldwin, A. S. Jr. (1995) Characterization of mechanisms involved in transrepression of NF-kappa B by activated glucocorticoid receptors. *Mol. Cell. Biol.* **15,** 943–953.

56. De Bosscher, K., Schmitz, M. L., Vanden Berghe, W., Plaisance, S., Fiers, W., and Haegeman, G. (1997) Glucocorticoid-mediated repression of nuclear factor-kappaB-dependent transcription involves direct interference with transactivation. *Proc. Natl. Acad. Sci. USA* **94,** 13504–13509.

57. Caldenhoven, E., Liden, J., Wissink, S., et al. (1995) Negative cross-talk between RelA and the glucocorticoid receptor: a possible mechanism for the antiinflamma-tory action of glucocorticoids. *Mol. Endocrinol.* **9,** 401–412.

58. Beauparlant, P. and Hiscott, J. (1996) Biological and biochemical inhibitors of the NF-kappa B/Rel proteins and cytokine synthesis. *Cytokine Growth Factor Rev.* **7,** 175–190.

59. Liden, J., Delaunay, F., Rafter, I., Gustafsson, J., and Okret, S. (1997) A new func-tion for the C-terminal zinc finger of the glucocorticoid receptor. Repression of RelA transactivation. *J. Biol. Chem.* **272,** 21467–21472.

60. Newton, R., Hart, L. A., Stevens, D. A., et al. (1998) Effect of dexamethasone on interleukin-1 beta-(IL-1beta)-induced nuclear factor-kappaB (NF-kappaB) and kappaB-dependent transcription in epithelial cells. *Eur. J. Biochem.* **254,** 81–89.

61. Heck, S., Bender, K., Kullmann, M., Gottlicher, M., Herrlich, P., and Cato, A. C. (1997) I kappaB alpha-independent downregulation of NF-kappaB activity by glucocorticoid receptor. *EMBO J.* **16,** 4698–4707.

62. Gveric, D., Kaltschmidt, C., Cuzner, M. L., and Newcombe, J. (1998) Transcription factor NF-kappaB and inhibitor I kappaBalpha are localized in macrophages in active multiple sclerosis lesions. *J. Neuropathol. Exp. Neurol.* **57,** 168–178.

63. Ray, K. P., Farrow, S., Daly, M., Talabot, F., and Searle, N. (1997) Induction of the E-selectin promoter by interleukin 1 and tumour necrosis factor alpha, and inhibition by glucocorticoids. *Biochem. J.* **328,** 707–715.

64. Arenzana-Seisdedos, F., Turpin, P., Rodriguez, M., et al. (1997). Nuclear localization of l kappa B alpha promotes active transport of NF-kappa B from the nucleus to the cytoplasm. *J. Cell Sci.* **110,** 369–378.

64a. Bethea, J. R., Castro, M., Keane, R. W., Lee, T. T., Dietrich, W. D. and Yezierski, R. P. (1998). Traumatic spinal cord injury induces nuclear factor-kappaB activation. *J. Neurosci.* **18,** 3251–3260.

65. Brattsand, R. and Linden, M. (1996) Cytokine modulation by glucocorticoids: mechanisms and actions in cellular studies. *Aliment. Pharmacol. Ther.* **10,** 81–90; discussion 91–82.

66. Hasegawa, T., Sorensen, L., Dohi, M., Rao, N. V., Hoidal, J. R., and Marshall, B. C. (1997) Induction of urokinase-type plasminogen activator receptor by IL-1 beta. *Am. J. Respir. Cell. Mol. Biol.* **16,** 683–692.

67. Kleinert, H., Euchenhofer, C., Ihrig-Biedert, I., and Forstermann, U. (1996) Glucocorticoids inhibit the induction of nitric oxide synthase II by down-regulating cytokine-induced activity of transcription factor nuclear factor-kappa B. *Mol. Pharmacol.* **49,** 15–21.

68. Nishida, K., Kitazawa, R., Mizuno, K., Maeda, S., and Kitazawa, S. (1997) Identification of regulatory elements of human alpha 6 integrin subunit gene. *Biochem. Biophys. Res. Commun.* **241,** 258–263.

69. Wissink, S., van de Stolpe, A., Caldenhoven, E., Koenderman, L., and van der Saag, P. T. (1997) NF-kappa B/Rel family members regulating the ICAM-1 promoter in monocytic THP-1 cells. *Immunobiology* **198,** 50–64.

70. van der Saag, P. T., Caldenhoven, E., and van de Stolpe, A. (1996) Molecular mechanisms of steroid action: a novel type of cross-talk between glucocorticoids and NF-kappa B transcription factors. *Eur. Respir. J. Suppl.* **22,** 146s–153s.

71. Brostjan, C., Anrather, J., Csizmadia, V., Natarajan, G., and Winkler, H. (1997) Glucocorticoids inhibit E-selectin expression by targeting NF-kappaB and not ATF/c-Jun. *J. Immunol.* **158,** 3836–3844.

72. Ray, K. P. and Searle, N. (1997) Glucocorticoid inhibition of cytokine-induced E-selectin promoter activation. *Biochem. Soc. Trans.* **25,** 189S.

73. Ohtsuka, T., Kubota, A., Hirano, T., et al. (1996) Glucocorticoid-mediated gene suppression of rat cytokine-induced neutrophil chemoattractant CINC/gro, a member of the interleukin-8 family, through impairment of NF-kappa B activation. *J. Biol. Chem.* **271,** 1651–1659.

74. Mukaida, N., Okamoto, S., Ishikawa, Y., and Matsushima, K. (1994) Molecular mechanism of interleukin-8 gene expression. *J. Leukoc. Biol.* **56,** 554–558.

75. Flavin, M. P., Ho, L. T., and Coughlin, K. (1997) Neurotoxicity of soluble macrophage products in vitro—influence of dexamethasone. *Exp. Neurol.* **145,** 462–470.

76. Gorovits, R., Ben-Dror, I., Fox, L. E., Westphal, H. M., and Vardimon, L. (1994) Developmental changes in the expression and compartmentalization of the glucocorticoid receptor in embryonic retina. *Proc. Natl. Acad. Sci. USA* **91,** 4786–4790.

77. Gorovits, R., Yakir, A., Fox, L. E., and Vardimon, L. (1996) Hormonal and non-hormonal regulation of glutamine synthetase in the developing neural retina. *Brain Res. Mol. Brain Res.* **43,** 321–329.

78. Pike, C. J., Ramezan-Arab, N., Miller, S., and Cotman, C. W. (1996). Beta-amyloid increases enzyme activity and protein levels of glutamine synthetase in cultured astrocytes. *Exp. Neurol.* **139,** 167–171.

79. Ben-Dror, I., Havazelet, N., and Vardimon, L. (1993) Developmental control of glucocorticoid receptor transcriptional activity in embryonic retina. *Proc. Natl. Acad. Sci. USA* **90,** 1117–1121.

80. Berko-Flint, Y., Levkowitz, G., and Vardimon, L. (1994) Involvement of c-Jun in the control of glucocorticoid receptor transcriptional activity during development of chicken retinal tissue. *EMBO J.* **13,** 646–654.

81. Vardimon, L., Ben-Dror, I., Havazelet, N., and Fox, L. E. (1993) Molecular control of glutamine synthetase expression in the developing retina tissue. *Dev. Dyn.* **196,** 276–282.

82. Watanabe, T. (1997) Apoptosis induced by glufosinate ammonium in the neuroepithelium of developing mouse embryos in culture. *Neurosci. Lett.* **222,** 17–20.

83. Gorovits, R., Avidan, N., Avisar, N., Shaked, I., and Vardimon, L. (1997) Glutamine synthetase protects against neuronal degeneration in injured retinal tissue. *Proc. Natl. Acad. Sci. USA* **94,** 7024–7029.

84. Farooque, M., Hillered, L., Holtz, A., and Olsson, Y. (1996) Effects of methylprednisolone on extracellular lactic acidosis and amino acids after severe compression injury of rat spinal cord. *J. Neurochem.* **66,** 1125–1130.

85. Newton, R., Adcock, I. M., and Barnes, P. J. (1996) Superinduction of NF-kappa B by actinomycin D and cycloheximide in epithelial cells. *Biochem. Biophy. Res. Commun.* **218,** 518–523.

86. Menegazzi, M., Guerriero, C., Carcereri de Prati, A., Cardinale, C., Suzuki, H., and Armato, U. (1996) TPA and cycloheximide modulate the activation of NF-kappa B and the induction and stability of nitric oxide synthase transcript in primary neonatal rat hepatocytes. *FEBS Lett.* **379,** 279–285.

87. Wang, C. Y., Mayo, M. W., and Baldwin, A. S., Jr. (1996) TNF- and cancer therapy-induced apoptosis: potentiation by inhibition of NF-kappaB [see comments]. *Science* **274,** 784–787.

88. Lee, S. Y., Kaufman, D. R., Mora, A. L., Santana, A., Boothby, M., and Choi, Y. (1998) Stimulus-dependent synergism of the antiapoptotic tumor necrosis factor

receptor-associated factor 2 (TRAF2) and nuclear factor kappaB pathways [In Process Citation]. *J. Exp. Med.* **188**, 1381–1384.

89. Lee, S. Y. and Choi, Y. (1997) TRAF-interacting protein (TRIP): a novel component of the tumor necrosis factor receptor (TNFR)- and CD30-TRAF signaling complexes that inhibits TRAF2-mediated NF-kappaB activation. *J. Exp. Med.* **185**, 1275–1285.

90. Lee, S. Y., Reichlin, A., Santana, A., Sokol, K. A., Nussenzweig, M. C., and Choi, Y. (1997) TRAF2 is essential for JNK but not NF-kappaB activation and regulates lymphocyte proliferation and survival. *Immunity* **7**, 703–713.

91. Takahashi, N., Takeuchi, K., Sugawara, A., et al. (1998) Structure and transcriptional function of the 5′-flanking region of rat thromboxane receptor gene. *Biochem. Biophys. Res. Commun.* **244**, 489–493.

92. Roy, S., Cain, K. J., Chapin, R. B., Charboneau, R. G., and Barke, R. A. (1998) Morphine modulates NF kappa B activation in macrophages. *Biochem. Biophys. Res. Commun.* **245**, 392–396.

93. Suh, N., Honda, T., Finlay, H. J., et al. (1998) Novel triterpenoids suppress inducible nitric oxide synthase (iNOS) and inducible cyclooxygenase (COX-2) in mouse macrophages. *Cancer Res.* **58**, 717–723.

94. Cai, J., Huang, Z. Z., and Lu, S. C. (1997) Differential regulation of gamma-glutamylcysteine synthetase heavy and light subunit gene expression. *Biochem. J.* **326**, 167–172.

95. Corsini, E., Terzoli, A., Bruccoleri, A., Marinovich, M., and Galli, C. L. (1997) Induction of tumor necrosis factor-alpha in vivo by a skin irritant, tributyltin, through activation of transcription factors: its pharmacological modulation by anti-inflammatory drugs. *J. Invest. Dermatol.* **108**, 892–896.

96. Ayyavoo, V., Mahboubi, A., Mahalingam, S., et al. (1997) HIV-1 Vpr suppresses immune activation and apoptosis through regulation of nuclear factor kappa B [see comments]. *Nature Med.* **3**, 1117–1123.

97. Kalkhoven, E., Wissink, S., van der Saag, P. T., and van der Burg, B. (1996) Negative interaction between the RelA(p65) subunit of NF-kappaB and the progesterone receptor. *J. Biol. Chem.* **271**, 6217–6224.

98. van der Burg, B. and van der Saag, P. T. (1996) Nuclear factor-kappa-B/steroid hormone receptor interactions as a functional basis of anti-inflammatory action of steroids in reproductive organs. *Mol. Hum. Reprod.* **2**, 433–438.

99. Adcock, I. M., Stevens, D. A., and Barnes, P. J. (1996) Interactions of glucocorticoids and beta 2-agonists. *Eur. Respir. J.* **9**, 160–168.

100. Lee, M. R., Liou, M. L., Yang, Y. F., and Lai, M. Z. (1993) cAMP analogs prevent activation-induced apoptosis of T cell hybridomas. *J. Immunol.* **151**, 5208–5217.

101. Passegue, E., Richard, J. L., Boulla, G., and Gourdji, D. (1995) Multiple intracellular signallings are involved in thyrotropin-releasing hormone (TRH)-induced c-fos and jun B mRNA levels in clonal prolactin cells. *Mol. Cell. Endocrinol.* **107**, 29–40.

102. Grandison, L., Nolan, G. P., and Pfaff, D. W. (1994) Activation of the transcription factor NF-KB in GH3 pituitary cells. *Mol. Cell. Endocrinol.* **106**, 9–15.

103. Howard, P. W. and Maurer, R. A. (1994) Thyrotropin releasing hormone stimulates transient phosphorylation of the tissue-specific transcription factor, Pit-1. *J. Biol. Chem.* **269,** 28662–28669.

104. Mason, M. E., Friend, K. E., Copper, J., and Shupnik, M. A. (1993) Pit-1/GHF-1 binds to TRH-sensitive regions of the rat thyrotropin beta gene. *Biochemistry* **32,** 8932–8938.

105. Inoue, H. and Tanabe, T. (1998) Transcriptional role of the nuclear factor kappa B site in the induction by lipopolysaccharide and suppression by dexamethasone of cyclooxygenase-2 in U937 cells. *Biochem. Biophys. Res. Commun.* **244,** 143–148.

106. Medvedev, A. E., Flo, T., Ingalls, R. R., et al. (1998) Involvement of CD14 and complement receptors CR3 and CR4 in nuclear factor-kappaB activation and TNF production induced by lipopolysaccharide and group B streptococcal cell walls. *J. Immunol.* **160,** 4535–4542.

107. Handel, M. L. (1997) Transcription factors AP-1 and NF-kappa B: where steroids meet the gold standard of anti-rheumatic drugs. *Inflamm. Res.* **46,** 282–286.

108. Liu, X. Z., Xu, X. M., Hu, R., et al. (1997) Neuronal and glial apoptosis after traumatic spinal cord injury. *J. Neurosci.* **17,** 5395–5406.

109. Kato, H., Kanellopoulos, G. K., Matsuo, S., et al. (1997) Protection of rat spinal cord from ischemia with dextrorphan and cycloheximide: effects on necrosis and apoptosis. *J. Thorac. Cardiovasc. Surg.* **114,** 609–618.

110. Ciutat, D., Caldero, J., Oppenheim, R. W., and Esquerda, J. E. (1996) Schwann cell apoptosis during normal development and after axonal degeneration induced by neurotoxins in the chick embryo. *J. Neurosci.* **16,** 3979–3990.

111. Yaginuma, H., Tomita, M., Takashita, N., et al. (1996) A novel type of programmed neuronal death in the cervical spinal cord of the chick embryo. *J. Neurosci.* **16,** 3685–3703.

112. Kato, H., Kanellopoulos, G. K., Matsuo, S., Wu, Y. J., Jacquin, M. F., Hsu, C. Y., Choi, D. W. and Kouchoukos, N. T. (1997). Protection of rat spinal cord from ischemia with dextrorphan and cycloheximide: effects on necrosis and apoptosis. *J. Thorac. Cardiovasc. Surg.* **114,** 609–618.

113. Tong, J. X., Vogelbaum, M. A., Drzymala, R. E., and Rich, K. M. (1997) Radiation-induced apoptosis in dorsal root ganglion neurons. *J. Neurocytol.* **26,** 771–777.

114. Li, Y. Q. and Wong, C. S. (1998) Apoptosis and its relationship with cell proliferation in the irradiated rat spinal cord [In Process Citation]. *Int. J. Radiat. Biol.* **74,** 405–417.

115. Guth, L., Zhang, Z., DiProspero, N. A., Joubin, K., and Fitch, M. T. (1994). Spinal cord injury in the rat: treatment with bacterial lipopolysaccharide and indomethacin enhances cellular repair and locomotor function. *Exp. Neurol.* **126,** 76–87.

116. Guth, L., Zhang, Z., and Roberts, E. (1994) Key role for pregnenolone in combination therapy that promotes recovery after spinal cord injury. *Proc. Natl. Acad. Sci. USA* **91,** 12308–12312.

117. Fujiki, M., Zhang, Z., Guth, L., and Steward, O. (1996) Genetic influences on cellular reactions to spinal cord injury: activation of macrophages/microglia and astrocytes is delayed in mice carrying a mutation (WldS) that causes delayed Wallerian degeneration. *J. Comp. Neurol.* **371,** 469–484.

118. Zhang, Z., Fujiki, M., Guth, L., and Steward, O. (1996). Genetic influences on cellular reactions to spinal cord injury: a wound-healing response present in normal mice is impaired in mice carrying a mutation (WldS) that causes delayed Wallerian degeneration. *J. Comp. Neurol.* **371,** 485–495.

119. Zhang, Z., Krebs, C. J., and Guth, L. (1997) Experimental analysis of progressive necrosis after spinal cord trauma in the rat: etiological role of the inflammatory response. *Exp. Neurol.* **143,** 141–152.

120. Winkler, T. (1994) Possibilities to evaluate and diminish the effects of the trauma in spinal cord lesions. An experimental study in the rat. *Scand. J. Rehabil. Med. Suppl.* **30,** 81–82.

121. Jacobs, T. P., Hallenbeck, J. M., Devlin, T. M., and Feuerstein, G. Z. (1987) Prostaglandin derivative PGBx improves neurologic recovery after ischemic spinal injury. *Pharm. Res.* **4,** 130–132.

122. Jacobs, T. P., Shohami, E., Baze, W., et al. (1987) Thromboxane and 5-HETE increase after experimental spinal cord injury in rabbits. *Cent. Nerv. Syst. Trauma* **4,** 95–118.

123. Shohami, E., Jacobs, T. P., Hallenbeck, J. M., and Feuerstein, G. (1987) Increased thromboxane A2 and 5-HETE production following spinal cord ischemia in the rabbit. *Prostaglandins Leukot. Med.* **28,** 169–181.

124. Cai, W., Hu, L., and Foulkes, J. G. (1996) Transcription-modulating drugs: mechanism and selectivity. *Curr. Opin. Biotechnol.* **7,** 608–615.

125. Siegal, T., Siegal, T., Shohami, E., and Shapira, Y. (1988) Comparison of soluble dexamethasone sodium phosphate with free dexamethasone and indomethacin in treatment of experimental neoplastic spinal cord compression. *Spine* **13,** 1171–1176.

126. Sharma, H. S., Olsson, Y., Nyberg, F., and Dey, P. K. (1993) Prostaglandins modulate alterations of microvascular permeability, blood flow, edema and serotonin levels following spinal cord injury: an experimental study in the rat. *Neuroscience* **57,** 443–449.

127. Winkler, T., Sharma, H. S., Stalberg, E., and Olsson, Y. (1993) Indomethacin, an inhibitor of prostaglandin synthesis attenuates alteration in spinal cord evoked potentials and edema formation after trauma to the spinal cord: an experimental study in the rat. *Neuroscience* **52,** 1057–1067.

128. Olsson, Y., Sharma, H. S., Pettersson, A., and Cervos-Navarro, J. (1992) Release of endogenous neurochemicals may increase vascular permeability, induce edema and influence cell changes in trauma to the spinal cord. *Prog. Brain Res.* **91,** 197–203.

129. Siegal, T., Siegal, T., and Lossos, F. (1990) Experimental neoplastic spinal cord compression: effect of anti-inflammatory agents and glutamate receptor antagonists on vascular permeability. *Neurosurgery* **26,** 967–970.

130. Yang, L. C., Marsala, M., and Yaksh, T. L. (1996) Effect of spinal kainic acid receptor activation on spinal amino acid and prostaglandin E2 release in rat. *Neuroscience* **75,** 453–461.

131. Bret-Dibat, J. L., Kent, S., Couraud, J. Y., Creminon, C., and Dantzer, R. (1994) A behaviorally active dose of lipopolysaccharide increases sensory neuropeptides levels in mouse spinal cord. *Neurosci. Lett.* **173,** 205–209.

132. Sharma, H. S., Olsson, Y., and Cervos-Navarro, J. (1993) Early perifocal cell changes and edema in traumatic injury of the spinal cord are reduced by indomethacin, an inhibitor of prostaglandin synthesis. Experimental study in the rat. *Acta Neuropathol.* **85,** 145–153.

133. Hall, E. D. and Wolf, D. L. (1986) A pharmacological analysis of the pathophysiological mechanisms of posttraumatic spinal cord ischemia. *J. Neurosurg.* **64,** 951–961.

134. Sharma, H. S. and Westman, J. (1997) Prostaglandins modulate constitutive isoform of heat shock protein (72 kD) response following trauma to the rat spinal cord. *Acta Neurochir. Suppl.* **70,** 134–137.

135. Hallenbeck, J. M., Jacobs, T. P., and Faden, A. I. (1983) Combined PGI2, indomethacin, and heparin improves neurological recovery after spinal trauma in cats. *J. Neurosurg.* **58,** 749–754.

136. Guth, L., Zhang, Z., and Roberts, E. (1994) Key role for pregnenolone in combination therapy that promotes recovery after spinal cord injury. *Proc. Natl. Acad. Sci. USA* **91,** 12308–12312.

137. Ghosh, P., Sica, A., Cippitelli, M., et al. (1996). Activation of nuclear factor of activated T cells in a cyclosporin A-resistant pathway. *J. Biol. Chem.* **271,** 7700–7704.

138. Wong, B. R., Rho, J., Arron, J., et al. (1997) TRANCE is a novel ligand of the tumor necrosis factor receptor family that activates c-Jun N-terminal kinase in T cells. *J. Biol. Chem.* **272,** 25190–25194.

139. Deguchi, K., Takeuchi, H., Miki, H., et al. (1991) Effects of FK 506 on acute experimental allergic encephalomyelitis. *Transplant. Proc.* **23,** 3360–3362.

140. Palladini, G., Caronti, B., Pozzessere, G., et al. (1996) Treatment with cyclosporine A promotes axonal regeneration in rats submitted to transverse section of the spinal cord—II—recovery of function. *J. Hirnforsch.* **37,** 145–153.

141. Steiner, J. P., Connolly, M. A., Valentine, H. L., et al. (1997) Neurotrophic actions of nonimmunosuppressive analogues of immunosuppressive drugs FK506, rapamycin and cyclosporin A. *Nature Med.* **3,** 421–428.

142. Chang, H. Y., Takei, K., Sydor, A. M., Born, T., Rusnak, F., and Jay, D. G. (1995) Asymmetric retraction of growth cone filopodia following focal inactivation of calcineurin. *Nature* **376,** 686–690.

143. Lyons, W. E., Steiner, J. P., Snyder, S. H., and Dawson, T. M. (1995) Neuronal regeneration enhances the expression of the immunophilin FKBP-12. *J. Neurosci.* **15,** 2985–2994.

144. Kar, S. and Quirion, R. (1995) Neuropeptide receptors in developing and adult rat spinal cord: an in vitro quantitative autoradiography study of calcitonin gene-related peptide, neurokinins, mu-opioid, galanin, somatostatin, neurotensin and vasoactive intestinal polypeptide receptors. *J. Comp. Neurol.* **354,** 253–281.

145. Dawson, T. M., Steiner, J. P., Lyons, W. E., Fotuhi, M., Blue, M., and Snyder, S. H. (1994) the immunophilins, FK506 binding protein and cyclophilin, are discretely localized in the brain: relationship to calcineurin. *Neuroscience* **62,** 569–580.

146. Renoir, J. M., Mercier-Bodard, C., Hoffmann, K., et al. (1995) Cyclosporin A potentiates the dexamethasone-induced mouse mammary tumor virus-chloram-

phenicol acetyltransferase activity in LMCAT cells: a possible role for different heat shock protein-binding immunophilins in glucocorticosteroid receptor-mediated gene expression. *Proc. Natl. Acad. Sci. USA* **92,** 4977–4981.

147. Sacerdote, P., di San Secondo, V. E., Sirchia, G., Manfredi, B., and Panerai, A. E. (1998) Endogenous opioids modulate allograft rejection time in mice: possible relation with Th1/Th2 cytokines. *Clin. Exp. Immunol.* **113,** 465–469.

148. Sacerdote, P., Manfredi, B., Mantegazza, P., and Panerai, A. E. (1997) Antinociceptive and immunosuppressive effects of opiate drugs: a structure-related activity study. *Br. J. Pharmacol.* **121,** 834–840.

149. Sowa, G., Gekker, G., Lipovsky, M. M., et al. (1997) Inhibition of swine microglial cell phagocytosis of *Cryptococcus neoformans* by femtomolar concentrations of morphine. *Biochem. Pharmacol.* **53,** 823–828.

150. Nair, M. P., Schwartz, S. A., Polasani, R., Hou, J., Sweet, A., and Chadha, K. C. (1997) Immunoregulatory effects of morphine on human lymphocytes. *Clin. Diagn. Lab. Immunol.* **4,** 127–132.

151. Ientile, R., Ginoprelli, T., Cannavo, G., Romeo, S., and Macaione, S. (1997) Beta-endorphin enhances polyamine transport in human lymphocytes. *Life Sci.* **60,** 1545–1551.

152. Coe, C. L. and Erickson, C. M. (1997) Stress decreases lymphocyte cytolytic activity in the young monkey even after blockade of steroid and opiate hormone receptors. *Dev. Psychobiol.* **30,** 1–10.

153. Whitlock, B. B., Liu, Y., Chang, S., et al. (1996) Initial characterization and autoradiographic localization of a novel sigma/opioid binding site in immune tissues. *J. Neuroimmunol.* **67,** 83–96.

154. Scifo, R., Cioni, M., Nicolosi, A., et al. (1996) Opioid-immune interactions in autism: behavioural and immunological assessment during a double-blind treatment with naltrexone. *Ann. 1st Super Sanita* **32,** 351–359.

155. Alicea, C., Belkowski, S., Eisenstein, T. K., Adler, M. W., and Rogers, T. J. (1996) Inhibition of primary murine macrophage cytokine production in vitro following treatment with the kappa-opioid agonist U50,488H. *J. Neuroimmunol.* **64,** 83–90.

156. Carr, D. J. (1991) The role of endogenous opioids and their receptors in the immune system. *Proc. Soc. Exp. Biol. Med.* **198,** 710–720.

157. Pruett, S. B., Han, Y. C., and Fuchs, B. A. (1992) Morphine suppresses primary humoral immune responses by a predominantly indirect mechanism. *J. Pharmacol. Exp. Ther.* **262,** 923–928.

158. Fecho, K., Maslonek, K. A., Dykstra, L. A., and Lysle, D. T. (1996) Assessment of the involvement of central nervous system and peripheral opioid receptors in the immunomodulatory effects of acute morphine treatment in rats. *J. Pharmacol. Exp. Ther.* **276,** 626–636.

159. Carr, D. J., Carpenter, G. W., Garza, H. H., Jr., Baker, M. L., and Gebhardt, B. M. (1995) Cellular mechanisms involved in morphine-mediated suppression of CTL activity. *Adv. Exp. Med. Biol.* **373,** 131–139.

160. Radulovic, J., Miljevic, C., Djergovic, D., et al. (1995) Opioid receptor-mediated suppression of humoral immune response in vivo and in vitro: involvement of kappa opioid receptors. *J. Neuroimmunol.* **57,** 55–62.

161. Minault, M., Lecron, J. C., Labrouche, S., Simonnet, G., and Gombert, J. (1995) Characterization of binding sites for neuropeptide FF on T lymphocytes of the Jurkat cell line. *Peptides* **16,** 105–111.

162. Sonetti, D., Ottaviani, E., Bianchi, F., et al. (1994) Microglia in invertebrate ganglia. *Proc. Natl. Acad. Sci. USA* **91,** 9180–9184.

163. Pryor, S. C. and Fascher, E. (1995) The effects of opiates and opioid neuropeptides on mosquito hemocytes. *Acta Biol. Hung.* **46,** 329–340.

164. Stefano, G. B., Casares, F., and Liu, U. (1995) Naltrindole sensitive delta 2 opioid receptor mediates invertebrate immunocyte activation. *Acta Biol. Hung.* **46,** 321–327.

165. Chao, C. C., Gekker, G., Sheng, W. S., Hu, S., Tsang, M., and Peterson, P. K. (1994) Priming effect of morphine on the production of tumor necrosis factor-alpha by microglia: implications in respiratory burst activity and human immunodeficiency virus-1 expression. *J. Pharmacol. Exp. Ther.* **269,** 198–203.

166. Chao, C. C., Molitor, T. W., Close, K., Hu, S., and Peterson, P. K. (1993) Morphine inhibits the release of tumor necrosis factor in human peripheral blood mononuclear cell cultures. *Int. J. Immunopharmacol.* **15,** 447–453.

167. Maity, R., Mukherjee, R., and Skolnick, P. (1995) Morphine inhibits the development of allogeneic immune responses in mouse lymph node. *Immunopharmacology* **29,** 175–183.

168. Naliboff, B. D., Solomon, G. F., Gilmore, S. L., Benton, D., Morley, J. E., and Fahey, J. L. (1995). The effects of the opiate antagonist naloxone on measures of cellular immunity during rest and brief psychological stress. *J. Psychosom. Res.* **39,** 345–359.

169. Bouix, O., el Mezouini, M., and Orsetti, A. (1995) Effects of naloxone opiate blockade on the immunomodulation induced by exercise in rats. *Int. J. Sports Med.* **16,** 29–33.

170. Lin, J., Lu, G., and Weng, J. (1994) Effects of beta-endorphin on phytohemagglutinin-induced lymphocyte proliferation and mouse plaque-forming cell response via an opioid receptor mechanism. *Chin. Med. Sci. J.* **9,** 245–247.

171. Freier, D. O. and Fuchs, B. A. (1994) A mechanism of action for morphine-induced immunosuppression: corticosterone mediates morphine-induced suppression of natural killer cell activity. *J. Pharmacol. Exp. Ther.* **270,** 1127–1133.

172. Schafer, M., Carter, L., and Stein, C. (1994) Interleukin 1 beta and corticotropin-releasing factor inhibit pain by releasing opioids from immune cells in inflamed tissue. *Proc. Natl. Acad. Sci. USA* **91,** 4219–4223.

173. Akira, S. and Kishimoto, T. (1997) NF-IL6 and NF-kappa B in cytokine gene regulation. *Adv. Immunol.* **65,** 1–46.

174. Auphan, N., Didonato, J. A., Helmberg, A., Rosette, C., and Karin, M. (1997) Immunoregulatory genes and immunosuppression by glucocorticoids. *Arch. Toxicol. Suppl.* **19,** 87–95.

175. Joyce, D. A., Steer, J. H., and Abraham, L. J. (1997) Glucocorticoid modulation of human monocyte/macrophage function: control of TNF-alpha secretion. *Inflamm. Res.* **46,** 447–451.

176. Okada, N., Kobayashi, M., Mugikura, K., et al. (1997) Interleukin-6 production in human fibroblasts derived from periodontal tissues is differentially regulated by cytokines and a glucocorticoid. *J. Periodont. Res.* **32,** 559–569.

177. Fernandez-Ruiz, E., Rebollo, A., Nieto, M. A., et al. (1989) IL-2 protects T cell hybrids from the cytolytic effect of glucocorticoids. Synergistic effect of IL-2 and dexamethasone in the induction of high-affinity IL-2 receptors. *J. Immunol.* **143,** 4146–4151.

178. Xie, H., Seward, R. J., and Huber, B. T. (1997) Cytokine rescue from glucocorticoid induced apoptosis in T cells is mediated through inhibition of IkappaBalpha. *Mol. Immunol.* **34,** 987–994.

179. Wissink, S., van Heerde, E. C., Schmitz, M. L., et al. (1997) Distinct domains of the ReIA NF-kappaB subunit are required for negative cross-talk and direct interaction with the glucocorticoid receptor. *J. Biol. Chem.* **272,** 22278–22284.

180. Wang, Y., Zhang, J. J., Dai, W., Lei, K. Y., and Pike, J. W. (1997) Dexamethasone potently enhances phorbol ester-induced IL-1beta gene expression and nuclear factor NF-kappaB activation. *J. Immunol.* **159,** 534–537.

181. de Lange, P., Koper, J. W., Huizenga, N. A., et al. (1997) Differential hormone-dependent transcriptional activation and repression by naturally occurring human glucocorticoid receptor variants. *Mol. Endocrinol.* **11,** 1156–1164.

182. Batuman, O. A., Ferrero, A. P., Diaz, A., Berger, B., and Pomerantz, R. J. (1994) Glucocorticoid-mediated inhibition of interleukin-2 receptor alpha and-beta subunit expression by human T cells. *Immunopharmacology* **27,** 43–55.

183. Kato, H., Kanellopoulos, G. K., Matsuo, S., et al. (1997) Neuronal apoptosis and necrosis following spinal cord ischemia in the rat. *Exp. Neurol.* **148,** 464–474.

184. Villarroya, H., Violleau, K., Ben Younes-Chennoufi, A., and Baumann, N. (1996) Myelin-induced experimental allergic encephalomyelitis in Lewis rats: tumor necrosis factor alpha levels in serum and cerebrospinal fluid immunohistochemical expression in glial cells and macrophages of optic nerve and spinal cord. *J. Neuroimmunol.* **64,** 55–61.

185. Dugandzija-Novakovic, S. and Shrager, P. (1995) Survival, development, and electrical activity of central nervous system myelinated axons exposed to tumor necrosis factor in vitro. *J. Neurosci. Res.* **40,** 117–126.

186. Taupin, V., Renno, T., Bourbonniere, L., Peterson, A. C., Rodriguez, M., and Owens, T. (1997) Increased severity of experimental autoimmune encephalomyelitis, chronic macrophage/microglial reactivity, and demyelination in transgenic mice producing tumor necrosis factor-alpha in the central nervous system. *Eur. J. Immunol.* **27,** 905–913.

187. Friedlander, R. M., Gagliardini, V., Hara, H., et al. (1997) Expression of a dominant negative mutant of interleukin-1 beta converting enzyme in transgenic mice prevents neuronal cell death induced by trophic factor withdrawal and ischemic brain injury. *J. Exp. Med.* **185,** 933–940.

188. Nuttin, B. (1995) Interleukin-6 and spinal cord damage in rats. *Verh. K. Acad. Geneeskd. Belg.* **57,** 315–349.

189. Klusman, I., and Schwab, M. E. (1997) Effects of pro-inflammatory cytokines in experimental spinal cord injury. *Brain Res.* **762,** 173–184.

190. Bethea, J. R., Castro, M., Bricenco, C., Gomez, F., Marcillo, A. E., and Dietrich, D. W. (1998) Systemically administered interleukin-10 (IL-10) attenuates injury induced inflammation and is neuroprotective following traumatic spinal cord injury. *J. Neurotrauma* **15,** 906.

191. Lezoualc'h, F., Sagara, Y., Holsboer, F., and Behl, C. (1998) High constitutive NF-kappaB activity mediates resistance to oxidative stress in neuronal cells. *J. Neurosci.* **18,** 3224–3232.

192. Lukaszewicz, G., Abcouwer, S. F., Labow, B. I., and Souba, W. W. (1997) Glutamine synthetase gene expression in the lungs of endotoxin-treated and adrenalectomized rats. *Am. J. Physiol.* **273,** L1182–1190.

193. Abcouwer, S. F., Lukaszewicz, G. C., and Souba, W. W. (1996) Glucocorticoids regulate glutamine synthetase expression in lung epithelial cells. *Am. J. Physiol.* **270,** L141–151.

194. Abcouwer, S. F., Lohmann, R., Bode, B. P., Lustig, R. J., and Souba, W. W. (1997) Induction of glutamine synthetase expression after major burn injury is tissue specific and temporally variable. *J. Trauma* **42,** 421–427; discussion 427–428.

195. Elgadi, K. M., Labow, B. I., Abcouwer, S. F., and Souba, W. W. (1998) Sepsis increases lung glutamine synthetase expression in the tumor-bearing host. *J. Surg. Res.* **78,** 18–22.

196. Wamil, A. W., Wamil, B. D., and Hellerqvist, C. G. (1998) CM101-mediated recovery of walking ability in adult mice paralyzed by spinal cord injury. *Proc. Natl. Acad. Sci. USA* **95,** 13188–13193.

197. Schwartz, M., Belkin, M., Yoles, E., and Solomon, A. (1996) Potential treatment modalities for glaucomatous neuropathy: neuroprotection and neuroregeneration. [Review]. *J. Glaucoma* **5,** 427–432.

198. Rapalino, O., Lazarov-Spiegler, O., Agranov, E., et al. (1998) Implantation of stimulated homologous macrophages results in partial recovery of paraplegic rats [In Process Citation]. *Nature Med.* **4,** 814–821.

199. Lazarov-Spiegler, O., Solomon, A. S., Zeev-Brann, A. B., Hirschberg, D. L., Lavie, V., and Schwartz, M. (1996) Transplantation of activated macrophages overcomes central nervous system regrowth failure. *FASEB J.* **10,** 1296–1302.

200. Rabchevsky, A. G. and Streit, W. J. (1997) Grafting of cultured microglial cells into the lesioned spinal cord of adult rats enhances neurite outgrowth. *J. Neurosci. Res.* **47,** 34–48.

201. Constantini, S. and Young, W. (1994) The effects of methylprednisolone and the ganglioside GM1 on acute spinal cord injury in rats. *J. Neurosurg.* **80,** 97–111.

202. Geisler, F. H., Dorsey, F. C., and Coleman, W. P. (1990) GM1 gangliosides in the treatment of spinal cord injury: report of preliminary data analysis. *Acta Neurobiol. Exp.* **50,** 515–521.

203. Geisler, F. H., Dorsey, F. C., and Coleman, W. P. (1991) Recovery of motor function after spinal-cord injury—a randomized, placebo-controlled trial with GM-1 ganglioside. *N. Engl. J. Med.* **324,** 1829–1838.

204. Geisler, F. H., Dorsey, F. C., and Coleman, W. P. (1992) GM-1 ganglioside in human spinal cord injury. *J. Neurotrauma* **9,** S517–S530.

205. Geisler, F. H. (1993) GM-1 ganglioside and motor recovery following human spinal cord injury. *J. Emerg. Med.* **1,** 49–55.

206. Geisler, F. H., Dorsey, F. C., and Coleman, W. P. (1993) Past and current clinical studies with GM-1 ganglioside in acute spinal cord injury [Review]. *Ann. Emerg. Med.* **22,** 1041–1047.

207. Schneider, J. S. (1998) GM1 ganglioside in the treatment of Parkinson's disease. *Ann. NY Acad. Sci.* **845,** 363–373.

208. Schneider, J. S., Roeltgen, D. P., Mancall, E. L., Chapas-Crilly, J., Rothblat, D. S., and Tatarian, G. T. (1998) Parkinson's disease: improved function with GM1 ganglioside treatment in a randomized placebo-controlled study. *Neurology* **50,** 1630–1636.

209. Baumgartner, W. A., Redmond, J. M., Zehr, K. J., et al. (1998) The role of the monosialoganglioside, GM1 as a neuroprotectant in an experimental model of cardiopulmonary bypass and hypothermic circulatory arrest. *Ann. NY Acad. Sci.* **845,** 382–390.

210. Tseng, E. E., Brock, M. V., Lange, M. S., et al. (1998) Monosialoganglioside GM1 inhibits neurotoxicity after hypothermic circulatory arrest. *Surgery* **124,** 298–306.

211. Lainetti, R. D., Pereira, F. C., and Da-Silva, C. F. (1998) Ganglioside GM1 potentiates the stimulatory effect of nerve growth factor on peripheral nerve regeneration in vivo. *Ann. NY Acad. Sci.* **845,** 415–416.

212. Ferrari, G., Minozzi, M. C., Zanellato, A. M., and Silvestrini, B. (1998) GM1, like IGF-I and GDNF, prevents neuronal apoptosis. *Ann. NY Acad. Sci.* **845,** 408.

213. Ferrari, G. and Greene, L. A. (1998) Promotion of neuronal survival by GM1 ganglioside. Phenomenology and mechanism of action. *Ann. NY Acad. Sci.* **845,** 263–273.

214. Rothblat, D. S. and Schneider, J. S. (1998) The effects of L-deprenyl treatment, alone and combined with GM1 ganglioside, on striatal dopamine content and substantia nigra pars compacta neurons. *Brain Res.* **779,** 226–230.

215. Saito, M., Guidotti, A., Berg, M. J., and Marks, N. (1998) The semisynthetic glycosphingolipid LIGA20 potently protects neurons against apoptosis. *Ann. NY Acad. Sci.* **845,** 253–262.

216. Farooqui, T. and Yates, A. J. (1998) Effect of GM1 on TrkA dimerization. *Ann. NY Acad. Sci.* **845,** 407.

217. Duchemin, A. M., Neff, N. H., and Hadjiconstantinou, M. (1998) Induction of Trk phosphorylation in rat brain by GM1 ganglioside. *Ann. NY Acad. Sci.* **845,** 406.

218. Ariga, T. and Yu, R. K. (1998) GM1 ganglioside inhibits amyloid beta-protein induced-cytokine release. *Ann. NY Acad. Sci.* **845,** 403.

219. Oderfeld-Nowak, B. and Zaremba, M. (1998) GM1 ganglioside potentiates trimethyltin-induced expression of interleukin-1 beta and the nerve growth factor in reactive astrocytes in the rat hippocampus: an immunocytochemical study. *Neurochem. Res.* **23,** 443–453.

220. Rampersaud, A. A., Van Brocklyn, J. R., and Yates, A. J. (1998) GM1 activates the MAP kinase cascade through a novel wortmannin-sensitive step upstream from c-Raf. *Ann. NY Acad. Sci.* **845,** 424.

221. Wang, L. J., Colella, R., and Roisen, F. J. (1998) Ganglioside GM1 alters neuronal morphology by modulating the association of MAP2 with microtubules and actin filaments. *Brain Res. Dev. Brain Res.* **105,** 227–239.

222. Wu, G., Fang, Y., Lu, Z. H., and Ledeen, R. W. (1998) Induction of axon-like and dendrite-like processes in neuroblastoma cells. *J. Neurocytol.* **27,** 1–14.

223. Wu, G., Lu, Z. H., Wei, T. J., Howells, R. D., Christoffers, K., and Ledeen, R. W. (1998) The role of GM1 ganglioside in regulating excitatory opioid effects. *Ann. NY Acad. Sci.* **845,** 126–138.

224. Crain, S. M. and Shen, K. F. (1998) GM1 ganglioside-induced modulation of opioid receptor-mediated functions. *Ann. NY Acad. Sci.* **845,** 106–125.

225. Avrova, N. F., Victorov, I. V., Tyurin, V. A., et al. (1998) Inhibition of glutamate-induced intensification of free radical reactions by gangliosides: possible role in their protective effect in rat cerebellar granule cells and brain synaptosomes [In Process Citation]. *Neurochem. Res.* **23,** 945–952.

226. Yates, A. J. and Rampersaud, A. (1998) Sphingolipids as receptor modulators. An overview. *Ann. NY Acad. Sci.* **845,** 57–71.

227. Faden, A. I., Jacobs, T. P., and Holaday, J. W. (1980) Opiate antagonist improves neurologic recovery after spinal injury. *Science* **211,** 493–494.

228. Faden, A. I., Jacobs, T. P., and Holaday, J. W. (1982) Comparison of early and late naloxone treatment in experimental spinal injury. *Neurology (NY)* **32,** 677–681.

229. Flamm, E. S., Young, W., Demopoulos, H. B., DeCrescito, V., and Tomasula, J. J. (1982) Experimental spinal cord injury: treatment with naloxone. *Neurosurgery* **10,** 227–231.

230. Flamm, E. S., Young, W., Collins, W. F., Piepmeier, J., Clifton, G. L., and Fischer, B. (1985) A phase I trial of naloxone treatment in acute spinal cord injury. *J. Neurosurg.* **63,** 390–397.

231. Young, W., DeCrescito, V., Flamm, E. S., Blight, A. R., and Gruner, J. A. (1988) Pharmacological therapy of acute spinal cord injury: studies of high dose methylprednisolone and naloxone. *Clin. Neurosurg.* **34,** 675–697.

232. Young, W., Flamm, E. S., Demopoulos, H. B., Tomasula, J. J., and DeCrescito, V. (1981) Effect of naloxone on posttraumatic ischemia in experimental spinal contusion. *J. Neurosurg.* **55,** 209–219.

233. Koreh, K., Seligman, M. L., Flamm, E. S., and Demopoulos, H. R. (1981) Lipid peroxidant properties of naloxone in vitro. *Biochem. Biophys. Res. Commun.* **102,** 1317–1322.

234. Faden, A. I. (1985) New pharmacologic approaches to spinal cord injury: opiate antagonists and thyrotropin-releasing hormone. *Cent. Nerv. Syst. Trauma* **2,** 5–8.

235. Benzel, E. C., Khare, V., and Fowler, M. R. (1992) Effects of naloxone and nalmefene in rat spinal cord injury induced by the ventral compression technique. *J. Spinal Disord.* **5,** 75–77.

236. Vink, R., McIntosh, T. K., Rhomhanyi, R., and Faden, A. I. (1990) Opiate antagonist nalmefene improves intracellular free Mg^{2+}, bioenergetic state, and neurologic outcome following traumatic brain injury in rats. *J. Neurosci.* **10,** 3524–3530.

237. Yum, S. W. and Faden, A. I. (1990) Comparison of the neuroprotective effects of the N-methyl-D-aspartate antagonist MK-801 and the opiate-receptor antagonist nalmefene in experimental spinal cord ischemia. *Arch. Neurol.* **47,** 277–281.

238. Puniak, M. A., Freeman, G. M., Agresta, C. A., Van, N. L., Barone, C. A., and Salzman, S. K. (1991) Comparison of a serotonin antagonist, opioid antagonist, and TRH analog for the acute treatment of experimental spinal trauma. *J. Neurotrauma* **8,** 193–203.

239. Arias, M. J. (1987) Treatment of experimental spinal cord injury with TRH, naloxone, and dexamethasone. *Surg. Neurol.* **28,** 335–338.

240. Faden, A. I., Jacobs, T. P., and Smith, M. T. (1983) Comparison of thyrotropin-releasing hormone (TRH), naloxone, and dexamethasone treatments in experimental spinal injury. *Neurology* **33,** 673–678.

241. Faden, A. I. and Jacobs, T. P. (1985) Effect of TRH analogs on neurologic recovery after experimental spinal trauma. *Neurology* **35,** 1331–1334.

242. Faden, A. I., Sacksen, I., and Noble, L. J. (1988) Structure-activity relationships of TRH analogs in rat spinal cord injury. *Brain Res.* **448,** 287–293.

243. Faden, A. I. (1989) TRH analog YM-14673 improves outcome following traumatic brain and spinal cord injury in rats: dose-response studies. *Brain Res.* **486,** 228–235.

244. Behrmann, D. L., Bresnahan, J. C., and Beattie, M. S. (1993) A comparison of YM-14673, U-50488H, and nalmefene after spinal cord injury in the rat. *Exp. Neurol.* **119,** 258–267.

245. Faden, A. I. and Jacobs, T. P. (1985) Opiate antagonist WIN 44, 441–3 stereospecifically improves neurologic recovery after ischemic spinal injury. *Neurology* **35,** 1311–1315.

246. Winkler, T., Sharma, H. S., Stalberg, E., Olsson, Y., and Nyberg, F. (1994) Opioid receptors influence spinal cord electrical activity and edema formation following spinal cord injury: experimental observations using naloxone in the rat. *Neurosci. Res.* **21,** 91–101.

247. Hall, E. D., Wolf, D. L., Althaus, J. S., and Von Voigtlander, P. F. (1987) Beneficial effects of the kappa opioid receptor agonist U-50488H in experimental acute brain and spinal cord injury. *Brain Res.* **435,** 174–180.

248. Qu, Z. X., Xu, J., Hogan, E. L., and Hsu, C. Y. (1993) Effect of U-50488h, a selective opioid kappa receptor agonist, on vascular injury after spinal cord trauma. *Brain Res.* **626,** 45–49.

249. Gomez-Pinilla, F., Tram, H., Cotman, C. W., and Nieto-Sampedro, M. (1989) Neuroprotective effect of MK-801 and U-50488H after contusive spinal cord injury. *Exp. Neurol.* **104,** 118–124.

250. Gannon, R. L. and Terrian, D. M. (1991) U-50,488H inhibits dynorphin and glutamate release from guinea pig hippocampal mossy fiber terminals. *Brain Res.* **548,** 242–247.

251. Faden, A. I. and Simon, R. P. (1988) A potential role for excitotoxins in the pathophysiology of spinal cord injury. *Ann. Neurol.* **23,** 623–626.

252. Faden, A. I. (1990) Opioid and nonopioid mechanisms may contribute to dynorphin's pathophysiological actions in spinal cord injury. *Ann. Neurol.* **27,** 67–74.

253. Bakshi, R. and Faden, A. I. (1990) Blockade of the glycine modulatory site of NMDA receptors modifies dynorphin-induced behavioral effects. *Neurosci. Lett.* **110,** 113–117.

254. Bakshi, R., Newman, A. H., and Faden, A. I. (1990) Dynorphin A-(1–17) induces alterations in free fatty acids, excitatory amino acids, and motor function through an opiate-receptor-mediated mechanism. *J. Neurosci.* **10,** 3793–3800.

255. Faden, A. I. (1992) Dynorphin increases extracellular levels of excitatory amino acids in the brain through a non-opioid mechanism. *J. Neurosci.* **12,** 425–429.

256. Bakshi, R. and Faden, A. I. (1990) Competitive and non-competitive NMDA antagonists limit dynorphin A-induced rat hindlimb paralysis. *Brain Res.* **507,** 1–5.

257. Choi, D. W. (1992) Excitotoxic cell death [Review]. *J. Neurobio.* **23,** 1261–1276.

258. Faden, A. I., Ellison, J. A., and Noble, L. J. (1990) Effects of competitive and non-competitive NMDA receptor antagonists in spinal cord injury. *Eur. J. Pharmacol.* **175,** 165–174.

259. Lucas, J. H., Wang, G. F., and Gross, G. W. (1990) NMDA antagonists prevent hypothermic injury and death of mammalian spinal neurons. *J. Neurotrauma* **7,** 229–236.

260. Martinez-Arizala, A., Rigamonti, D. D., Long, J. B., Kraimer, J. M., and Holaday, J. W. (1990) Effects of NMDA receptor antagonists following spinal ischemia in the rabbit. *Exp. Neurol.* **108,** 232–240.

261. Agrawal, S. K. and Fehlings, M. G. (1997) Role of NMDA and non-NMDA ionotropic glutamate receptors in traumatic spinal cord axonal injury. *J. Neurosci.* **17,** 1055–1063.

262. Craenen, G., Jeftinija, S., Grants, I., and Lucas, J. H. (1996) The role of excitatory amino acids in hypothermic injury to mammalian spinal cord neurons. *J. Neurotrauma* **13,** 809–818.

263. Haghighi, S. S., Johnson, G. C., de Vergel, C. F., and Vergel Rivas, B. J. (1996) Pretreatment with NMDA receptor antagonist MK801 improves neurophysiological outcome after an acute spinal cord injury. *Neurol. Res.* **18,** 509–515.

264. Liu, S., Ruenes, G. L., and Yezierski, R. P. (1997) NMDA and non-NMDA receptor antagonists protect against excitotoxic injury in the rat spinal cord. *Brain Res.* **756,** 160–167.

265. Holtz, A. and Gerdin, B. (1991) MK 801, an OBS N-methyl-D-aspartate channel blocker, does not improve the functional recovery nor spinal cord blood flow after spinal cord compression in rats. *Acta Neurol. Scand.* **84,** 334–338.

266. von Euler, M., Li-Li, M., Whittemore, S., Seiger, A., and Sundstrom, E. (1997) No protective effect of the NMDA antagonist memantine in experimental spinal cord injuries. *J. Neurotrauma* **14,** 53–61.

267. Simpson, R., Jr., Robertson, C. S., and Goodman, J. C. (1990) Spinal cord ischemia-induced elevation of amino acids: extracellular measurement with microdialysis. *Neurochem. Res.* **15,** 635–639.

268. Panter, S. S. and Faden, A. I. (1992) Pretreatment with NMDA antagonists limits release of excitatory amino acids following traumatic brain injury. *Neurosci. Lett.* **136,** 165–168.

269. Panter, S. S., Yum, S. W., and Faden, A. I. (1990) Alteration in extracellular amino acids after traumatic spinal cord injury [see comments]. *Ann. Neurol.* **27**, 96–99.

270. Liu, D., Thangnipon, W., and McAdoo, D. J. (1991) Excitatory amino acids rise to toxic levels upon impact injury to the rat spinal cord. *Brain Res.* **547**, 344–348.

271. Liu, D. and McAdoo, D. J. (1993) An experimental model combining microdialysis with electrophysiology, histology, and neurochemistry for exploring mechanisms of secondary damage in spinal cord injury: effects of potassium. *J. Neurotrauma* **10**, 349–362.

272. Liu, D. (1994) An experimental model combining microdialysis with electrophysiology, histology, and neurochemistry for studying excitotoxicity in spinal cord injury. Effect of NMDA and kainate. *Mole. Chem. Neuropathol.* **23**, 77–92.

273. Marsala, M., Sorkin, L. S., and Yaksh, T. L. (1994) Transient spinal ischemia in rat: characterization of spinal cord blood flow, extracellular amino acid release, and concurrent histopathological damage. *J. Cereb. Blood Flow Metab.* **14**, 604–614.

274. Yezierski, R. P., Liu, S., Ruenes, G. L., Kajander, K. J., and Brewer, K. L. (1998) Excitotoxic spinal cord injury: behavioral and morphological characteristics of a central pain model. *Pain* **75**, 141–155.

275. Wrathall, J. R., Teng, Y. D., Choiniere, D., and Mundt, D. J. (1992) Evidence that local non-NMDA receptors contribute to functional deficits in contusive spinal cord injury. *Brain Res.* **586**, 140–143.

276. Wrathall, J. R., Teng, Y. D., Choiniere, D., and Mundt, D. J. (1992) Evidence that local non-NMDA receptors contribute to functional deficits in contusive spinal cord injury. *Brain Res.* **586**, 140–143.

277. Wrathall, J. R., Bouzoukis, J., and Choiniere, D. (1992) Effect of kynurenate on functional deficits resulting from traumatic spinal cord injury. *Eur. J. Pharmacol.* **14**, 1–4.

278. Wrathall, J. R., Choiniere, D., and Teng, Y. D. (1994) Dose-dependent reduction of tissue loss and functional impairment after spinal cord trauma with the AMPA/kainate antagonist NBQX. *J. Neurosci.* **14**, 6598–6607.

279. Wrathall, J. R., Teng, Y. D., and Choiniere, D. (1996) Amelioration of functional deficits from spinal cord trauma with systemically administered NBQX, an antagonist of non-N-methyl-D-aspartate receptors. *Exp. Neurol.* **137**, 119–126.

280. Wrathall, J. R., Teng, Y. D., and Marriott, R. (1997) Delayed antagonism of AMPA/kainate receptors reduces long-term functional deficits and tissue loss resulting from experimental spinal cord injury. *J. Neurosci.* **17**, 4359–4366.

281. von Euler, M., Seiger, A., Holmberg, L., and Sundstrom, E. (1994) NBQX, a competitive non-NMDA receptor antagonist, reduces degeneration due to focal spinal cord ischemia. *Exp. Neurol.* **129**, 163–168.

282. Follesa, P., Wrathall, J. R., and Mocchetti, I. (1998) 2,3-Dihydroxy-6-nitro-7-sulfamoyl-benzo(F)-quinoxaline (NBQX) increases fibroblast growth factor mRNA levels after contusive spinal cord injury. *Brain Res.* **782**, 306–309.

283. Teng, Y. D. and Wrathall, J. R. (1996) Evaluation of cardiorespiratory parameters in rats after spinal cord trauma and treatment with NBQX, an antagonist of excitatory amino acid receptors. *Neurosci. Lett.* **209**, 5–8.

284. Bowes, M. P., Swanson, S., and Zivin, J. A. (1996) The AMPA antagonist LY293558 improves functional neurological outcome following reversible spinal cord ischemia in rabbits. *J. Cereb. Blood Flow Metab.* **16,** 967–972.

285. Blight, A. R., Leroy, E. C. Jr., and Heyes, M. P. (1997) Quinolinic acid accumulation in injured spinal cord: time course, distribution, and species differences between rat and guinea pig. *J. Neurotrauma* **14,** 89–98.

286. Popovich, P. G., Reinhard, J. F. Jr., Flanagan, E. M., and Stokes, B. T. (1994) Elevation of the neurotoxin quinolinic acid occurs following spinal cord trauma. *Brain Res.* **633,** 348–352.

287. Popovich, P. G., Reinhard, J. F. Jr., Flanagan, E. M., and Stokes, B. T. (1994) Elevation of the neurotoxin quinolinic acid occurs following spinal cord trauma. *Brain Res.* **633,** 348–352.

288. Sundstrom, E., Mo, L. L., and Seiger, A. (1995) In vivo studies on NMDA-evoked release of amino acids in the rat spinal cord. *Neurochem. Int.* **27,** 185–193.

289. Liu, D. (1994) An experimental model combining microdialysis with electrophysiology, histology, and neurochemistry for studying excitotoxicity in spinal cord injury. Effect of NMDA and kainate. *Mol. Chem. Neuropathol.* **23,** 77–92.

290. Rokkas, C. K., Helfrich, L. R. Jr., Lobner, D. C., Choi, D. W., and Kouchoukos, N. T. (1994) Dextrorphan inhibits the release of excitatory amino acids during spinal cord ischemia. *Ann. Thorac. Surg.* **58,** 312–319; discussion 319–320.

291. Sluka, K. A. and Westlund, K. N. (1993) Spinal cord amino acid release and content in an arthritis model: the effects of pretreatment with non-NMDA, NMDA, and NK1 receptor antagonists. *Brain Res.* **627,** 89–103.

292. Paleckova, V., Palecek, J., McAdoo, D. J., and Willis, W. D. (1992) The non-NMDA antagonist CNQX prevents release of amino acids into the rat spinal cord dorsal horn evoked by sciatic nerve stimulation. *Neurosci. Lett.* **148,** 19–22.

293. Radhakrishnan, V. and Henry, J. L. (1993) Excitatory amino acid receptor mediation of sensory inputs to functionally identified dorsal horn neurons in cat spinal cord. *Neuroscience* **55,** 531–544.

294. Regan, R. F. and Choi, D. W. (1991) Glutamate neurotoxicity in spinal cord cell culture. *Neuroscience* **43,** 585–591.

295. Rusin, K. I., Jiang, M. C., Cerne, R., and Randic, M. (1993) Interactions between excitatory amino acids and tachykinins in the rat spinal dorsal horn. *Brain Res. Bull.* **30,** 329–338.

296. Takeda, H., Caiozzo, V. J., and Gardner, V. O. (1993) A functional in vitro model for studying the cellular and molecular basis of spinal cord injury. *Spine* **18,** 1125–1133.

297. Tymianski, M., Charlton, M. P., Carlen, P. L., and Tator, C. H. (1993) Source specificity of early calcium neurotoxicity in cultured embryonic spinal neurons. *J. Neurosci.* **13,** 2085–2104.

298. Smith, D. O., Franke, C., Rosenheimer, J. L., Zufall, F., and Hatt, H. (1991) Glutamate-activated channels in adult rat ventral spinal cord cells. *J. Neurophysio.* **66,** 369–378.

299. Sakatani, K., Chesler, M., Hassan, A. Z., Lee, M., and Young, W. (1993) Non-synaptic modulation of dorsal column conduction by endogenous GABA in neonatal rat spinal cord. *Brain Res.* **622,** 43–50.

300. Sakatani, K., Hassan, A., Lee, M., Chesler, M., and Young, W. (1993) Non-synaptic modulation of dorsal column conduction by endogenous GABA in neonatal rat spinal cord. *Brain Res.* **662,** 43–50.

301. Saruhashi, Y., Young, W., Sugimori, M., Abrahams, J., and Sakuma, J. (1996) Evidence of GABA and 5-HT sensitivity in adult rat spinal axons: studies with randomized double pulse stimulation. *Neuroscience,* in press.

302. Ciporen, J. N., Sakuma, J., and Young, W. (1996) Pregnenolone blocks GABA-A receptor induced axonal depression in rat dorsal columns. *Neuroscience* submitted.

303. Saruhashi, Y., Young, W., Hassan, A. Z., and Park, R. (1994) Excitatory and inhibitory effects of serotonin on spinal axons. *Neuroscience* **61,** 645–653.

304. Honmou, O. and Young, W. (1995) Norepinephrine modulates excitability of neonatal rat optic nerves through calcium-mediated mechanisms. *Neuroscience* **65,** 241–251.

305. Lee, M., Sakatani, K., and Young, W. (1993) A role of GABAA receptors in hypoxia-induced conduction failure of neonatal rat spinal dorsal column axons. *Brain Res.* **601,** 14–19.

306. Lee, M., Sakatani, K., and Young, W. (1993) A role of GABA-A receptors in hypoxia induced conduction failure of neonatal rat spinal dorsal column axons. *Brain Res.* **601,** 14–19.

307. Saruhashi, Y., Young, W., and Hassan, A. Z. (1997) Calcium-mediated intracellular messengers modulate the serotonergic effects on axonal excitability. *Neuroscience* **81,** 959–965.

308. Saruhashi, Y., Young, W., Sugimori, M., Abrahams, J., and Sakuma, J. (1997) Evidence for serotonin sensitivity of adult rat spinal axons: studies using randomized double pulse stimulation. *Neuroscience* **80,** 559–566.

309. Sakuma, J., Ciporen, J., Abrahams, J., and Young, W. (1996) Independent depressive mechanisms of GABA and (+/–)-8-hydroxy-dipropylaminotetralin hydrobromide on young rat spinal axons. *Neuroscience* **75,** 927–938.

310. Farooque, M., Hillered, L., Holtz, A., and Olsson, Y. (1996) Changes of extracellular levels of amino acids after graded compression trauma to the spinal cord: an experimental study in the rat using microdialysis. *J. Neurotrauma* **13,** 537–548.

311. Farooque, M., Hillered, L., Holtz, A., and Olsson, Y. (1997) Effects of moderate hypothermia on extracellular lactic acid and amino acids after severe compression injury of rat spinal cord. *J. Neurotrauma* **14,** 63–69.

312. McAdoo, D. J., Hughes, M. G., Xu, G. Y., Robak, G., and de Castro, R. Jr. (1997) Microdialysis studies of the role of chemical agents in secondary damage upon spinal cord injury. *J. Neurotrauma* **14,** 507–515.

313. Rokkas, C. K., Helfrich, L. R. Jr., Lobner, D. C., Choi, D. W., and Kouchoukos, N. T. (1994) Dextrorphan inhibits the release of excitatory amino acids during spinal cord ischemia. *Ann. Thorac. Surg.* **58,** 312–319; discussion 319–320.

314. Liu, D. X., Valadez, V., Sorkin, L. S., and McAdoo, D. J. (1990) Norepinephrine and serotonin release upon impact injury to rat spinal cord. *J. Neurotrauma* **7,** 219–227.

315. Baykal, S., Ceylan, S., Usul, H., Akturk, F., and Deger, O. (1996) Effect of thyrotropin-releasing hormone on Na(+)-K(+)-adenosine triphosphatase activity following experimental spinal cord trauma. *Neurol. Med. Chir.* **36,** 296–299.

316. Gonzalez, S., Grillo, C., Gonzalez Deniselle, M. C., Lima, A., McEwen, B. S., and De Nicola, A. F. (1996) dexamethasone up-regulates mRNA for Na$^+$, K$^{(+)}$-ATPase in some spinal cord neurones after cord transection. *Neuroreport* **7,** 1041–1044.

317. Schwab, M. E. and Bartholdi, D. (1996) Degeneration and regeneration of axons in the lesioned spinal cord [Review]. *Physiol. Rev.* **76,** 319–370.

318. Bregman, B. S., Diener, P. S., McAtee, M., Dai, H. N., and James, C. (1997) Intervention strategies to enhance anatomical plasticity and recovery of function after spinal cord injury. [Review]. *Adv. Neurol.* **72,** 257–275.

319. Spillmann, A. A., Bandtlow, C. E., Lottspeich, F., Keller, F., and Schwab, M. E. (1998) Identification and characterization of a bovine neurite growth inhibitor (bNI-220). *J. Biol. Chem.* **273,** 19283–19293.

320. Ye, J. H. and Houle, J. D. (1997) Treatment of the chronically injured spinal cord with neurotrophic factors can promote axonal regeneration from supraspinal neurons. *Exp. Neurol.* **143,** 70–81.

321. Oudega, M. and Hagg, T. (1996) Nerve growth factor promotes regeneration of sensory axons into adult rat spinal cord. *Exp. Neurol.* **140,** 218–229.

322. Guest, J. D., Hesse, D., Schnell, L., Schwab, M. E., Bunge, M. B., and Bunge, R. P. (1997) Influence of IN-1 antibody and acidic FGF-fibrin glue on the response of injured corticospinal tract axons to human Schwann cell grafts. *J. Neurosci. Res.* **50,** 888–905.

323. Kalderon, N. and Fuks, Z. (1996) Severed corticospinal axons recover electrophysiologic control of muscle activity after x-ray therapy in lesioned adult spinal cord. *Proc. Natl. Acad. Sci. USA* **93,** 11185–11190.

324. Kalderon, N. and Fuks, Z. (1996) Structural recovery in lesioned adult mammalian spinal cord by x-irradiation of the lesion site. *Proc. Natl. Acad. Sci. USA* **93,** 11179–11184.

325. Chen, A., Xu, X. M., Kleitman, N., and Bunge, M. B. (1996) Methylprednisolone administration improves axonal regeneration into Schwann cell grafts in transected adult rat thoracic spinal cord. *Exp. Neurol.* **138,** 261–276.

326. Guest, J. D., Rao, A., Olson, L., Bunge, M. B., and Bunge, R. P. (1997) The ability of human Schwann cell grafts to promote regeneration in the transected nude rat spinal cord. *Exp. Neurol.* **148,** 502–522.

327. Xu, X. M., Chen, A., Guenard, V., Kleitman, N., and Bunge, M. B. (1997) Bridging Schwann cell transplants promote axonal regeneration from both the rostral and caudal stumps of transected adult rat spinal cord. *J. Neurocytol.* **26,** 1–16.

328. Li, Y. and Raisman, G. (1997) Integration of transplanted cultured Schwann cells into the long myelinated fiber tracts of the adult spinal cord. *Exp. Neurol.* **145,** 397–411.

329. Menei, P., Montero-Menei, C., Whittemore, S. R., Bunge, R. P., and Bunge, M. B. (1998) Schwann cells genetically modified to secrete human BDNF promote enhanced axonal regrowth across transected adult rat spinal cord. *Eur. J. Neurosci.* **10,** 607–621.

330. Li, Y., Field, P. M., and Raisman, G. (1997) Repair of adult rat corticospinal tract by transplants of olfactory ensheathing cells [see comments]. *Science* **277,** 2000–2002.

331. Ramon-Cueto, A., Plant, G. W., Avila, J., and Bunge, M. B. (1998) Long-distance axonal regeneration in the transected adult rat spinal cord is promoted by olfactory ensheathing glia transplants. *J. Neurosci.* **18,** 3803–3815.

332. Nakahara, Y., Gage, F. H., and Tuszynski, M. H. (1996) Grafts of fibroblasts genetically modified to secrete NGF, BDNF, NT-3, or basic FGF elicit differential responses in the adult spinal cord. *Cell Transplant.* **5,** 191–204.

333. Blesch, A. and Tuszynski, M. H. (1997) Robust growth of chronically injured spinal cord axons induced by grafts of genetically modified NGF-secreting cells. *Exp. Neurol.* **148,** 444–452.

334. McTigue, D. M., Horner, P. J., Stokes, B. T., and Gage, F. H. (1998) Neurotrophin-3 and brain-derived neurotrophic factor induce oligodendrocyte proliferation and myelination of regenerating axons in the contused adult rat spinal cord. *J. Neurosci.* **18,** 5354–5365.

335. Tessler, A., Fischer, I., Giszter, S., Himes, B. T., Miya, D., Mori, F., and Murray, M. (1997) embryonic spinal cord transplants enhance locomotor performance in spinalized newborn rats [Review]. *Adv. Neurol.* **72,** 291–303.

336. Zompa, E. A., Cain, L. D., Everhart, A. W., Moyer, M. P., and Hulsebosch, C. E. (1997) transplant therapy: recovery of function after spinal cord injury. *J. Neurotrauma* **14,** 479–506.

337. Falci, S., Holtz, A., Akesson, E., et al. (1997) Obliteration of a posttraumatic spinal cord cyst with solid human embryonic spinal cord grafts: first clinical attempt. *J. Neurotrauma* **14,** 875–884.

338. Olby, N. J. and Blakemore, W. F. (1996) Primary demyelination and regeneration of ascending axons in the dorsal funiculus of the rat spinal cord following photochemically induced injury. *J. Neurocytol.* **25,** 465–480.

339. Cheng, H., Cao, Y., and Olson, L. (1996) Spinal cord repair in adult paraplegic rats: partial restoration of hind limb function [see comments]. *Science* **273,** 510–513.

340. Asada, Y., Nakamura, T., and Kawaguchi, S. (1998) Peripheral nerve grafts for neural repair of spinal cord injury in neonatal rat: aiming at functional regeneration. *Transplant. Proc.* **30,** 147–148.

341. Gauthier, P., Decherchi, P., Rega, P., and Lammari-Barreault, N. (1996) Nerve transplantation in the central nervous system: a strategy for inducing nerve fiber regeneration in lesions of the brain and spinal cord. *Rev. Neurol.* **152,** 106–115.

342. Decherchi, P., Lammari-Barreault, N., and Gauthier, P. (1996) Regeneration of respiratory pathways within spinal peripheral nerve grafts. *Exp. Neurol.* **137,** 1–14.

343. Kikukawa, S., Kawaguchi, S., Mizoguchi, A., Ide, C., and Koshinaga, M. (1998) Regeneration of dorsal column axons after spinal cord injury in young rats. *Neurosci. Lett.* **249,** 135–138.

344. Inoue, T., Kawaguchi, S., and Kurisu, K. (1998) Spontaneous regeneration of the pyramidal tract after transection in young rats. *Neurosci. Lett.* **247,** 151–154.

345. Davies, S. J., Fitch, M. T., Memberg, S. P., Hall, A. K., Raisman, G., and Silver, J. (1997) Regeneration of adult axons in white matter tracts of the central nervous system. *Nature* **390,** 680–683.

346. Davies, S., Illis, L. S., and Raisman, G. (1995) Regeneration in the central nervous system and related factors. Summary of the Bermuda Paraplegia Conference, April 1994 (International Spinal Research Trust). *Paraplegia* **33,** 10–17.

347. Davies, S. J., Goucher, D. R., Doller, C. and Silver, J. (1999). Robust regeneration of adult sensory axons in degenerating white matter of the adult rat spinal cord. *J. Neurosci.* **19,** 5810–5822.

347a. Davies, S. J. and Silver, J. (1998). Adult axon regeneration in adult CNS white matter. *Trends Neurosci.* **21,** 515.

348. Z'Graggen, W. J., Metz, G. A., Kartje, G. L., Thallmair, M., and Schwab, M. E. (1998) Functional recovery and enhanced corticofugal plasticity after unilateral pyramidal tract lesion and blockade of myelin-associated neurite growth inhibitors in adult rats. *J. Neurosci.* **18,** 4744–4757.

Index

Acidosis, induction by calcium in cell death, 41

Actin, filament dynamics in axons, 138, 139

γ-Aminobutyric acid (GABA), antagonist neuroprotection, 249, 250

Animal models, *see also* Cat models; Rat models,
    clinical therapy testing, 13, 14
    requirements, 1
    utility, v, vi

Apoptosis,
    endonuclease activation by calcium, 40
    inducing factors,
        afferent activity in cat models, 5
        nerve growth factor, 4
        tumor necrosis factor-α, 4
    oligodendrocytes following contusion injury, 3, 4

Astrocyte,
    activation in spinal cord injury, 121, 122, 170, 251, 252
    axon regeneration effects, 122, 123, 170, 171, 195, 251
    markers, 202
    proteoglycan synthesis, 122, 123
    transplantation with Schwann cells, 230, 231

Axon,
    actin filament dynamics, 138, 139
    age effects on regrowth, 175—177
    astrocyte effects on regeneration, 121—123
    central versus peripheral neuron capacity for growth, 169, 170, 178, 179

Axon *(cont.)*,
    demyelination, *see* Demyelination, axons
    growth modulators,
        assays, 116, 133, 134
        axonal guidance molecules, 117, 118, 133—135
        cell adhesion molecules, 117, 118, 131, 132
        inhibitors, overview, 133, 169, 170
        overview, vii, viii
        trophic factors, 115—117, 131
    growth requirements in regeneration and sprouting, 174
    membrane turnover in growth, 139
    microtubule formation and transport, *see* Microtubule, axonal arrays
    myelin,
        regeneration inhibition,
            chick embryo development studies, 119
            immune system role, 120, 121
            mature nervous system studies, 120, 121
            myelin-associated glycoprotein, 135, 144
            myelin-free lamprey regeneration studies, 120
            NI-35, 135, 143, 144
            rodent embryo studies, 119, 120
        stabilization of projections, 118, 119
    neurotrophic rescue from retrograde death following axotomy, 171, 172

From: *Neurobiology of Spinal Cord Injury*
Edited by: R. G. Kalb and S. M. Strittmatter © Humana Press Inc., Totowa, NJ

DATE DUE